Introduction to X-Ray Spectrometric Analysis

Wilhelm Conrad Röntgen
(1845-1923)
Discoverer of x-rays

Henry Gwyn Jeffreys Moseley
(1887-1915)
Founder of x-ray spectrometry

Introduction to X-Ray Spectrometric Analysis

Eugene P. Bertin

RCA Laboratories
David Sarnoff Research Center
Princeton, New Jersey

Plenum Press · New York and London

Library of Congress Cataloging in Publication Data

Bertin, Eugene P   1921-
    Introduction to X-ray spectrometric analysis.

    Includes bibliographies and index.
    1. X-ray spectroscopy. I. Title.
QD96.X2B46   543'.085   77-27244
ISBN 0-306-31091-0

Photograph of W. C. Röntgen from Otto Glasser, *Dr. W. C. Röntgen* (2d ed.; Springfield, Illinois: Charles C Thomas, 1958); courtesy of Mrs. Otto Glasser and the publisher.

Photograph of H. G. J. Moseley from Mary Elvira Weeks, *Discovery of the Elements* (6th ed.; Easton, Pennsylvania: Journal of Chemical Education, 1960); courtesy of the publisher.

© 1978 Plenum Press, New York
A Division of Plenum Publishing Corporation
227 West 17th Street, New York, N.Y. 10011

Printed in the United States of America

Preface

X-ray fluorescence spectrometry has been an established, widely practiced method of instrumental chemical analysis for about 30 years. However, although many colleges and universities offer full-semester courses in optical spectrometric methods of instrumental analysis and in x-ray diffraction, very few offer full courses in x-ray spectrometric analysis. Those courses that are given are at the graduate level. Consequently, proficiency in this method must still be acquired by: self-instruction; on-the-job training and experience; "workshops" held by the x-ray instrument manufacturers; the one- or two-week summer courses offered by a few universities; and certain university courses in analytical and clinical chemistry, metallurgy, mineralogy, geology, ceramics, etc. that devote a small portion of their time to applications of x-ray spectrometry to those respective disciplines.

Moreover, with all due respect to the books on x-ray spectrometric analysis now in print, in my opinion none is really suitable as a text or manual for beginners in the discipline. In 1968, when I undertook the writing of the first edition of my previous book, *Principles and Practice of X-Ray Spectrometric Analysis*,* my objective was to provide a student text. However, when all the material was compiled, I decided to provide a more comprehensive book, which was also lacking at that time. Although that book explains principles, instrumentation, and methods at the beginner's level, this material is distributed throughout a mass of detail and more advanced material. That 679-page book and its 1079-page second edition are simply too voluminous and comprehensive for novices. *X-Rays, Electrons, and Analytical Chemistry*, by Liebhafsky, Pfeiffer, Winslow,

---

* The publication details for this and the other books referred to here are given in the Suggested Reading list at the end of Chapter 1.

and Zemany (566 pages), is, by the authors' own expressed intent, a
graduate-level book and contains much advanced material. Probably the
best beginners' book, Adler's *X-Ray Emission Spectrography in Geology*,
devotes only 165 of its 258 pages to x-ray fluorescence spectrometry, the
remainder to electron-probe microanalysis. Other books on *general* x-ray
spectrometric analysis are unsuitable *for beginners* for one or more of the
following reasons: (1) they may contain material that is either extraneous
or too advanced; (2) they may not give elementary explanations of principles, instrumentation, and methods; (3) they may lack practical laboratory
material, such as specimen preparation and operation of the pulse-height
selector; (4) they may devote undue space to specialized topics of particular
interest to their author(s); and/or (5) they may lack adequate treatment
of energy dispersion. Of course, books on *limited* aspects of the field, such
as quantitative analysis or energy dispersion, are unsuitable for beginners.

This book is intended to be a general-purpose text for all types of
instruction: (1) self-instruction for technicians new to x-ray spectrometry,
and for technicians having practical experience in x-ray spectrometry but
little or no formal instruction and therefore lacking a basic understanding
of the field; (2) workshops and short courses; (3) university courses partially
devoted to x-ray spectrometry; and (4) full-semester undergraduate courses.
The book also constitutes a concise summary of all aspects of the subject
for scientists and technicians in other disciplines who want to evaluate the
applicability of the method to their own fields. The book is intended particularly for those who are not content simply to operate the instrument
and perform the analyses, but who want explanations of the principles,
instrument, and method.

This book has sprung from two sources: my previous book, both
editions of which were very well received, and more important, from my
lectures, since its inception in 1966, at Professor Henry Chessin's annual
summer short course on x-ray spectrometry at the State University of New
York at Albany. Attendees at this course come from private, institutional,
university, industrial, and government laboratories throughout the United
States and Canada, and frequently from foreign countries. They represent
the full range of experience from complete novices to highly competent
analysts. I have benefited greatly from my association with these people.

The book presupposes no knowledge of x-ray spectrometry, but does
require a familiarity with the language of physical science and with simple
algebra. The emphasis is on x-ray fluorescence spectrometry for *chemical
analysis* (as distinguished from x-ray spectrometry for atomic and molecular
structure), particularly as applied on standard commercial wavelength-

dispersive and energy-dispersive instruments. Full consideration is given to energy dispersion, and a chapter is devoted to electron-probe microanalysis.

In addition to providing *answers*—or how to find them—an important objective of a book of this type is to raise *questions* that otherwise might not occur. For example, the novice is not likely to realize that his analyses may be affected by the orientation of grind marks on solid specimens, or by particle size and distribution in powders, or by the kind of acid used in preparation of liquids!

The principles underlying instrumentation, methods, and techniques are given in elementary descriptive, rather than mathematical terms. Elementary explanations or definitions are given of such subjects as atomic structure, radioisotopes, crystal structure and Miller indexes, ion exchange, and cathode-ray tube displays. Essential working mathematical equations are included, but are not derived. Essential laboratory procedures are given in detail sufficient to serve as instructions. Operation of instrumentation is described in some detail. Specific instruments or applications are not described except when they illustrate general instrumentation, methods, or techniques. An understanding of *general* principles and instrumentation should lead to proficiency in *specific* instruments and applications.

Much of the text and many of the figures and tables are taken from my *Principles and Practice of X-Ray Spectrometric Analysis*. However, much new material is added to provide an overall book much more concise and better suited to pedagogy than the previous book.

The book is divided into 11 chapters: Chapters 1 and 2 describe the nature and properties of x-rays. Chapter 3 gives a general introduction to x-ray spectrometric analysis, and Chapters 4–6 go into the details (excitation, wavelength dispersion, detection and readout, x-ray intensity measurement, energy-dispersive x-ray spectrometry). Chapter 7 discusses qualitative and semiquantitative analysis, Chapter 8 deals with performance criteria, and Chapter 9 takes up quantitative analysis. Specimen preparation and presentation are discussed in Chapter 10, and Chapter 11 describes electron-probe microanalysis. The chapters vary widely in length. By far the most detailed treatment is given to explanation of basic concepts for which comprehensive explanations are not given in other books on x-ray spectrochemistry.

At the end of each chapter are some selected references for further study, mostly to other books. I also recommend that the reader have access to my *Principles and Practice . . .* for further study. Moreover, many laboratory procedures are described in sufficient detail to serve as instructions for self-training experiments.

It is my objective—and hope—that the reader who carefully studies this book will benefit in three ways. He/she should acquire a working proficiency permitting intelligent operation of the instrument and performance of analyses previously developed by a competent x-ray spectrochemist. He/she should be able to "graduate" profitably to more comprehensive and advanced books. Finally, he/she should acquire a basic foundation of knowledge that will serve as a mental "framework" or "file" to which knowledge can be added systematically as it is acquired by laboratory experience, reading books and journals, and attending technical meetings.

Eugene P. Bertin
Harrison, New Jersey

Contents

Chapter 2. Properties of X-Rays

Chapter 3. X-Ray Emission Spectrometric Analysis: General Introduction

Chapter 4. Excitation; Wavelength Dispersion

Chapter 5. Detection and Readout; X-Ray Intensity Measurement

Chapter 9. Quantitative Analysis

Chapter 10. Specimen Preparation and Presentation

Chapter 11. Electron-Probe Microanalysis

Chapter 1

Excitation and Nature
of X-Rays and X-Ray Spectra

Before undertaking the study of x-ray spectrometric analysis, we shall devote two chapters to the origin and nature of x-rays and x-ray spectra.

1.1. HISTORY

Wilhelm Conrad Roentgen (see Frontispiece) discovered x-rays on 8 November 1895 at the University of Würzburg in Germany.

Henry Gwyn Jeffreys Moseley (see Frontispiece) laid the foundation of x-ray spectrometric analysis in 1913 at the University of Manchester in England. His first published x-ray spectra are shown in Figure 1.1. They illustrate the basis of qualitative x-ray spectrometric analysis by showing the relationship of x-ray spectral-line wavelength and atomic number, and the contamination of cobalt by iron and nickel; they illustrate the basis of quantitative analysis by showing the greater intensity of copper lines than zinc lines in the x-ray spectrum of brass (70Cu–30Zn). Moseley foresaw the application of x-ray spectrometry to trace analysis when he noted the unexpectedly high intensities of x-ray lines of elements at low concentration. He also foresaw the use of x-ray spectra in the discovery of then unknown chemical elements by prediction of the wavelengths of their characteristic x-ray lines.

Other outstanding contributions in the development of x-ray spectrometry are listed chronologically in Table 1.1. Note that the Friedman–Birks prototype of modern commercial flat-crystal spectrometers did not appear until 1948.

TABLE 1.1. Chronological Development of X-Ray Spectrometry

1895	W. C. Roentgen[a,b] discovered x-rays.
1896	J. Perrin measured x-ray intensity, using an air-ionization chamber.
1909	C. G. Barkla noted evidence of absorption edges.
1911	C. G. Barkla noted evidence of emission-line series, which he designated *K*, *L*, *M*, *N*, etc.
1912	M. von Laue, W. Friedrich, and E. P. Knipping demonstrated diffraction of x-rays by crystals.
1913	W. L. and W. H. Bragg built the Bragg x-ray spectrometer.
1913	H. G. J. Moseley[a] showed the relationship between wavelength of x-ray spectral lines and atomic number, noted that copper lines are stronger than zinc lines in the x-ray spectrum of brass, and thereby established the basis of qualitative and quantitative x-ray spectrochemical analysis.
1913	W. D. Coolidge introduced the hot-filament, high-vacuum x-ray tube.
1913	J. Chadwick first observed excitation of characteristic x-ray spectra by ions (α-particles), laying the foundation of ion-induced x-ray spectrometry. Since Chadwick's α-ray source was a radioisotope—radon, $^{222}_{86}$Rn—he may also be credited with the first observation of radioisotope excitation of x-ray spectra.
1913–23	M. Siegbahn did his classic work of measuring wavelengths of x-ray spectra of the chemical elements.
1922	A. Hadding first applied x-ray spectra specifically to chemical analysis (of minerals).
1923	D. Coster and G. von Hevesy discovered hafnium, the first element to be identified by its x-ray spectrum.
1923	G. von Hevesy proposed quantitative analysis by secondary excitation of x-ray spectra.
1923	R. Glocker and W. Frohnmeyer applied x-ray absorption-edge spectrometry.
1924	W. Soller constructed an x-ray spectrometer using parallel-foil collimators.
1925	D. Coster and Y. Nishina applied x-ray secondary-emission (fluorescence) spectrometry at von Hevesy's suggestion (above).
1928	R. Glocker and H. Schreiber[c] applied x-ray secondary-emission (fluorescence) spectrometry.
1928	H. Geiger and W. Müller developed the gas-filled detector tube to a high degree of reliability.
1938	Hilger and Watts, Ltd. offered the first commercial x-ray spectrometer; the instrument was designed by T. H. Laby.

TABLE 1.1 *(continued)*

1948	H. Friedman and L. S. Birks built the prototype of the first commercial x-ray secondary-emission spectrometer having a sealed-off x-ray tube; the instrument was a modified version of a North American Philips (Norelco) diffractometer for orientation of quartz oscillator-crystal plates.
1949	R. Castaing and A. Guinier built the first electron-probe x-ray primary-emission spectrometer (electron-probe microanalyzer).
1964	A. A. Sterk first applied ion (proton) excitation of x-ray spectra to actual chemical analysis.

---

[a] See Frontispiece.
[b] In 1901, Roentgen received the first Nobel Prize in Physics for his discovery.
[c] Schreiber coined the term x-ray "fluorescence" in 1929.

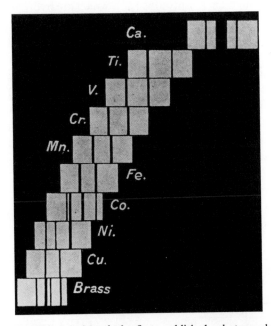

FIGURE 1.1. Moseley's first published photograph of x-ray spectra, showing lines of calcium through zinc. [H. G. J. Moseley, *Philosophical Magazine* **26**, 1024 (1913); Taylor and Francis Ltd., London; courtesy of the publisher.]

1.2. DEFINITION

Radiation may be defined as energy, in the form of waves or particles, emanating from its source through space in straight lines; however, radiation in the form of charged particles may be deflected from its linear path by electric and/or magnetic fields.

All types of radiation have a dual nature in that they exhibit some properties best explained in terms of particles and other properties best explained in terms of waves. Nevertheless, most radiation has either *predominantly* corpuscular or wave properties, and it is convenient to classify all types of radiation in these two categories. The elemental units ("atoms") of either corpuscular or wave energy are known as *quanta*. When predominantly wave radiation exhibits the corpuscular aspect of its dual nature, the individual elemental "particles" are also known as *photons*.

The primarily corpuscular radiation includes alpha (α) rays, or helium nuclei (He^{2+}); beta (β^-) rays, or electrons (e^-); positrons (β^+), or positively charged electrons (e^+); neutrons (n); and the primary cosmic rays, which consist mostly of high-energy protons (p^+), which are hydrogen nuclei (H^+).

The primarily wave radiation comprises the electromagnetic spectrum, which is divided into overlapping regions, as shown in Figure 1.2. The visible region (4000–7500 Å) is defined by human visual response. The gamma (γ) region comprises high-energy electromagnetic radiation originating in the nuclei of atoms undergoing radioactive decay. The secondary cosmic radiation results from interaction of the primary cosmic radiation (corpuscular) with terrestrial matter. The other spectral regions are defined somewhat arbitrarily on the basis of wavelength or, more realistically, the technology used to generate, transmit, detect, and apply the radiation in each region.

X-rays may be defined as electromagnetic radiation of wavelength $\sim 10^{-5}$ to ~ 100 Å produced by deceleration of high-energy electrons and/or by electron transitions in the inner orbits of atoms. Both these processes are discussed in detail below. The 10^{-5}-Å radiation is produced in betatrons

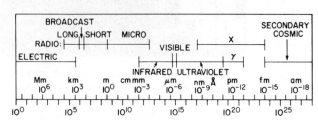

FIGURE 1.2. Electromagnetic spectrum.

operating at $\sim$1 GV (10^9 V); 100 Å represents the K-band spectra of the lightest elements. In conventional x-ray spectrometry, the spectral region of interest is $\sim$0.1 Å (U $K\alpha$) to $\sim$20 Å (F $K\alpha$). In ultrasoft x-ray spectrometry, the region of interest is $\sim$10 Å to $\sim$100 Å (Be $K\alpha$).

1.3. PROPERTIES

Following are the principal properties of x-rays and the application(s) of each in x-ray spectrochemistry.

X-rays are emitted on bombardment of matter by sufficiently energetic electrons, protons ($^1H^+$), deuterons ($^2H^+$), α-particles ($^4He^{2+}$), or heavier ions, and by irradiation of matter by sufficiently energetic electromagnetic radiation, particularly x- and γ-rays. It follows that an important property of x-rays is their ability to excite secondary characteristic x-ray line spectra (see the next paragraph). All these modes of excitation are applied in x-ray spectrochemical analysis, and excitation by x-rays, the basis of x-ray "fluorescence" spectrometry, is the most widely applied mode.

X-rays are emitted as noncharacteristic continuous spectra or as line and band spectra having wavelengths characteristic of the emitting element; they undergo differential absorption in matter and thereby produce characteristic absorption spectra. Continuous spectra from x-ray tubes are largely responsible for excitation of characteristic secondary ("fluorescence") spectra. Continuous spectra scattered from the specimen are responsible for most of the background in x-ray fluorescence spectra. Characteristic line spectra are the basis of all x-ray emission spectrometric methods. Characteristic absorption spectra are the basis of certain x-ray absorption analytical methods.

X-rays propagate at the velocity of light ($c = 3\times10^8$ m/s), without transference of mass, in straight lines undeflected by electric or magnetic fields. The linear propagation is the basis of x-ray collimation and "focusing."

On encountering matter, depending on conditions, x-rays may pass through unaffected or undergo reflection, refraction, diffraction (by pinholes, slits, gratings, or crystals), polarization, coherent or incoherent scatter, photoelectric absorption, or, at energy >1.02 MeV, electron–positron pair formation. Crystal diffraction is the basis of wavelength dispersion. Scatter is the principal source of background in secondary spectra and the basis of several analytical methods. Photoelectric absorption is the basis of x-ray secondary (fluorescence) excitation.

X-rays may cause ionization in gases, liquids, and solids, thereby altering their electrical properties. Gas ionization is the basis of gas-filled x-ray detectors—ionization chambers, Geiger counters, and proportional counters. Ionization of silicon is the basis of lithium-drifted, solid-state semiconductor x-ray detectors, without which energy dispersion would be impractical.

X-rays may induce photolysis (radiation-induced chemical change), including temporary or permanent color change, in matter. Photolysis causes bubble formation and precipitation in liquid specimens and contributes to aging of solid, briquet, and powder standards.

X-rays may cause permanent physical damage (radiation damage)—cracks, fissures, voids, scaling, etc.—in solids; this also contributes to aging of standards and causes deterioration of thin plastic specimen-cell covers and detector windows on prolonged use.

X-rays expose photographic materials; this is the basis of photographic recording of x-ray spectra and of film-badge dosimetry.

X-rays excite visible and ultraviolet luminescence in certain materials; this is the basis of scintillation counters and fluorescent screens.

X-rays excite photo-, Auger, and Compton-recoil electrons in matter. This is the basis of x-ray induced photo- and Auger-electron spectrometry, two forms of "electron spectrometry for chemical analysis" (ESCA).

X-rays may kill, damage, and/or cause genetic change in biological tissue—the basis of the x-ray health hazard. This hazard is insidious because x-rays are invisible and otherwise undetected by human senses.

X-rays exhibit the particle-wave duality mentioned above. The corpuscular (photon) properties include photoelectric absorption, incoherent scatter, gas ionization, and scintillation production. The wave properties include velocity, reflection, refraction, diffraction, polarization, and coherent scatter.

1.4. UNITS OF MEASUREMENT

In practice, the x-ray spectrochemist is concerned with three parameters for x-ray measurement: wavelength, photon energy, and intensity.

Wavelength λ (lambda) is a measure of the wave aspect of x-rays and is used in wavelength-dispersive x-ray spectrometry. Figure 1.3 represents a train of 10 waves. Wavelength is the distance between any two successive peaks or "valleys," or between any two successive corresponding points. In x-ray spectrochemistry, the unit of wavelength is the angstrom (Å):

FIGURE 1.3. Representation of waves, showing wavelength λ and amplitude. If the wave train is 3×10^{10} cm long and the lead wave travels this distance in 1 s, frequency ν is $(3 \times 10^{10}\ \text{cm/s})/\lambda_{\text{cm}} = 10\ \text{s}^{-1}$ or 10 Hz.

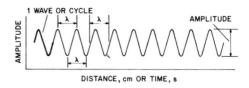

$1\ \text{Å} \equiv 10^{-10}\ \text{m}$. The angstrom is a nonsystematic unit, and wavelength should be expressed in nanometers (nm): $1\ \text{nm} \equiv 10^{-9}\ \text{m} = 10\ \text{Å}$. However, the angstrom is so well established that its abandonment for the nanometer seems unlikely. In this book, λ indicates wavelength in angstroms except where another unit is indicated by subscript, for example, λ_{cm}. X-ray wavelength regions are sometimes designated *ultrahard* ($<0.1\ \text{Å}$), *hard* (0.1–1 Å), *soft* (1–10 Å), and *ultrasoft* ($>10\ \text{Å}$).

A less commonly used wave parameter is frequency ν (nu). The unit of frequency is the hertz (Hz): $1\ \text{Hz} \equiv 1$ wave per second (s^{-1}). In Figure 1.3, if it is assumed that the length of the 10-wave train is 3×10^{10} cm and that the leading wave travels this distance in 1 s, it is evident that wavelength and frequency are related as follows:

$$\lambda_{\text{cm}} = c/\nu \qquad \text{cm} \tag{1.1}$$

$$\lambda_{\text{Å}} = c/\nu \times 10^8 \quad \text{Å} \tag{1.2}$$

$$\nu = c/\lambda_{\text{cm}} \qquad \text{Hz, s}^{-1} \tag{1.3}$$

where c is the velocity of light, 3×10^{10} cm/s.

Photon energy E_x is a measure of the photon nature of x-rays and is used in energy-dispersive x-ray spectrometry. In x-ray spectrochemistry, the unit of photon energy is the electron volt (eV) or kilo-electron-volt (keV). Photon energy and wavelength are related as follows:

$$E_x = hc/\lambda e \tag{1.4}$$

where substitution of values for Planck's constant h (6.6×10^{-27} erg s), the velocity of light c (3×10^{10} cm/s), the electron charge e (4.8×10^{-10} esu), and conversion factors 10^{-8} cm/Å and 1/300 esu/V gives

$$E_x = \frac{(6.6 \times 10^{-27})(3 \times 10^{10})}{(4.8 \times 10^{-10})(10^{-8})(1/300)\lambda}$$

$$= 12{,}396/\lambda \tag{1.5}$$

$$\lambda = 12{,}396/E_x \tag{1.6}$$

where E_x and λ are in electron volts and angstroms, respectively. If E_x is in kilo-electron-volts, the constant is 12.396. Notice that photon energy increases as wavelength decreases, and *vice versa*. In this book, E_x indicates photon energy in eV or keV, as indicated where used. Incidentally, some writers use the symbol E for electric potential ("voltage"); in this book, V (italic) indicates potential, V (roman) the unit of potential, the volt.

   Intensity I of an x-ray beam is defined in physics in terms of energy per unit area per unit time. However, in x-ray spectrochemistry, a more useful concept is the number of effective x-ray photons entering the x-ray detector per unit time, that is, the number of photons counted per unit time. Since the effective detector area is constant, it may be disregarded, and the unit is simply counts per second, counts per minute, or counts in a preset counting time. Many writers use the symbol R (for count rate) for intensity. Incidentally, some writers use the symbol I for electric current; in this book, i indicates current.

1.5. X-RAY TUBES

1.5.1. Conventional High-Power X-Ray Spectrometer Tubes

   The most widely applied source of x-rays is the x-ray tube, which has many forms for various applications. The basic structure and components of the tubes used in x-ray fluorescence spectrometers are shown in Figure 1.4. The external features are a heavy rayproof metal head c that confines the x-rays, a thin beryllium window d that permits emergence of the useful beam, and a reentrant glass envelope i that retains the vacuum and provides a long electrical insulation path. This vacuum enclosure houses the actual x-ray generating structure: A helical tungsten filament g is heated to in-candescence by an electric current passed through the external filament terminals h. The hot filament emits electrons (*thermionic emission*), which are roughly focused by a concave focusing electrode f and attracted and accelerated to the target or anode b, which is operated at a positive potential of up to 100 kV. Let it suffice here to state that the high-speed electrons bombarding the target excite x-ray emission.

   The target consists of a thin disk or plating of the target metal b imbedded in or plated on a heavy, hollow, water-cooled copper block a, which conducts heat away from the electron-bombarded target area. The filament-target space is surrounded by a cylindrical metal internal shield e

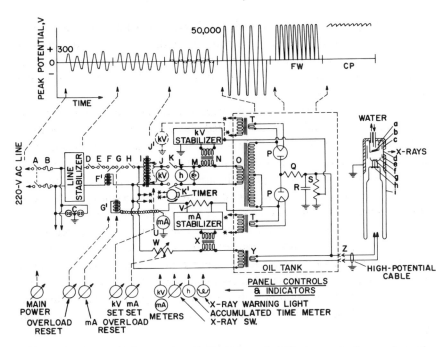

FIGURE 1.4. X-ray generator. Panel controls and indicators are shown along the bottom. Wave forms (potential *versus* time) at selected points are shown along the top.

X-ray power supply:

A. Main power switch
B. Fuses
C. Accessory outlets
D. Water switch, high-pressure
E. Water switch, low-pressure
F. Line-overload relay
F'. Solenoid coil for F
G. mA-overload relay
G'. Solenoid coil for G
H. Door interlocks
I. Autotransformer, kV selector
J. kV meter
J'. kV meter, alternate position
K. X-ray switch
K'. Time-delay switch

L. Accumulated-time meter
M. X-ray warning light
N. kV stabilizer reactor
O. High-potential transformer
PP. High-potential rectifiers
Q. Resistor ⎱ Constant-potential filter
R. Capacitor ⎰
S. Potential-divider resistor
TT. Rectifier-filament transformers
U. mA meter
V. Resistor for mA stabilizer
W. Filament-control rheostat
X. mA stabilizer reactor
Y. X-ray tube filament transformer
Z. High-potential socket

X-ray tube:

a. Water-cooled target block
b. Target
c. Rayproof head
d. Beryllium window
e. Internal shield

f. Focusing electrode
g. Filament
h. Filament terminals
i. Glass envelope

having an opening in register with the beryllium window *d*. This shield intercepts improperly focused electrons from the filament and electrons scattered from the target, and it minimizes metallization of the window by tungsten sublimed from the filament and/or metal sputtered from the target and target block.

In practice, the target, which is integral with the tube head, is operated at ground potential and the well-insulated filament at up to 100 kV negative. This mode of operation permits safe contact with the tube head and specimen chamber, and water cooling without precaution against grounding through the cooling-water line. A rayproof housing must be provided to prevent access to the glass end of the operating tube, both because of the poor x-ray shielding at this end and because of the electric shock hazard.

The x-ray tube is extremely inefficient. Only $\sim1\%$ of the target power (~2.5 kW at 50 kV, 50 mA) is converted to x-rays, the other 99% to heat. Consequently, without water cooling, operation at high power for only a few seconds may ruin the tube. Unless the target metal is itself a good heat conductor, it must be very thin, either a thin disk or plating. This is feasible because 50–100-kV electrons penetrate only ~1–3 μm into a metal surface.

The target area actually bombarded by the filament electrons is the *focal spot*; it is the source of primary x-rays, and target-to-window and target-to-specimen distances are measured from the center of the focal spot. X-rays are emitted from the focal spot in all directions, but emerge from the metal head *c* only through the beryllium window *d*. The angle between the normal to the window and the target surface is the *target angle* or *anode angle*; it is typically $\sim20°$, so that the solid angle of the useful emergent cone of x-rays is $\sim40°$. The focal spot in an x-ray spectrometer, unlike that in a diffractometer, is not itself a part of the x-ray optical system and therefore may be relatively large—10×5 mm, compared with 10×1 mm in a diffraction tube. This larger area permits greater power dissipation— 2.5–3.5 kW (or kVA), compared with ~1 kW in a diffraction tube.

Nevertheless, the focal spot is relatively small, so the emergent x-ray beam is substantially spherically divergent and therefore approximately "obeys" the inverse-square law (Section 2.1.7). It follows that if a specimen just outside the window is to receive high primary intensity, the target-to-specimen distance—and therefore the target-to-window distance—must be as small as possible. Different x-ray tube manufacturers accomplish this by reducing the diameter of the tube head, flattening the window side of the head, and placing the filament and target asymmetrically off-axis toward the window. Incidentally, the smaller the target-to-window distance, the more severe are variations in primary intensity on the specimen arising from minor variations in the specimen plane.

High-power x-ray spectrometer tubes are available commercially having targets of chromium, copper, molybdenum, rhodium, silver, tungsten, platinum, and gold. Other targets are available on special order. Tubes are available having beryllium windows 1, 0.25, and 0.125 mm (0.040, 0.010, and 0.005 inch) thick. The applications of these various targets and windows are considered in appropriate subsequent sections.

1.5.2. Other X-Ray Tubes

The type of x-ray tube described above is typical of the conventional high-power tubes used for excitation on wavelength-dispersive x-ray fluorescence spectrometers. These tubes operate at maximum ratings of 50, 75, or 100 kV, 2.5–3.5 kW (kVA). Several other types of x-ray tube are used in x-ray spectrochemical analysis.

The *dual-target x-ray tube* has essentially the same construction as the conventional tube except for two features: it has two adjacent targets—one tungsten, platinum, or molybdenum, the other chromium—and two filaments, each having its focal spot on one of the targets. The appropriate filament is energized by an external selector switch. This single tube permits efficient excitation of both short- and long-wavelength spectral regions.

The *end-window x-ray tube* has the x-ray window in the end (top) of the head, normal to the tube axis. This tube is advantageous for multichannel spectrometers (Section 3.3.3.3).

Continuously pumped demountable x-ray tubes have several advantages: Burned-out filaments are easily replaced, and targets are easily cleaned, resurfaced, and interchanged. There is no limitation on the target element because gas-free operation and long life are not required; therefore, reasonably stable compounds can be used when pure metals are unsuitable.

Low-power x-ray tubes operate at a maximum of about 50 kV, 1 mA ($\lesssim$100 W), require no water cooling, and can be energized by small, well-regulated, solid-state power supplies. These tubes are used for excitation on energy-dispersive x-ray spectrometers. Some low-power tubes are simply scaled-down conventional tubes, but others are based on other principles.

X-ray tubes for ultralong-wavelength spectrometry are demountable, have interchangeable targets having long-wavelength spectral lines, and operate at relatively low potential but high current, for example, $\lesssim$15 kV, $\lesssim$200 mA.

1.6. THE PRIMARY X-RAY GENERATOR

The x-ray tube energized by the x-ray tube power supply constitutes the x-ray generator. Although low-power ($\lesssim$100 W) x-ray tubes are energized by small, light-weight, solid-state power supplies, high-power tubes for wavelength-dispersive x-ray fluorescence spectrometers require large, elaborate supplies. A typical high-power supply is shown in Figure 1.4. It must provide up to 50-, 75-, or 100-kV, 50–75-mA pulsating or constant-potential dc target power and up to 15-V, 5-A ac filament power. Both these types of power must be derived from the 110- or 220-V, 60-Hz ac line. The ac line potential is increased or decreased by transformers, converted to dc by rectifiers, and smoothed by filters. The nature of alternating current is considered briefly in Section 1.6.1. Transformers, rectifiers, and filters are considered in Section 1.6.2. The x-ray tube power supply and its operation are described in Section 1.6.3.

1.6.1. Alternating Current

Direct current (dc) is an electron current that flows always in the same direction. Alternating current (ac) is an electron current that periodically reverses its direction, that is, alternates.

Ac *potential* is represented by a sine wave on a plot of potential *versus* time, as shown by the first four wave forms along the top of Figure 1.4. The interpretation is as follows. Assume that you are facing the switch box to which your x-ray power supply is connected. The box may be regarded as the power *source* for which the power supply is the *load*. Consider an instant at which the left and right terminals are positive and negative, respectively. The wave forms in Figure 1.4 indicate that the ac potential across the terminals starts at zero (volt), rises to a maximum (*peak*), and falls to zero; the polarity then reverses, so the left and right terminals are negative and positive, respectively; the potential across the terminals then rises to a peak and falls to zero, as before. This is one *cycle* of ac potential, and in most power lines in the United States, the ac cycle repeats at a rate of 60 times a second; that is, the *frequency f* is 60 hertzes (Hz). The positive and negative half-cycles are *alternations*.

Ac *current*, like ac potential, is also represented by a sine wave, but on a plot of current *versus* time. Consider the current flowing from the source (switch box) to the load (x-ray power supply). The negative terminal of a source is negative because it has an excess of (negative) electrons; the positive terminal is positive because it has a deficiency of electrons.

Consequently, electron current always flows from the negative terminal of a source, through the load, back to the positive terminal. So, at an instant when the left and right terminals of the ac source are positive and negative, respectively, the ac current flows out of the right (negative) terminal, through the load, into the left terminal (positive), starting at zero (ampere), rising to a peak, and falling to zero. Then the terminals reverse their polarity and the current reverses its direction and flows out of the left terminal into the right, starting at zero, rising to a peak, and falling to zero, as before. The ac current repeats this cycle at a frequency of 60 Hz.

In the wave forms in Figure 1.4, it is evident that the mean value of an ac potential or current is zero because positive and negative alternations cancel arithmetically. Consequently, ac potential and current are stated in *peak* or in *root-mean-square* (rms) or *effective* values. The rms value is the square root of the sum of the squares of the instantaneous values of potential or current summed over the entire cycle:

$$V_{\text{rms}} = 0.707 V_{\text{peak}}; \qquad i_{\text{rms}} = 0.707 i_{\text{peak}} \qquad (1.7)$$

$$V_{\text{peak}} = 1.414 V_{\text{rms}}; \qquad i_{\text{peak}} = 1.414 i_{\text{rms}} \qquad (1.8)$$

where V and i are potential (V) and current (A), respectively; $1.414 = 2^{1/2}$ and $0.707 = 1/2^{1/2}$. For example, in the 110-V and 220-V ac line, the peak potentials are 155 and 310 V, respectively.

1.6.2. Transformers, Rectifiers, Filters

Transformers (Figure 1.5) are inductive devices for increasing or decreasing an ac potential without substantial loss of power. A transformer consists of two separate coils of wire on a laminated iron core. If an ac potential V_{pri} is applied to the primary (input) winding, an induced ac potential V_{sec} of the same frequency appears across the secondary (output)

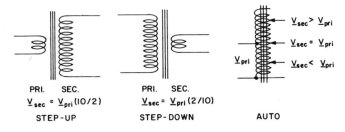

FIGURE 1.5. Transformers: step-up (left), step-down (center), auto (right). The ratios are the secondary/primary turns ratios.

winding:

$$V_{sec}/V_{pri} = n_{sec}/n_{pri} \qquad (1.9)$$

$$V_{sec} = V_{pri}(n_{sec}/n_{pri}) \qquad (1.10)$$

where n_{pri} and n_{sec} are the numbers of turns of wire in the primary and secondary windings, respectively, and n_{sec}/n_{pri} is the *turns ratio*. If $n_{sec} > n_{pri}$ (turns ratio > 1), the device is a *step-up* transformer and $V_{sec} > V_{pri}$; if $n_{sec} < n_{pri}$ (turns ratio < 1), the device is a *step-down* transformer and $V_{sec} < V_{pri}$.

An *autotransformer* (Figure 1.5) is a transformer in which a single winding serves as both primary and secondary. The input is applied across part or all of the winding. The output is taken from the bottom of the coil and a movable contact, which may be set to give output potential less than, equal to, or greater than the input potential. The turns-ratio principle [Equation (1.10)] applies as in transformers having separate windings. Variacs (trade name, General Radio Co.) and Powerstats (trade name, Superior Electric Co.) are continuously variable autotransformers.

The transformer violates no conservation laws because, disregarding minor losses, the primary and secondary *powers* are the same:

$$(Vi)_{sec} = (Vi)_{pri} \qquad (1.11)$$

Let us consider a practical application of transformers. Suppose that an x-ray generator operating from a 220-V ac line is to have a maximum output of 50 kV, 50 mA. What must be the turns ratio of the high-potential transformer, and what current will the primary winding draw from the ac line? An autotransformer can be used to increase the 220-V line potential to 250 V (see below). Then, from Equation (1.9), $50,000/250 = 200/1$. Disregarding minor power losses, a rearranged Equation (1.11) gives

$$i_{pri} = (Vi)_{sec}/V_{pri} \qquad (1.12)$$

Then, $(50,000 \times 0.050)/250 = 10$ A.

Rectifiers are devices for converting alternating current to direct current, which always has the same polarity and always flows in the same direction. One form of rectifier is shown in Figure 1.4 (*PP*). These are the electron-tube type and consist of a thermionic-emissive tungsten filament surrounded by a cylindrical metal anode in a glass vacuum enclosure. Electron current can flow from the filament to the anode only when the anode is positive. Thus, the tube functions as a one-way electronic valve, passing only the positive alternations of an ac current, and this process is known as *half-wave rectification* (HW). However, two or four rectifiers can be arranged to reverse and pass the negative alternations as well, and this process is known as *full-wave rectification* (FW). Full-wave rectification is shown in the fifth wave form at the top of Figure 1.4. Half-wave rectification would be represented by the fourth wave form with the negative (lower) alternations omitted. Both these modes of rectification are explained briefly in Section 1.6.3.4.

Two full-wave rectifier circuits are commonly used in high-power generators for x-ray spectrometers. One circuit is shown in Figure 1.4. This circuit requires a high-potential transformer *O* having a center-tapped secondary winding delivering *twice* the x-ray tube potential, and two rectifiers *PP*, each having its own filament transformer *TT*. This circuit is unsuitable for 75- and 100-kV units because of the size and expense of transformers insulated for 150–200 kV. The other circuit, the bridge circuit, requires a transformer delivering only the x-ray tube potential, but four rectifiers, two with their own filament transformers, two with a common transformer.

The rectifier output, the fifth wave form at the top of Figure 1.4, is pulsating dc. It is true dc because, although it varies from zero to a peak, the potential always has the same polarity and the current always flows in the same direction.

The principal disadvantages of electron-tube rectifiers are that they require a filament transformer insulated for high potential, produce heat, and may emit x-rays. (The basic similarity of these tubes to x-ray tubes is evident.) Another form of rectifier, the solid-state rectifier, consists of a semiconductor having high resistance to current passing through it in one direction, low resistance in the other direction. These rectifiers are very compact and do not have the disadvantages cited above.

Filters are devices that reduce the fluctuation (*ripple*) in a pulsating direct current. The resistor–capacitor combination *QR* in Figure 1.4 constitutes a filter. The capacitor charges, that is, stores electrons, during the peaks and discharges during the "valleys," tending to keep the wave form smooth. Another form of filter uses an inductor (instead of a capacitor), which stores energy in a magnetic field during the peaks; during the valleys, this field collapses, inducing a current in the circuit, again tending to keep the wave form smooth. The output of the filter is shown as the sixth wave form in Figure 1.4 and is known as *constant-potential* dc (CP). The wave form shows a slight residual ripple, given by

$$V_{\text{ripple}} = (i/f)/C \qquad (1.13)$$

where V_{ripple} is the "peak-to-valley" residual ripple potential (V), i is the current (A) drawn from the power supply, f is the ripple frequency (120 Hz for full-wave rectified 60-Hz ac), and C is the capacitance (farads, F). For example, for a full-wave rectified 60-Hz constant-potential supply using a 0.2-μF capacitor and delivering 50 mA, the ripple is $(0.050/120)/(2 \times 10^{-7}) = 2100$ V, or $(2100/50,000) \times 100 = 4.2\%$.

We can now consider how transformers, rectifiers, and a filter are combined with other components to provide a practical x-ray power supply.

1.6.3. X-Ray Tube Power Supply

The x-ray tube power supply is shown in Figure 1.4 and consists of: (1) a high-potential dc target supply; (2) a low-potential ac filament supply; (3) line, mA, and kV stabilizers; and (4) safety and protective devices. Panel controls and indicators are shown along the bottom of the figure. Wave forms of the current at six key points are shown along the top.

1.6.3.1. AC-Line Input

In Figure 1.4, when the main power switch A is closed, the 220-V ac line passes through the switch and fuses B to the accessory outlets C and any auxiliary equipment not requiring regulated (stabilized) power, and to the line stabilizer. The stabilizer maintains constant 220-V output for input fluctuations of 190–250 V. Thus, the input line is divided into unregulated and regulated branches. The auxiliary equipment on the unregulated line includes the vacuum pump, water circulating pump and cooler, spectro-goniometer motor drive, specimen rotator, and semiautomatic or automatic drive mechanisms. If all the safety and protective devices D–H (Section 1.6.3.6) are closed, which they normally are, the regulated line energizes the autotransformer I in the high-potential supply, x-ray tube filament transformer Y, rectifier-tube filament transformers TT, mA and kV stabilizers, and time-delay switch K'.

1.6.3.2. High-Potential Power Supply

In this section, we describe a 50-kV supply, but the discussion applies also to the 75- and 100-kV supplies also available commercially.

The regulated 220-V line goes to the input of the autotransformer I, which permits the potential applied to the primary winding of the high-potential transformer O to be varied from zero to ~250 V. The output is taken from the bottom of the winding and the movable contact. When the contact is set near the bottom of the winding (dashed arrow), the output is near 0 V. As the contact is moved upward (solid arrow), output potential increases until at the point where the input is connected, output and input are equal at 220 V. As the contact is moved still higher, output exceeds input up to ~250 V.

The movable contact on the autotransformer is operated by the kV-set knob, ·and the potential across the output is indicated on the meter J. However, the scale of this meter is graduated to indicate the x-ray tube operating potential, 0–50 (or 75 or 100) kV, rather than 0–250 V, as one would expect. How this can be done is explained below. The output of the autotransformer is applied to the input of the high-potential transformer O, but not until both the manual x-ray switch K and the 30–60-s time-delay switch K' are closed. The x-ray tube is energized simultaneously with the transformer O, so the x-ray warning light M and x-ray tube accumulated time meter L are also energized.

The high-potential transformer is a step-up transformer and increases the output of the autotransformer (0–250 V) to the x-ray tube target potential (0–50, 75, or 100 kV). It is shown above that if 50 kV is required from a 250-V input, a turns ratio of 50,000/250 = 200/1 is required. Then for autotransformer output 50 V, the high potential is $50 \times 200 = 10,000$ V (10 kV), for 125 V it is 25 kV, etc., until for 250 V it is 50 kV. It is now evident why it is the kV-set knob that operates the autotransformer. It is also evident why the panel meter J can indicate kilovolts, since the high-potential output is always 200 times the autotransformer output. Thus, a small, rugged, inexpensive, and easily insulated 0–250-V ac voltmeter at the autotransformer output serves the same function as a bulky, expensive, 0–50-kV ac kilovolt-meter at the output of transformer O. A preferable way to indicate the x-ray tube potential J' is described in Section 1.6.3.5.

In Figure 1.4, transformer O actually has a turns ratio of 400/1 and delivers 100 kV maximum for a 50-kV generator. The secondary coil is "center-tapped" and actually consists of two 50-kV windings. This arrangement is required for full-wave rectification, as explained below.

The ac output of the transformer is rectified—converted to pulsating dc—by the two rectifiers PP, each having its own step-down filament transformer TT, similar to the x-ray tube filament transformer Y. The ac input to the primary windings of these transformers is constant and in no way affected by the autotransformer I or the rheostat W for the x-ray tube filament transformer. Electrons can flow from filament to anode in these rectifiers only when the filament is negative with respect to the anode.

In full-wave supplies, the pulsating dc output of the rectifiers is connected directly to the high-potential socket Z. In constant-potential supplies, the full-wave rectified current is smoothed by an electric filter consisting of a resistor Q (or an inductor) and capacitor R. The potential divider S is a high resistance—10 MΩ or more—and allows the charge on the capacitor to leak off when the supply is turned off. Otherwise, the capacitor would retain its charge and present an electric-shock hazard, even when the supply is turned off. Other uses of the potential divider are given below.

The high-potential transformer O, rectifiers PP, constant-potential filter QR, potential divider S, rectifier filament transformers TT, and x-ray tube filament transformer Y (see below) all require high electrical insulation —up to 100 kV for a 50-kV supply. Consequently, all these components are mounted in an oil-filled tank, as shown in Figure 1.4. Incidentally, even though the three filament transformers TTY have primary and secondary potentials of only 220 and $\sim$10 V, respectively, they must be insulated to withstand the full 100-kV potential of transformer O. Otherwise, the high-potential could arc through the filament transformers to the ac line and ground.

1.6.3.3. X-Ray Tube Filament Supply

The x-ray tube filament transformer Y is a step-down transformer similar to the rectifier filament transformers TT and reduces the stabilized 220-V ac line to 5–10 V to heat the x-ray tube filament. The rheostat W varies the primary potential in much the same way as the autotransformer I varies the primary potential to transformer O. In fact, a second autotransformer can be used in place of the rheostat. The secondary of the filament transformer is connected to the high-potential socket Z, thence through the high-potential cable to the x-ray tube filament terminals h. If the resistance of the rheostat W is decreased, the potential applied to the primary winding, and therefore the output of the secondary winding, increases. As a result, more current flows through the x-ray tube filament, increasing its temperature and electron emission. Thus, the x-ray tube current and x-ray output increase. It is now evident why the knob that operates the rheostat is the mA-set knob, as shown.

1.6.3.4. Operation

We may now consider the operation of the full-wave rectified high-potential system. On positive ac alternations, the top of the secondary winding of O is positive and the bottom is negative; that is, the top and bottom have a deficiency and excess of electrons, respectively. Electron current always flows from the negative terminal of a source, so current flows to the filament of the lower rectifier P, across the vacuum to the anode, through the resistor Q (if present) to the socket Z, thence through the cable to the x-ray tube filament g. The x-ray tube filament electrons flow across the vacuum to the target b, through the copper block a and tube head c to ground. The return current flows through the ground to the milliammeter U, through winding G' and resistance V (both considered later), to the center tap of the transformer O.

On negative alternations, the polarity of the secondary winding of O is reversed, with the top of the winding negative. Now, the same thing happens, except that the electron current flows from the top of the winding through the upper rectifier. Thus, successive half-cycles flow from alternate halves of the transformer and through alternate rectifiers, but the current applied to the x-ray tube is ʳlways in the same direction. It is now evident why for full-wave rectified 50-kV operation (with two rectifier units), transformer O must, in effect, have two 50-kV windings.

To explain half-wave rectifier operation, we may imagine that, in Figure 1.4, the top half of the secondary coil of transformer O, the top rectifier P, and the top rectifier filament transformer T are absent. The center tap then becomes the top of the high-potential secondary, which develops only 50 kV maximum. On positive ac alternations, the top (formerly center tap) of the secondary is positive, the bottom negative. Current flows from the bottom of the winding, through the bottom rectifier P, to the socket Z, through the cable and x-ray tube (filament g to target b), to ground, through the milliammeter U, back to the top ("center tap") of the secondary of O. On negative alternations, the top of the winding is negative. Current cannot flow from anode to filament in either x-ray tube or rectifier, so no current flows. Incidentally, if there were no rectifier at all, the x-ray tube would act as its own half-wave rectifier; this is *self-rectification*. In this arrangement, the bottom of the high-potential secondary would be connected directly to the socket Z. Current could flow from the x-ray tube filament to target only during the alternations when the bottom of the transformer is negative. No modern spectrometer power unit operates on half-wave rectification or self-rectification; all are either full-wave or constant-potential.

The electron current referred to in the preceding discussion is the x-ray *tube current* and is responsible for the generation of the x-rays; it is indicated on the milliammeter U, and is usually of the order 50 mA or less. This current must be distinguished from the x-ray tube *filament current*, which flows in the circuit consisting of the x-ray tube filament and the secondary of the filament transformer Y, is responsible only for heating the x-ray tube filament, and is usually of the order 5 A or less.

The wave forms of the potential at key components in the system are shown at the top of Figure 1.4 on a scale of potential *versus* time. The 220-V ac line is represented by the first wave form, where line fluctuations are indicated—rather naively—by the unequal amplitudes of the alternations. The stabilized 220-V ac is represented by the second wave form, where all the alternations are of equal amplitude. The 0–250-V ac output of the autotransformer I is represented by the third drawing. When the kV knob is set with the contact low on the autotransformer (dashed arrow), the output potential is low (dashed wave); when the knob is near the top of the winding (solid arrow), the output potential is high (solid wave) and may be made even higher than the 220-V input. The 50-kV ac output of the high-potential transformer O is shown next, followed by the 50-kV full-wave rectified pulsating output of the rectifiers. For 60-Hz ac, the full-wave pulse rate is 120 per second. In a full-wave generator, this is substantially the wave form applied to the x-ray tube. In a constant-potential generator, the full-wave pulsating current is smoothed by an electric filter QR, providing nearly pure direct current, as shown by the last wave form. Half-wave rectification would be represented by the fourth wave form (50-kV ac) with the negative (bottom) alternations omitted, or by the fifth wave form (50-kV FW) with every other alternation omitted.

It is customary to state ac and pulsating dc potentials in volts *rms* (root mean square), also known as volts *effective*. This convention is followed in the x-ray generator except for the ac output of the high-potential transformer and the pulsating dc output of the rectifiers. These potentials are stated in kilovolts *peak*—KV(P)—for the following reasons. It is the peak x-ray tube potential that determines the wavelength profile of the continuous x-ray spectrum (Section 1.7), and it is the peak potential that must exceed the excitation potential of an element if its characteristic spectrum is to be excited (Section 1.8.6). The relationship of peak and rms potentials is given by Equations (1.7) and (1.8).

In constant-potential generators, both the kilovoltmeter and milliammeter indicate substantially continuous dc, and x-ray tube power is given by

$$P_{CP} = Vi \qquad (1.14)$$

where V is x-ray tube potential (V), i is x-ray tube current (A), and P_{CP} is power (watts, W). In full-wave generators, the kilovoltmeter indicates

peak kilovolts and the milliammeter rms current; x-ray tube power is
given by

$$P_{\mathrm{FW}} = 0.707 V_{\mathrm{peak}} i_{\mathrm{rms}} \qquad (1.15)$$

For example, an x-ray tube operating at 50 kV, 50 mA from a constant-
potential generator dissipates $50{,}000 \times 0.050 = 2500$ W; the same tube
operating at the same conditions from a full-wave generator dissipates
$(0.707 \times 50{,}000)(0.050) = 1768$ W. It follows that, at the same nominal
operating conditions (kV, mA), the x-ray output is substantially lower with
full-wave than with constant-potential operation. About 1% of the x-ray
tube target power is converted to x-rays.

1.6.3.5. Stabilization

Because primary x-ray intensity is proportional to iV^2 [Equation
(1.18)], it is essential that the x-ray tube current and, especially, potential
be highly regulated, $\pm 0.05\%$ or less. All commercial x-ray spectrometers
have an ac line stabilizer and an "mA stabilizer," and some also have a
"kV stabilizer." These three types of stabilizer are shown schematically
in Figure 1.4, and are considered below in order.

A simple "constant-potential" or "isolation" transformer (not shown)
may be used between the ac line and the input terminals of the x-ray generator in laboratories where extremely wide line fluctuations are encountered. This is a transformer having a 1-to-1 turns ratio and a heavy
iron core. However, these transformers used alone provide only about
$\pm 1\%$ output regulation for $\pm 15\%$ input fluctuation. A good line stabilizer
holds the output constant at 220 V $\pm$ 0.1% or less for an input fluctuation
of 190–250 V or more, and may be electronic, electromagnetic, or a combination of both.

The "mA stabilizer" functions substantially as follows. The x-ray tube
current returning to the center tap of the high-potential transformer O
through the milliammeter U passes through the resistor V and develops a
potential across it. Fluctuations in the tube current cause corresponding
fluctuations in this potential, which are amplified and, through transformer
X (or a reactor), induce potentials in the primary circuit of the x-ray tube
filament transformer Y in such phase as to counteract the fluctuations.

The "kV stabilizer" functions similarly. The high potential output of
the power supply appears across the resistive potential divider S. A certain
small fraction of this total potential is applied to the kV stabilizer, where
it is compared with a fixed reference potential, thereby generating a dif-

ference potential. Fluctuations in the high potential across S cause corresponding fluctuations in the difference potential, which are amplified and, through transformer N (or a reactor), induce potentials in the primary circuit of the high-potential transformer in such phase as to counteract the fluctuations.

Incidentally, the fraction of the x-ray tube potential taken from the divider S may also be indicated on a dc panel voltmeter J' graduated in kV, providing an alternative means of indicating x-ray tube potential. This alternative is advantageous in that the potential indicated on J' is derived directly from the potential actually applied to the x-ray tube.

1.6.3.6. Safety and Protective Devices

In Figure 1.4, K' is a 30–60-s time-delay switch that opens when the power supply is turned off and is closed by a bimetallic strip heated by a resistor or, preferably, by a small timer motor. When the main power switch A is turned on, the bimetallic strip heater or timer motor is energized and, after 30–60 s, the time-delay switch is closed. The time-delay switch prevents the application of power to the primary of the high-potential transformer O until the x-ray tube and rectifier filaments and the mA and kV stabilizers have had time to warm up. Closing the x-ray switch K cannot energize transformer O unless switch K' has closed, and not even then unless all the other safety and protective devices D–H are closed.

Provision must be made to shut off the instrument if any difficulty develops that might present a hazard to equipment or personnel. Safety and protective devices are indicated schematically by switches D–H, shown open in Figure 1.4, but normally closed. D and E represent high- and low-pressure switches in the cooling water line of the x-ray tube. F represents the line overload relay; if any of the circuits connected to the stabilized 220-V line draw excessive current, the solenoid F' is magnetized sufficiently to open F. G represents the x-ray tube current overload relay; if the x-ray tube current becomes excessive, the solenoid G' opens G. The overload relays F and G have reset buttons on the panel, as shown. H represents the several interlocks on cabinet doors and the x-ray tube housing; if any of these are disturbed while the equipment is in operation, H opens and turns off the equipment. The fuses B will "blow" if excessive current is drawn by the line stabilizer or by any equipment connected directly to the 200-V ac line or to the outlets C, or if the protective devices fail.

Suppose that an x-ray generator like that described above, but having 100-kV instead of 50-kV output, were fitted with an x-ray tube having a target of high atomic number, say tungsten (Z 74). Suppose further that the emergent x-rays were dispersed into a spectrum, just as white light is dispersed into a spectrum by a glass prism. Suppose finally that the intensity of each wavelength present in the dispersed spectrum were measured with an x-ray detector and intensity meter. A plot of intensity *versus* wavelength would appear as shown in Figure 1.6. This spectrum is seen to consist of two superimposed spectra of distinctly different types—a continuous band of wavelengths, the *continuous spectrum*, and three series of discrete wavelengths, the *characteristic line spectrum*. These two types of spectra are considered in Sections 1.7 and 1.8, respectively.

1.7. THE CONTINUOUS SPECTRUM

The *continuous spectrum*, also known as the *continuum, white* spectrum, *general* spectrum, and *Brehmsstrahlung*, is characterized by four major features: a continuous range of wavelengths (analogous to white light), having an abrupt short-wavelength limit λ_{min}, rising to a maximum intensity $\lambda_{I_{max}}$, then gradually decreasing in intensity at longer wavelengths. $\lambda_{I_{max}}$ occurs at $\sim 1.5\lambda_{min}$ and, for most practical purposes, may be regarded

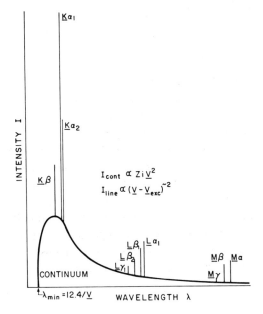

FIGURE 1.6. Typical x-ray spectrum of a heavy element, showing the countinuous spectrum and K, L, and M characteristic line spectra.

as the *effective* wavelength λ_{eff} of the continuum, that is, the single wavelength having substantially the same absorption in a given absorber as the continuum.

The continuum profile is seen in Figure 1.6 and approximated by the Kramers formula,

$$I_\lambda \propto iZ\left(\frac{1}{\lambda_{\min}} - \frac{1}{\lambda}\right)\frac{1}{\lambda^2} \tag{1.16}$$

where I_λ is the intensity at wavelength λ, $\lambda_{\min}$ is the short-wavelength limit, i is x-ray tube current, and Z is the atomic number of the x-ray tube target.

The continuum may be recorded by directing the primary x-ray beam into an x-ray spectrometer. This can be done by rotating the x-ray tube about its axis, or, more conveniently, by placing a piece of paraffin or polyethylene, or a cell of water, in the specimen chamber to scatter some of the primary x-rays into the spectrometer.

The continuum arises when high-speed electrons undergo stepwise deceleration in matter (hence the appropriate German term, *Brehmsstrahlung*, which means, literally, "braking radiation"). The principal source of high-intensity continuum is the x-ray tube. Consider an electron moving from filament to target in an x-ray tube operating at potential V. The shortest wavelength $\lambda_{\min}$ that this electron can possibly generate is emitted if on striking the target the electron decelerates to zero velocity in a single step, giving up all its energy as one x-ray photon. This wavelength is given by an equation analogous to Equation (1.6),

$$\lambda_{\min} = 12{,}396/V \tag{1.17}$$

When this equation is applied to x-ray tubes operating from full-wave supplies, V is *peak* potential (V). However, most electrons give up their energy in numerous unequal decrements ΔV, each resulting in an x-ray photon of some wavelength $\lambda = 12{,}396/\Delta V > \lambda_{\min}$. The net effect of all the electrons bombarding the target undergoing such stepwise deceleration is the generation of a continuous spectrum of wavelengths $\gtrless \lambda_{\min}$.

From Equation (1.17), the minimum potential V at which an x-ray tube can be operated if it is to generate x-rays of a specified wavelength λ is $V = 12{,}396/\lambda$. However, if λ is to be generated at high intensity, the x-ray tube must be operated at substantially higher potential.

The effect of x-ray tube target atomic number Z, current i, and potential V on the integrated continuum intensity I_{int} is shown in Figure 1.7 and expressed by the Beatty formula,

$$I_{\text{int}} = (1.4 \times 10^{-9})ZiV^2 \tag{1.18}$$

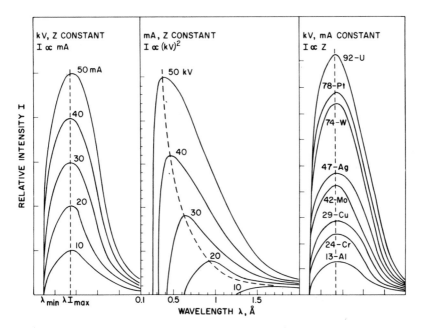

FIGURE 1.7. Effect of x-ray tube current (mA), potential (kV), and target atomic number (Z) on the continuous x-ray spectrum.

where I, i, and V are in watts, amperes, and volts, respectively. It is evident that none of these parameters substantially affects the general profile of the continuum, and *only* V affects the short-wavelength limit. A change in x-ray tube current causes a proportional change in continuum intensity because the number of electrons arriving at the target is directly proportional to the current. The atomic number of the target has much the same effect because the number of orbital electrons in each target atom is proportional to Z. X-ray tube potential has by far the most marked effect. As V increases: (1) the intensity at any wavelength already present increases because the filament electrons are accelerated to higher velocity and can undergo more decelerations; (2) λ_{min} and therefore $\lambda_{I_{max}}$ are displaced to progressively shorter wavelengths in accordance with Equation (1.17); and (3) the intensity at and near $\lambda_{I_{max}}$ increases rapidly. With full-wave rectification, the continuum pulsates forth and back in wavelength and up and down in intensity with each half-cycle; it attains its shortest λ_{min} and $\lambda_{I_{max}}$ and its highest intensity only at the potential peaks. With constant potential, the continuum remains substantially constant with time.

Incidentally, on some x-ray generators, particularly old ones, λ_{min} may *appear* to increase slightly on increase in x-ray tube current. However, this is actually caused by a slight decrease in x-ray tube potential on increase in current due to "loading" or poor regulation of the high-potential supply.

For practical purposes, continuum is excited only by electrons. Continuum excited by ions is weaker than by electrons by a factor equal to the square of the ratio of the electron and ion masses. For example, for protons, the continuum intensity is $\sim(1/1800)^2$ that for electrons, and is successively less for deuterons, α-particles, and heavier ions. Continuum is not generated by secondary excitation, that is, by x-rays or γ-rays, because photons do not undergo stepwise loss of energy (Section 2.1.2). However, continuum may appear in secondary x-ray spectra due to scatter of incident primary continuum by the specimen.

In x-ray fluorescence (that is, x-ray-tube-excited) spectrometric analysis, the continuum provides the principal source of specimen excitation and (scattered) background. In electron-excited x-ray spectrometry, the much higher continuum intensity is a serious limitation on minimum detection limit. The continuum may provide incident radiation for polychromatic and monochromatic x-ray absorption methods, including absorption-edge spectrometry. A selected continuum wavelength may be used for the background-ratio technique. These applications are discussed in later chapters.

1.8. THE CHARACTERISTIC LINE SPECTRUM

1.8.1. Atomic Structure

X-ray spectrometry is a method of analysis for chemical elements. A *chemical element* is a substance that cannot be separated into simpler substances by ordinary chemical means; that is, an element is not "made out of" any other substances. At present, 103 chemical elements are known. Just as all Arabic numbers consist of various combinations of the 10 digits and all words in the English language consist of various combinations of the 26 letters of the alphabet, all substances in the universe consist of various combinations of the 100-odd chemical elements.

The characteristic x-ray spectral lines of an element arise from electron transitions within the atoms of that element. An *atom* is the smallest particle of an element that can undergo—at least in principle—all the chemical actions characteristic of that element. It follows that an understanding of the nature and origin of x-ray spectra requires some knowledge of atomic structure.

For our purpose, atoms may be regarded as consisting of various numbers of three subatomic particles—electrons, protons, and neutrons.

An *electron e⁻* may be regarded as an "atom" of electric charge and has a rest mass m_e of 9.1×10^{-28} g and *negative* electric charge e of 4.8×10^{-10} esu. A *proton p⁺* has a mass of 1.67×10^{-24} g, ~1800 times the electron mass, and a *positive* charge equal to the electron charge. A *neutron n* may be regarded as a combined electron–proton pair; it has a mass about the same as the proton and no charge.

Each of the 100-odd elements is characterized by the number of protons in its atoms. This number varies from one for the lightest element (hydrogen) to 103 for the heaviest presently known (lawrencium) and is known as the *atomic number Z* of the element. Atoms, as distinguished from ions, are electrically neutral, so it follows that the atoms of each element also contain Z electrons. However, although all atoms of a specified element contain exactly Z protons and Z electrons, the number of neutrons in the atoms of an element varies from ~Z for the lightest elements to ~$1.5Z$ for the heaviest, and atoms of the same element may contain different numbers of neutrons (see the discussion of isotopes below).

Because the mass of the electron is very small (~1/1800) relative to that of the proton, the total number of protons and neutrons in an atom is a measure of its relative mass and is known as the *mass number M*. The *atomic weight A* of an element is the exact mass of its atom relative to the mass of a specific isotope of carbon, $^{12}_{6}C$ (see below), assigned the mass 12.0000. Atomic weights, unlike mass numbers, are nonintegral numbers for three reasons: (1) They include the mass of the electrons. (2) Most naturally occurring elements are mixtures of two or more isotopes (for most elements, in constant proportions), each having its own atomic weight; the atomic weight of the element is the weighted mean of the atomic weights of its isotopes. (3) Finally, the mass of an atom is slightly less than the sum of the masses of its protons, neutrons, and electrons. The mass defect is conserved as the binding energy holding the subatomic particles together.

Structurally, every atom consists of a dense central *nucleus* containing all of its Z protons and all of its $M - Z$ neutrons, with all of its Z electrons revolving around this nucleus in orbits, like planets around the sun. The electron orbits are grouped in *shells* designated K, L, M, N, etc., in order of increasing distance from the nucleus. The closer a shell lies to the positive nucleus, the more tightly are its negative electrons bound in the atom. The electrons in each shell are classified further with respect to angular momentum and direction of spin. Each of these parameters—shell, momentum, spin—is designated by a *quantum number*, which may have only certain values, and no two electrons in an atom may have an identical set of quantum numbers (Pauli exclusion principle).

TABLE 1.2. Electron–Orbital Configurations of the Atoms of the Chemical Elements

Period	Atomic number	Element	Shell:	K	L		M			N				O	
										Number of electrons in each nl orbital					
			nl:	$1s$	$2s$	$2p$	$3s$	$3p$	$3d$	$4s$	$4p$	$4d$	$4f$	$5s$	$5p$
1	1	H		1											
	2	He		2											
2	3	Li		K (2)	1										
	4	Be			2										
	5	B			2	1									
	6	C			2	2									
	7	N			2	3									
	8	O			2	4									
	9	F			2	5									
	10	Ne			2	6									
3	11	Na		K (2)	L (8)		1								
	12	Mg					2								
	13	Al					2	1							
	14	Si					2	2							
	15	P					2	3							
	16	S					2	4							
	17	Cl					2	5							
	18	Ar					2	6							
4A	19	K		K (2)	L (8)		2	6		1					
	20	Ca					2	6		2					
	21	Sc					2	6	1	2					
	22	Ti					2	6	2	2					
	23	V					2	6	3	2					

Group	Z	Element	K (2)	L (8)	3s	3p	3d	M (18)	4s	4p	4d	5s	5p
4B	24	Cr	2	8	2	6	4		2				
	25	Mn	2	8	2	6	5		2				
	26	Fe	2	8	2	6	6		2				
	27	Co	2	8	2	6	7		2				
	28	Ni	2	8	2	6	8		2				
	29	Cu	2	8	2	6	9		2				
	30	Zn	2	8	2	6	10		2				
	31	Ga	2	8	2	6	10		2	1			
	32	Ge	2	8	2	6	10		2	2			
	33	As	2	8	2	6	10		2	3			
	34	Se	2	8	2	6	10		2	4			
	35	Br	2	8	2	6	10		2	5			
	36	Kr	2	8	2	6	10		2	6			
5A	37	Rb	2	8				(18)	2	6		1	
	38	Sr	2	8				(18)	2	6		2	
	39	Y	2	8				(18)	2	6	1	2	
	40	Zr	2	8				(18)	2	6	2	2	
	41	Nb	2	8				(18)	2	6	4	1	
	42	Mo	2	8				(18)	2	6	5	1	
	43	Tc	2	8				(18)	2	6	5	2	
	44	Ru	2	8				(18)	2	6	7	1	
	45	Rh	2	8				(18)	2	6	8	1	
	46	Pd	2	8				(18)	2	6	10		
5B	47	Ag	2	8				(18)	2	6	10	1	
	48	Cd	2	8				(18)	2	6	10	2	
	49	In	2	8				(18)	2	6	10	2	1
	50	Sn	2	8				(18)	2	6	10	2	2
	51	Sb	2	8				(18)	2	6	10	2	3
	52	Te	2	8				(18)	2	6	10	2	4
	53	I	2	8				(18)	2	6	10	2	5
	54	Xe	2	8				(18)	2	6	10	2	6

TABLE 1.2 *(continued)*

Period	Atomic number	Element	Shell: K (2)	L (8)	M (18)	4s	4p	4d	4f	5s	5p	5d	5f	6s	6p	6d	7s
6A	55	Cs				2	6	10		2	6			1			
	56	Ba				2	6	10		2	6			2			
	57	La				2	6	10		2	6	1		2			
	58	Ce				2	6	10	2	2	6			2			
	59	Pr				2	6	10	3	2	6			2			
	60	Nd				2	6	10	4	2	6			2			
	61	Pm				2	6	10	5	2	6			2			
	62	Sm				2	6	10	6	2	6			2			
	63	Eu				2	6	10	7	2	6			2			
	64	Gd				2	6	10	7	2	6	1		2			
	65	Tb				2	6	10	9	2	6			2			
	66	Dy				2	6	10	10	2	6			2			
	67	Ho				2	6	10	11	2	6			2			
	68	Er				2	6	10	12	2	6			2			
	69	Tm				2	6	10	13	2	6			2			
	70	Yb				2	6	10	14	2	6			2			
	71	Lu				2	6	10	14	2	6	1		2			
	72	Hf				2	6	10	14	2	6	2		2			
	73	Ta				2	6	10	14	2	6	3		2			
	74	W				2	6	10	14	2	6	4		2			
	75	Re				2	6	10	14	2	6	5		2			
	76	Os				2	6	10	14	2	6	6		2			
	77	Ir				2	6	10	14	2	6	7		2			
	78	Pt				2	6	10	14	2	6	9		1			

Number of electrons in each *nl* orbital

Group	Z	Element	K (2)	L (8)	M (18)	N (32)	5s	5p	5d	5f	6s	6p	6d	7s
6B	79	Au	2	8	18	32	2	6	10		1			
	80	Hg	2	8	18	32	2	6	10		2			
	81	Tl	2	8	18	32	2	6	10		2	1		
	82	Pb	2	8	18	32	2	6	10		2	2		
	83	Bi	2	8	18	32	2	6	10		2	3		
	84	Po	2	8	18	32	2	6	10		2	4		
	85	At	2	8	18	32	2	6	10		2	5		
	86	Rn	2	8	18	32	2	6	10		2	6		
7A	87	Fr	2	8	18	32	2	6	10		2	6		1
	88	Ra	2	8	18	32	2	6	10		2	6		2
	89	Ac	2	8	18	32	2	6	10		2	6	1	2
	90	Th	2	8	18	32	2	6	10		2	6	2	2
	91	Pa	2	8	18	32	2	6	10	2	2	6	1	2
	92	U	2	8	18	32	2	6	10	3	2	6	1	2
	93	Np	2	8	18	32	2	6	10	4	2	6	1	2
	94	Pu	2	8	18	32	2	6	10	6	2	6		2
	95	Am	2	8	18	32	2	6	10	7	2	6		2
	96	Cm	2	8	18	32	2	6	10	7	2	6	1	2
	97	Bk	2	8	18	32	2	6	10	8	2	6	1	2
	98	Cf	2	8	18	32	2	6	10	10	2	6		2
	99	Es	2	8	18	32	2	6	10	11	2	6		2
	100	Fm	2	8	18	32	2	6	10	12	2	6		2
	101	Md	2	8	18	32	2	6	10	13	2	6		2
	102	No	2	8	18	32	2	6	10	14	2	6		2
	103	Lr	2	8	18	32	2	6	10	14	2	6	1	2
7B	104													
	105													

Table 1.2 gives the electron–orbital configurations of the atoms of all known elements. Table 1.3 lists the quantum numbers, their significance, the allowed values that each may assume in an atom, the allowed changes that each may undergo during an intraatomic transition (*selection rules*), and the total number of electrons in an atom that may have that quantum number. The magnetic quantum number m has no significance for our purposes and is included only for completeness. Figure 1.10 lists the n, l, and j quantum numbers for all possible electrons in the first four atom shells (K, L, M, N) and in the first five subgroups of the O shell. It is evident that on the basis of these n, l, and j quantum numbers, atomic electrons fall into *subgroups*, *orbitals*, or *energy levels*—one K, three L, five M, seven N, etc. The number of subgroups in each shell corresponds to the allowed values of j in that shell—1/2, 3/2, 5/2, 7/2, etc. (Table 1.3). Figure 1.10 also lists the number of electrons in each subgroup for gold atoms, and these numbers happen also to be the maximum numbers of electrons allowed in each subgroup in any atom (Table 1.3, column 6). Gold atoms also have one electron in the PI subgroup (not shown).

An element is a substance all atoms in which contain the same number of protons in their nuclei and therefore have the same atomic number Z. However, as mentioned above, atoms of the same element may contain different numbers of neutrons and therefore have different mass numbers M. An *isotope* is a form of a chemical element all atoms of which contain both the same number of protons and the same number of neutrons in their nuclei, and therefore have both the same atomic number and mass number. A *radioactive isotope* or *radioisotope* is an isotope that undergoes radioactive decay, that is, that spontaneously changes, at a certain rate, into an isotope of a different element or into a different isotope of the same element. For example, carbon (atomic number 6) has eight known isotopes: $^{9}_{6}C$, $^{10}_{6}C$, $^{11}_{6}C$, $^{12}_{6}C$, $^{13}_{6}C$, $^{14}_{6}C$, $^{15}_{6}C$, and $^{16}_{6}C$. All these isotopes have 6 protons in their atomic nuclei, but 3, 4, 5, 6, 7, 8, 9, and 10 neutrons, respectively. Of these isotopes, $^{12}_{6}C$ and $^{13}_{6}C$ are stable, the others radioactive; $^{12}_{6}C$, $^{13}_{6}C$, and $^{14}_{6}C$ occur in nature, the others are produced only by artificial transmutation.

The correct symbol for an isotope gives the symbol of the element, atomic number Z, and mass number M in the form $^{M}_{Z}El$, for example, $^{80}_{35}Br$. Unfortunately, the older notation $_{Z}El^{M}$ ($_{35}Br^{80}$) is also still commonly used. The former symbol is preferable because it leaves the right superscript and subscript positions open for, respectively, ionic charge and number of atoms in a molecule or radical; for example, a singly charged molecule of the bromine isotope above would be $^{80}_{35}Br^{+}_{2}$. When the discussion warrants, either the atomic or mass number may be omitted—^{80}Br or $_{35}Br$; the old

TABLE 1.3. Quantum Numbers

Symbol	Name	Significance	Allowed values, Common designations	Selection rules	Maximum number of electrons
n	Principal	Principal binding energy; indicates shell	1, 2, ..., n $K, L, M, N, \ldots$ [a]	$\Delta n \neq 0$	$2n^2$
l	Azimuthal	Orbital angular momentum; determines shape of orbital	0, 1, ..., $(n-1)$ $s, p, d, f, g, h, \ldots$ [b]	$\Delta l = \pm 1$	$2(2l+1)$
m	Magnetic	Projection of angular momentum (l) on magnetic field; indicates orientation of orbital in magnetic field	$-l, \ldots, 0, \ldots, +l$	—	
s	Spin	Direction of spin—clockwise or counterclockwise	$\pm \frac{1}{2}$	—	
j	Inner precession	Vector sum of l and s	$l \pm \frac{1}{2}$, except $j \neq 0 - \frac{1}{2}$	$\Delta j = \pm 1$ or 0	$2j + 1$

[a] So designated by Barkla (1911) to allow for possible shells inside the innermost shell known at the time.
[b] From the terminology of optical spectroscopy: *sharp, principal, diffuse, fundamental*; azimuthal quantum numbers 5 (*g*) and 6 (*h*) will occur only in superheavy elements as yet undiscovered, that is, not yet produced in the laboratory.

form of the former symbol is Br^{80}. Also, for simplicity, some authors use symbols of the form 35-Br and Br-80. In this book, the preferred notation noted above is used.

1.8.2. Origin and Nature

The characteristic x-ray line spectrum consists of series of discrete wavelengths characteristic of the emitting element. These x-ray *spectral lines* should record in x-ray spectrometers as extremely sharp lines, but actually have considerable breadth. However, this is due to imperfections in the dispersion, detection, and readout components.

The x-ray line spectrum of an element arises when electrons are expelled from the inner orbits (*K, L, M*) of its atoms and electrons from orbits

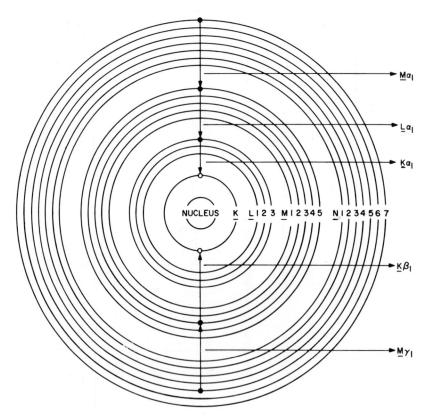

FIGURE 1.8. Origin of x-ray spectra; two of many possible series of spontaneous electron transitions that may follow creation of a *K*-shell vacancy.

farther out fall into the vacancies. Each such transition constitutes an energy loss, which appears as an x-ray photon. For example, on creation of a *K*-shell vacancy in a heavy atom, a succession of spontaneous electron transitions follows; each fills a vacancy in a lower level with resultant emission of an x-ray photon, but also creates a vacancy in a level farther out. Figure 1.8 shows two of the many possible series of transitions. The result of such processes in large numbers of atoms is the simultaneous emission of the *K*, *L*, and *M* series (see below) of x-ray spectra of that element.

Figure 1.9 shows the electron transitions that give rise to the principal x-ray spectral lines. Figure 1.10 shows schematically all *K*, *L*, *M*, and *N* levels and the first five *O* levels, giving for each the *n*, *l*, and *j* quantum numbers and their optical and x-ray notation. The up and down arrows show the electron transitions giving rise, respectively, to vacancies in each level and to all *K*, *L*, and *M* lines of relative intensity ∼1 or more. Each of these transitions is labeled with the symbol of the corresponding absorption edge (see below) or resulting line. The weakest of these lines are seldom observed in x-ray fluorescence spectrometry. However, in electron-probe microanalysis, where electron excitation and curved-crystal optics combine to give high efficiency, some of these lines may appear. Since these electron transitions correspond precisely to the difference in energy between two atomic orbitals, the emitted x-ray photon has energy characteristic of this

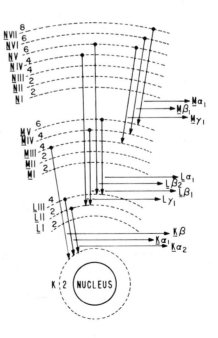

FIGURE 1.9. Origin of x-ray spectra; intraatom electron transitions responsible for the principal x-ray spectral lines of analytical interest. The numeral following each orbital symbol (*K*, *L*I, *L*II, etc.) is the number of electrons in that orbital when full.

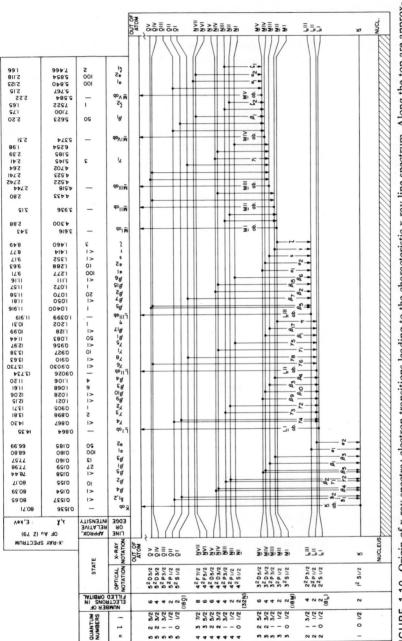

FIGURE 1.10. Origin of x-ray spectra; electron transitions leading to the characteristic x-ray line spectrum. Along the top are approximate relative spectral-line intensities and, for gold, wavelengths (Å) of lines and absorption edges, photon energies (keV) of lines, and excitation potentials (kV). If the atomic energy levels (K, LI, LII, LIII, MI, etc.) were plotted on a vertical logarithmic energy scale graduated in kilo-electron-volts, this figure would be an *energy-level diagram* for gold. On such a diagram, photon energies of lines and excitation potentials could be read from the arrows indicating electron transitions. $L\beta_9$, $L\beta_{10}$, $L\beta_{17}$, Ls, and Lt are forbidden lines. Gold has one electron in the *PI* orbital (not shown).

difference and thereby of the atom and element. At the top of the figure are approximate relative intensities, wavelengths, and photon energies of the lines for gold. The line transitions are substantially instantaneous, occurring within 10^{-12}–10^{-14} s of the creation of the electron vacancy.

Incidentally, physicists would prefer Figure 1.10 with the arrows pointing in the opposite directions. For example, the figure shows the $K\alpha_1$ line as originating from an *electron* transition from the LIII orbital to the K orbital; physicists regard it as originating from an *atom* transition from the K state (or K quantum state) to the LIII state.

X-ray spectral lines are grouped in series K, L, M, N; all lines in a series result from electron transitions *from* various higher orbitals *to* the indicated shell (physicists would say *atom* transitions from the same initial state to various final states).

Electron transitions cannot occur from *any* higher to *any* lower orbital. Only certain transitions are "permitted" by the selection rules, which are given in column 5 of Table 1.3: $\Delta n \neq 0$, $\Delta l = \pm 1$, $\Delta j = \pm 1$ or 0. Thus, an $LIII \rightarrow LI$ transition would violate the first rule, an $MIV(l = 2) \rightarrow LI(l = 0)$ transition would violate the second rule, and an $MV(j = \frac{5}{2}) \rightarrow LII(j = \frac{1}{2})$ transition would violate the third rule. X-ray spectral lines that "obey" the selection rules are known as *diagram* lines, those that don't as *forbidden* lines. An example of a forbidden line in Figure 1.10 is $L\beta_9$, $MV \rightarrow LI$ ($l = 2$ to $l = 0$, and $j = \frac{5}{2}$ to $j = \frac{1}{2}$).

A third type of x-ray line—*satellite* lines or *nondiagram* lines—arise in atoms having two or more inner-shell vacancies. A specified electron transition emits a slightly different wavelength in such atoms than in singly ionized atoms. An example is the Si $SK\alpha_3$ line, which arises from an $LIII \rightarrow K$ transition in an atom having both K and L vacancies; its wavelength is 7.077 Å, compared with 7.125 Å for diagram Si $K\alpha$. The origin of doubly ionized atoms is discussed in Section 2.4.

A final type of x-ray spectral "line" is the spectral *band*. A specified orbital in a specified atom usually has an extremely sharp energy level, and electron transitions originating from such orbitals emit sharp lines having breadth at half-height $\sim$0.001 Å. However, the energy levels of the outermost orbitals of atoms in solids and liquids may be broadened under the influence of ligand (combined) and other neighboring atoms. Electron transitions arising from such orbitals give rise to broad bands rather than sharp lines. The K spectra of the lightest elements (atomic number 4–9, beryllium to fluorine) are of this type. Because band spectra arise from electron transitions from outer orbitals, they may be influenced by the chemical environment of the atoms of their origin, especially by oxidation number and by identity and bond type of ligand atoms and radicals. Consequently, band spectra provide a valuable means for investigation of chemical bonding.

1.8.3. Notation

In the Siegbahn notation system, the symbol of an x-ray spectral line (for example, Ni $K\alpha_1$, Au $L\beta_2$, W $L\eta$, U Ll) consists of (1) the symbol of the chemical element; (2) the symbol of the series (K, L, M, N, etc.; the line originates from the filling of a vacancy in, or an electron transition to, the indicated shell); and (3) a lowercase Greek letter, usually with a numerical subscript, or a lowercase italic letter denoting the particular line in the series.

The α_1 line is usually the strongest in the series, and α lines arise from $\Delta n = 1$ transitions. The β_1 line is usually the second strongest, except that $K\alpha_2$ is stronger than $K\beta_1$. The β and γ lines usually arise from $\Delta n = 1$ or 2 transitions. Aside from these conventions, the notation is not systematic. However, the line resulting from a given transition in any element is always given the same symbol.

The symbol $K\alpha$, sometimes $K\alpha_{1,2}$, indicates the $K\alpha_1\alpha_2$ pair and has wavelength equal to the average of the individual lines weighted for their relative intensities, $K\alpha_1 : K\alpha_2 = 2 : 1$,

$$\lambda K\alpha = \frac{2\lambda K\alpha_1 + \lambda K\alpha_2}{3} \tag{1.19}$$

The symbol $K\beta$ indicates the $K\beta_1\beta_3$ pair.

Diffraction orders (Section 2.3.1) are indicated in any of several ways; for example, second-order Cu $K\alpha$ may be indicated Cu $K\alpha(2)$ (the notation used in this book), Cu $K\alpha(n = 2)$, Cu $2K\alpha$, or Cu $K\alpha$ II.

Another system of notation for x-ray spectral lines is the level-designation system, which gives the *atomic* transitions resulting in the lines. Thus, the $K\alpha_1$ line may be designated $K,LIII$ (or $K,L3$), indicating that it arises from an *atom* transition from the K to the LIII *state*—that is, an *electron* transition from the LIII to the K *orbital*. Similarly, the $L\alpha_1$ line may be designated $LIII,MV$ (or $L3,M5$), etc. Some of the less prominent L lines and most of the M and N lines have not been assigned Greek or italic symbols. These lines are indicated only by their level designations.

The prefix S before the symbol of an x-ray spectral line, for example, Al $SK\alpha_3$, indicates that it is a satellite or nondiagram line rather than a diagram line (Section 1.8.2).

Taking note of the "archaic and unsystematic" state of the present system of notation for x-ray spectral lines, Jenkins has proposed what appears to be a simple, logical notation system. For the series, he would retain the K, L, M, etc., designations for all lines arising from electron transitions to orbitals having principal quantum numbers $n = 1, 2, 3$, etc., respectively. Within a series, he would designate as α, β, γ, δ, etc., all lines arising from electron transitions for which $\Delta n = 1, 2, 3, 4$, etc., respectively. For lines having the same n and Δn transitions, he would add numerical subscripts 1, 2, 3, etc., for electron transitions from the I, II, III, etc., levels, respectively, of the orbital involved. An additional subscript s or f would indicate satellite and forbidden lines, respectively.

1.8.4. Wavelength and Photon Energy

The photon energy of an x-ray spectral line is the difference in energy ΔE between the initial and final energy levels involved in the intraatomic electron transition from which that line originates. For example, in Figure 1.10, for gold:

$$K \text{ state} \;\rightarrow\; L\text{III state} \;+\; K\alpha_1$$
$$80.71 \qquad\qquad 11.92 \qquad 68.80 \text{ keV}$$

$$K \text{ state} \;\rightarrow\; M\text{III state} \;+\; K\beta_1$$
$$80.71 \qquad\qquad 2.74 \qquad 77.98 \text{ keV}$$

$$L\text{III state} \;\rightarrow\; M\text{V state} \;+\; L\alpha_1$$
$$11.92 \qquad\qquad 2.22 \qquad 9.71 \text{ keV}$$

The slight discrepancies in the energies are not real and arise from small errors in rounding off the table values. Because of such relationships, x-ray spectrometry is an extremely valuable method for the study of atomic structure.

Wavelength is inversely proportional to photon energy and is calculated by Equation (1.6). Energy differences among the K, L, M, N, and O shells of atoms are such that the K, L, and M x-ray spectra of the elements lie in the photon energy region 0.1–100 keV and the wavelength region 100–0.1 Å. Figure 1.11 shows the strongest K, L, and M lines of all the elements. The absorption edges and higher orders also shown are discussed later.

For a specified element, wavelengths of lines in the same series decrease (photon energies increase) as ΔE (the "length") of the energy transition increases. For example, in Figure 1.10, compare the wavelengths and photon energies of Au K lines α_2 to $\delta_{1,2}$ and Au LIII lines l to β_5.

For a specified element, wavelengths of lines decrease (photon energies increase) from series M to L to K because electrons must "fall" farther to fill vacancies successively closer to the nucleus. However, there is overlap of the wavelength (photon energy) regions of lines of the LI, LII, and LIII subseries because these orbitals lie closer in energy than do the various orbitals from which electron transitions into them originate. The same applies to lines of the MI–MV subseries. Figure 1.10 illustrates all these features for gold lines. Figure 1.15 illustrates L-subseries overlap for gold lines.

For a specified spectral line, wavelength decreases (photon energy increases) as atomic number increases because positive nuclear charge, binding of orbital electrons, and interorbital energy differences all increase with increasing atomic number. Figure 1.11 and Table 1.4, columns 2–7, illustrate this phenomenon.

However, the variation of x-ray spectral-line wavelength with atomic number is not periodic, as in optical spectroscopy, but uniformly progressive, as shown in Figure 1.11. This is because x-ray spectra originate in *inner* atomic orbitals, which are substantially—but not entirely—independent of the chemical state of the atom (see the last paragraph of Section 1.8.2). The proximity of the atom nucleus to these inner orbitals explains the strong dependence of wavelength on atomic number, as defined by the Moseley law,

$$\nu^{1/2} = (c/\lambda_{\mathrm{cm}})^{1/2} = k_1(Z - k_2) \qquad (1.20)$$

where ν is frequency (Hz), c is the velocity of light (cm/s), λ is wavelength (cm), Z is atomic number, and k_1 and k_2 are constants different for each x-ray line. For practical purposes, this equation reduces to

$$\lambda \propto 1/Z^2; \qquad Z \propto 1/\lambda^{1/2} \qquad (1.21)$$

Figure 1.12 is a Moseley law plot for some prominent lines. The curves are linear except for small deviations at high Z.

Since spectral-line wavelength varies inversely as the square of the atomic number [Equation (1.21)], the difference in wavelength between the same lines, say $K\alpha$, of adjacent elements increases as atomic number decreases; it follows that the difference in photon energy decreases as atomic number decreases. This is shown in Table 1.5. (Adjacent elements are elements that are side by side in the periodic table; that is, that differ by 1 in atomic number.) This means that wavelength (crystal) dispersion and resolution of x-ray lines of adjacent elements become easier as atomic number decreases, and energy dispersion and resolution become more difficult.

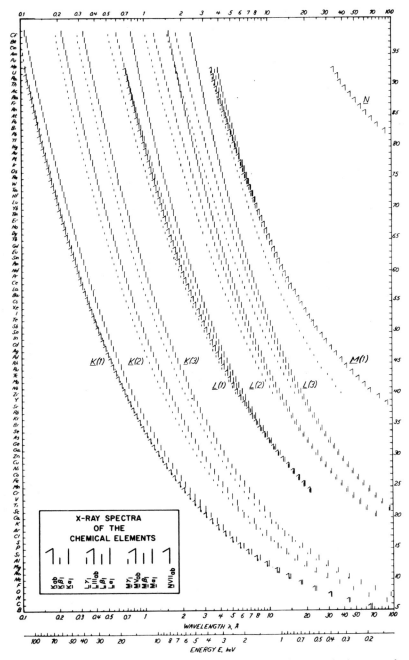

FIGURE 1.11. X-ray spectra of the chemical elements. The parenthetic numerals are diffraction orders (Section 2.3.1).

TABLE 1.4. Relationship of Atomic Number and X-Ray Spectral-Line Wavelength and Photon Energy, Absorption-Edge Wavelength, Excitation Potential, Fluorescent Yield, and Mass-Absorption Coefficient[a]

1	2	3	4	5	6	7	8	9	10	11	12	13	14	15	16	17	18	19
Atomic number Element	Spectral-line wavelength, λ (Å)			Spectral-line photon energy, E_x (keV)			Absorption-edge wavelength, λ_{ab} (Å)			Excitation potential, V_{exc} (kV)			Fluorescent yield, ω			Mass-absorption coefficient, μ/ϱ (cm²/g)		
	$K\alpha$	$L\alpha$	$M\alpha$	$K\alpha$	$L\alpha$	$M\alpha$	K	LIII	MV	K	LIII	MV	K	LIII	MV	0.5 Å	1 Å	10 Å
₄Be	113	—	—	0.11	—	—	107	—	—	0.12	—	—	—	—	—	0.2	0.5	312
₉F	18.3	—	—	0.68	—	—	17.9	—	—	0.69	—	—	0.005	—	—	0.7	5	3029
₂₀Ca	3.36	36.4	—	3.69	0.34	—	3.07	36.2	—	4.04	0.35	—	0.14	0.001	—	6	45	3028
₃₀Zn	1.44	12.3	—	8.63	1.01	—	1.28	12.0	—	9.66	1.02	—	0.46	0.01	—	20	136	10862
₄₀Zr	0.79	6.07	—	15.74	2.04	—	0.69	5.56	67.18	18.00	2.22	0.18	0.72	0.03	0.001	43	41	3305
₅₀Sn	0.49	3.60	—	25.19	3.44	—	0.42	3.15	24.90	29.19	3.93	0.50	0.85	0.08	0.002	12	80	7654
₆₀Nd	0.33	2.37	12.68	37.19	5.23	0.98	0.28	2.01	13.09	43.57	6.22	0.95	0.91	0.16	0.006	19	136	8404
₇₀Yb	0.24	1.67	8.15	52.02	7.41	1.67	0.20	1.39	8.16	61.30	8.94	1.52	0.94	0.26	0.013	31	214	3687
₈₀Hg	0.18	1.24	5.66	70.18	9.99	2.19	0.15	1.01	5.41	83.11	12.28	2.29	0.95	0.37	0.028	46	258	6784
₉₂U	0.13	0.91	3.92	97.17	13.61	3.16	0.11	0.72	3.51	115.6	17.16	3.53	0.96	0.48	0.064	72	102	9091

[a] Not shown are many other parameters related to atomic number, including absorption-edge jump ratio and difference, scatter, and coherent/incoherent scatter ratio. The lines in the mass-absorption coefficient columns indicate where discontinuities occur due to absorption edges.

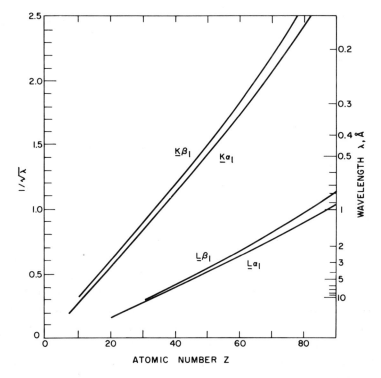

FIGURE 1.12. Moseley law diagram for selected x-ray spectral lines.

1.8.5. Intensity

At the outset, we must distinguish *relative* and *absolute* intensities of x-ray lines; this section is concerned only with the former. We must then distinguish theoretical and observed (measured) relative intensities.

The theoretical relative intensities of lines *within* a series depend on

TABLE 1.5. Wavelength and Photon-Energy Differences for $K\alpha$ Lines of Adjacent Chemical Elements

Elements	$\Delta\lambda$ (Å)	ΔE_x (keV)
$_{46}$Pd, $_{47}$Ag	0.026	~1
$_{28}$Ni, $_{29}$Cu	0.117	~0.5
$_{13}$Al, $_{14}$Si	1.213	~0.25
$_{6}$C, $_{7}$N	13	~0.1
$_{3}$Li, $_{4}$Be	127	~0.05

the relative probabilities of their respective intraatomic electron transitions. The relative intensities of the principal lines of analytical interest are typically as follows:

$K\alpha_1$	100	$L\alpha_1$	100	$M\alpha_1$	50
$K\alpha_2$	50	$L\beta_1$	75	$M\alpha_2$	50
$K\alpha_{1,2}$	150	$L\beta_2$	30	$M\alpha_{1,2}$	100
$K\beta_1$	20	$L\beta_3$	5	$M\beta_1$	100
		$L\beta_4$	3	$M\gamma_1$	5
		$L\gamma_1$	10		
		Ll	3		

In general, if a spectral line arises from an electron transition *from* an unfilled orbital, the transition probability and relative intensity of the resulting line are likely to vary with atomic number. Two examples follow. In Figure 1.10, it is seen that the $K\alpha_1$ and $K\alpha_2$ lines arise from LIII → K and LII → K transitions, respectively; these orbitals hold a maximum of 4 and 2 electrons, respectively. So one would expect the $K\alpha_1$ and $K\alpha_2$ transition probabilities and relative intensities to be 2 : 1. The L shell is filled for all elements having atomic number $\lesssim 10$ (neon) (Table 1.2), and in fact, the $K\alpha_1$: $K\alpha_2$ ratio is 2 : 1 for all elements having $Z \lesssim 10$. However, the $K\beta$ line (actually the unresolved $K\beta_1$–$K\beta_3$ lines) arises from (MIII,MII) → K transitions (Figure 1.10). The $K\alpha$: $K\beta$ ratio is 25 : 1 for $_{13}$Al, 5 : 1 for $_{29}$Cu, and 3 : 1 for $_{50}$Sn. This is because these M orbitals are not filled in the lighter elements, so the transition probability and line intensity are reduced. For the same reasons, L lines do not occur for light elements, or M lines for light and intermediate elements.

The theoretical relative intensities of the strongest lines (α_1) in the K, L, and M series of an element depend on the relative probabilities of expulsion of electrons from the respective shells of atoms of that element. In the next sections, we learn that these vacancies are usually created by high-energy electrons or x-rays, and excitation is considered in detail. For our purposes here, let the following brief summary suffice.

Suppose that a specified x-ray line series (K, L, M) of a specified element is to be excited. The exciting electrons or x-rays must be sufficiently energetic to expel electrons from that shell of that element. The closer the shell lies to the atom nucleus, and the higher the atomic number of the element, the greater is the energy required. For electron excitation, each element has K, L, and M *excitation potentials*, $V_K > V_L > V_M$, which are the minimum electron energies able to expel electrons from the respective atom shells and excite the corresponding x-ray line series of that element. For x-ray excitation, each element has K, L, and M *absorption edges*, $\lambda_K < \lambda_L < \lambda_M$, which are the longest (least energetic) x-ray wavelengths able to expel electrons from the respective shells and excite the corresponding line series of that element. The analogous parameters in terms of x-ray photon energies are the excitation potentials defined above.

If the primary (incident) x-radiation is of sufficiently short wavelength (high photon energy), it can expel electrons from any shell of atoms of the irradiated element and thereby simultaneously excites all its x-ray line series. However, such expulsions are most probable from the innermost shell that the primary x-rays can excite, the K shell in this case, so the K lines are produced at highest intensity. The K-shell excitation probability and K-line intensities increase as the primary wavelength increases (photon energy decreases) and approaches the short-wavelength side of the K-absorption edge; then both fall to zero at primary wavelength just longer than the K edge. Thereupon, L-shell excitation probability and L-line intensities predominate and increase as primary wavelength approaches the L edge, beyond which the M series is excited the same way.

If the K lines of a heavy element are being excited efficiently, the strongest lines in its three series have approximate intensity ratios $K\alpha_1 : L\alpha_1 : M\alpha \approx 100 : 10 : 1$. In the same spectral region, these first-order lines have about this same ratio for elements having the same concentration at the same excitation conditions.

For electron excitation, for a specified series of a specified element, line intensities are proportional to the $\sim$1.7th power of the difference between the electron energy and the series excitation potential of that element [Equation (1.22)].

The foregoing considerations apply to theoretical relative intensities based on electron expulsion (series) and transition (lines) probabilities. The relative intensities actually measured depend on many factors, including excitation conditions, dispersion crystal, radiation path, detector, electronic readout components, and specimen matrix and surface.

Theoretical and measured relative intensities differ for x-ray and electron excitation.

In this chapter, only general principles of excitation of characteristic x-ray spectra are considered. Excitation as applied specifically to x-ray spectrochemical analysis is considered in later chapters.

We have seen that x-ray line spectra arise when vacancies occur in the inner electron orbitals of atoms. X-ray line excitation, then, consists in creation of these vacancies in adequate numbers; this is done most commonly by irradiation of matter with sufficiently energetic electrons (*primary excitation*) or x-rays (*secondary excitation* or "fluorescence"). These two excitation modes are considered in Sections 1.8.6 and 1.8.7, respectively. Other modes are described briefly in Section 1.8.8.

1.8.6. Electron (Primary) Excitation

In this mode of excitation, the specimen to be excited is bombarded by sufficiently energetic electrons. This is the mode applied in x-ray tubes, electron-probe microanalyzers, and spectrometers having electron-gun or β-radioisotope excitation. The measure of the energy (velocity) the bombarding electrons must have if they are to expel orbital electrons from atoms of an element is the x-ray *excitation potential, critical potential, critical energy,* or *critical absorption energy.*

In this book, the general symbol for excitation potential is V_{exc}; the symbols for excitation potentials of specific orbitals have the form V_K, V_{LIII}, etc., and V_{CuK}, V_{AgLIII}, etc. The unit is the volt (V) or kilovolt (kV). Excitation potentials (V, kV) are numerically equal to electron energies in electron volts (eV) and kilo-electron-volts (keV).

The excitation potential is the minimum bombarding electron energy (keV) that can expel an electron from a specified orbital in an atom of a specified element. It is the minimum electron energy (keV) that can excite a specified x-ray line series of a specified element. In practical terms, it is the minimum x-ray tube operating potential that can accelerate electrons from the x-ray tube filament to target with sufficient velocity to expel electrons from a specified orbital in the target atoms. For full-wave rectification, it is the peak potential that must exceed the excitation potential. The K, L, and M excitation potentials of gold are represented in Figure 1.10 by the arrows directed upward from each level, and the numerical values (kV) are given at the top of the figure.

Each element has as many excitation potentials as it has subgroups: one K (V_K), three L (V_{LI}, V_{LII}, V_{LIII}), five M (V_{MI}, V_{MII}, etc.). For each element, the excitation potential increases for levels progressively closer to the nucleus: $V_K > V_{LI} > V_{LII} > V_{LIII} > V_{MI} > \cdots$. This is because the closer a negative electron lies to the positive nucleus, the more firmly it is bound and the greater is the energy required to expel it. For each level, say V_K, excitation potential increases with increasing atomic number. This is

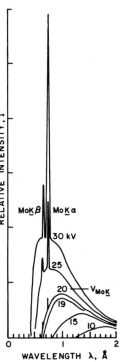

FIGURE 1.13. Primary excitation of the Mo K spectrum illustrating the Mo K excitation potential.

because as atomic number increases, the number of protons in the atom nucleus increases, and all electrons are more firmly bound. These phenomena are illustrated in Table 1.4, columns 11–13.

Figure 1.13 illustrates primary excitation using the Mo K spectrum as an example, but the principles are general. The figure shows spectra from an x-ray tube having a molybdenum target and operating at various potentials. Below the Mo K excitation potential, $V_{MoK} = 20$ kV, only continuum is generated. At 20 kV, very weak Mo K lines appear superimposed on the continuum. At higher potentials, Mo K-line intensity increases sharply.

In general,

$$I_K \propto i(V - V_K)^{\sim 1.7} \qquad (1.22)$$

where I_K is the intensity of a specific K line, V_K is K-excitation potential (kV), V is x-ray tube potential (kV, peak or constant potential), and i is x-ray tube current (mA). The quantity $(V - V_K)$ is the *overpotential* (or "overvoltage"); however, this term is also applied to the ratio V/V_K. For full-wave rectification, at $V \gtrsim 3V_K$, the exponent becomes ~ 1. Analogous proportionalities apply to L and M series lines.

In practice, for high line intensity, the x-ray tube should be operated at $V \lesssim 1.5V_{exc}$, and preferably $2\text{–}3V_{exc}$, insofar as is feasible.

An electron that has enough energy to expel a Z K electron obviously can also expel any Z L or Z M electron. However, an electron having just enough energy to expel, say, Z LII electrons, can also expel LIII, M, and N electrons, but not LI or K electrons. It follows that all spectral lines appear simultaneously that result from electron transitions to the innermost level excited and to all levels farther out. Even if only, say, LII vacancies were created by the incident electrons, vacancies farther out would result from electron transitions into the LII level (Figures 1.8 and 1.10), insofar as permitted by the selection rules.

In general, all lines of a given *level* appear at once. The exception is that lines arising from transitions to that level from levels very far from the nucleus do not appear if the atom is too small to have any electrons in that outer level.

Ordinarily, the L lines of the target would be substantially less intense than the K lines. However, the overpotential (above) can drastically alter this relationship. For example, in a thin-window rhodium-target tube operating at 45 kV(CP), the K lines (V_K 23.2 kV, $V/V_K \approx 2$) contribute $\sim 10\%$ of the total emitted flux, the L lines (V_{LIII} 3.0 kV, $V/V_{LIII} \approx 15$) contribute $\sim 25\%$.

1.8.7. X-Ray (Secondary, "Fluorescence") Excitation

1.8.7.1. Principles

In this mode of excitation, the specimen to be excited is irradiated by sufficiently energetic x-ray (or γ-ray) photons. This is the mode applied in x-ray fluorescence spectrometry, where the excitation x-rays come from x-ray tubes, secondary radiators ("fluorescers"), or radioisotope sources. The measure of the energy the x-rays must have if they are to expel orbital electrons from atoms of an element is the x-ray *absorption-edge* wavelength.

In this book, the general symbol for absorption-edge wavelength is λ_{ab}; the symbols for absorption edges of specific orbitals have the form $\lambda_{K_{ab}}$, $\lambda_{LIII_{ab}}$, etc., and $\lambda_{CuK_{ab}}$, $\lambda_{AgLIII_{ab}}$, etc. The unit is the angstrom (Å).

X-ray absorption and absorption edges are discussed in detail in Section 2.1. For our purpose here, let the following brief summary suffice. Figure 1.14 shows the x-ray absorption curve of uranium. The curve shows that, in general, the mass-absorption coefficient—that is, the x-ray "stopping power"—of a specified absorber (uranium) decreases as wavelength de-

FIGURE 1.14. X-ray absorption curve for uranium, showing x-ray "stopping power" of uranium as a function of x-ray wavelength. As wavelength decreases (photon energy increases) x-ray penetration increases and the stopping power of the absorber (uranium) decreases, except at the absorption edges.

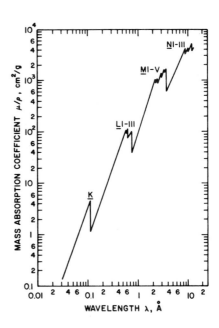

creases. This is to be expected because the shorter the wavelength, the greater are the photon energy and, therefore, the penetrating power. However, the curve shows several abrupt discontinuities; these are known as—and look like!— absorption *edges* and occur at those wavelengths just capable of expelling electrons from the several orbitals of uranium atoms, as indicated on the figure. Only x-rays having wavelength shorter (photon energy greater) than a specified absorption edge are sufficiently energetic to expel electrons from the corresponding orbital. X-rays having wavelength in the region just shorter than the absorption edge are most efficient in exciting the corresponding x-ray line series.

The absorption edge is the longest x-ray wavelength (minimum photon energy) that can expel an electron from a specified orbital in an atom of a specified element. It is the longest x-ray wavelength that can excite a specified x-ray line series of a specified element. The K, L, and M absorption edges of gold are represented in Figure 1.10 by the arrows directed upward from each level, and the numerical values (Å) are given at the top of the figure.

Each element has as many absorption edges as it has subgroups (and as it has excitation potentials): one K ($\lambda_{K_{ab}}$), three L ($\lambda_{LI_{ab}}$, $\lambda_{LII_{ab}}$, $\lambda_{LIII_{ab}}$), five M ($\lambda_{MI_{ab}}$, $\lambda_{MII_{ab}}$, etc.). For each element, the absorption-edge wavelength decreases for levels progressively closer to the nucleus: $\lambda_{K_{ab}} < \lambda_{LI_{ab}} < \lambda_{LII_{ab}} < \lambda_{LIII_{ab}} < \lambda_{MI_{ab}} < \lambda_{MII_{ab}} < \cdots$. For each level, say $\lambda_{K_{ab}}$, absorption-edge wavelength decreases with increasing atomic number. These relationships are analogous to those cited for excitation potentials (Section 1.8.6) and have the same explanation. These phenomena are illustrated in Table 1.4, columns 8–10.

An x-ray photon that has enough energy to expel a ZK electron obviously can also expel any ZL or ZM electron. However, a photon having wavelength between $\lambda_{ZLI_{ab}}$ and $\lambda_{ZLII_{ab}}$ can expel LII, LIII, M, and N electrons from atoms of element Z, but not LI or K electrons. The third and second paragraphs from the end of Section 1.8.6 regarding just which line series do and do not appear simultaneously apply here as well.

1.8.7.2. Relationship of Absorption Edges and Spectral-Line Series

Figure 1.15, which complements Figure 1.10, shows all K, L, and M absorption edges of gold and the principal lines associated with each. The figure is largely self-explanatory, but some features warrant discussion.

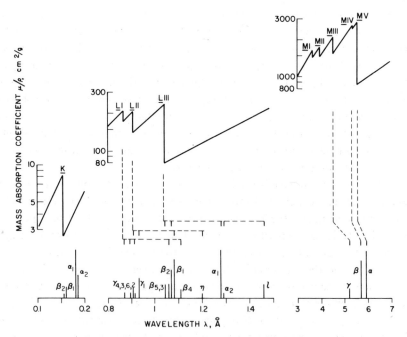

FIGURE 1.15. Relationship of x-ray absorption edges and spectral-line series for gold (correlate with Figure 1.10).

It must be reemphasized that only photons having wavelength equal to or less than that of an absorption edge can excite the associated lines.

Figure 1.15 shows that the lines of each series of an element occur on the long-wavelength side of the corresponding absorption edge where absorption is relatively low. It follows that an element is relatively transparent to its own spectral lines. It also follows that lines of a given series of an element cannot excite that series of that element, although they can excite a series having an absorption edge at longer wavelength. Thus, barium K lines excite barium L lines, but not barium K lines.

No line in a series can have wavelength shorter than that of the series edge. For example, in Figure 1.15, although some LI and LII lines are shorter than the LIII edge, no LIII line is shorter than the LIII edge. Also, in Figure 1.11, the $L\gamma_1$ lines on the short side of the LIII edges might be construed as an error in the figure. However, the $L\gamma_1$ line is associated with the LII edge (not shown), not the LIII edge.

The reason lines must have less energy—longer wavelength—than their absorption edges is as follows. The wavelength of an absorption edge is equivalent to the energy *absorbed* in creating an inner-orbital vacancy by lifting an electron from that level clear out of the atom to the conduction band or free state. The wavelength of a spectral line is equivalent to the energy *emitted* on filling that vacancy; this is effected by an electron transition, not from outside the atom, but from an outer orbital. Thus, less energy is emitted than was absorbed. The farther out the orbital from which the electron falls, the more nearly equivalent the transition is to the expulsion, and the closer the line lies to the absorption edge. This can be shown in Figure 1.10 by comparing in each series the transition for the shortest line with those for the other lines. The comparison of x-ray and optical absorption in Section 2.1.5 is relevant to this discussion.

1.8.7.3. Excitation with Polychromatic X-Rays

In an x-ray fluorescence spectrometer, the primary beam is not monochromatic, but consists of the continuum and the line spectrum of the x-ray tube target element. Only that part of the primary beam having wavelength shorter than the absorption edge of an element contributes to the excitation of the spectrum of that element. Figure 1.16 shows primary x-rays from an x-ray tube irradiating a molybdenum specimen external to the tube. Figure 1.17 shows the primary continuous x-ray spectrum at each of five x-ray tube potentials; the molybdenum K absorption edge, $\lambda_{\mathrm{Mo}K_{\mathrm{ab}}} = 0.62$ Å, is also shown.

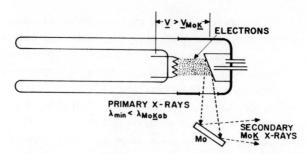

FIGURE 1.16. Arrangement for excitation of the molyb-denum K spectrum. For primary (electron) excitation, the Mo specimen is placed on the x-ray tube target. For secondary (x-ray) excitation ("fluorescence"), the Mo specimen is placed outside the window of the x-ray tube, which may have any target.

If the molybdenum K spectrum is to be excited, the primary beam must contain x-rays having wavelength $\lesssim \lambda_{\mathrm{MoK_{ab}}}$. This means that the short-wavelength limit of the continuum $\lambda_{\mathrm{min}} \lesssim \lambda_{\mathrm{MoK_{ab}}}$. Consequently, by Equation (1.17), the tube must be operated at a potential

$$V \gtrsim \frac{12{,}396}{\lambda_{\mathrm{MoK_{ab}}}} \gtrsim V_{\mathrm{MoK}} \tag{1.23}$$

For $\lambda_{\mathrm{MoK_{ab}}} = 0.62$ Å, $V = V_{\mathrm{MoK}} = 20{,}000$ V, or 20 kV. Figure 1.17 shows that at potentials < 20 kV, the primary spectrum contains no x-radiation having wavelength $\lesssim \lambda_{\mathrm{MoK_{ab}}}$, and even at 20 kV, there is very little. How-ever, as V is increased, the amount of useful primary radiation (shaded) increases sharply, and the molybdenum K spectrum is excited with great efficiency. In general, it is desirable to operate the tube at a potential such that the "hump" of the continuum $(\lambda_{I_{\mathrm{max}}})$ occurs at a wavelength shorter than that of the absorption edge to be excited.

So far, it has been tacitly assumed that only primary continuum excites the molybdenum K lines. However, the primary beam also contains target lines superimposed on the continuum. If the target is tungsten, and the tube is operated at $\gtrsim V_{\mathrm{WLII}}$ ($\sim$11.5 kV), tungsten $L\alpha_1$, $L\beta_1$, and $L\gamma_1$ lines appear. These lines, having wavelengths $> \lambda_{\mathrm{MoK_{ab}}}$, cannot excite the Mo K spectrum. However, if the target is silver, and the tube is operated at $\gtrsim V_{\mathrm{AgK}}$ ($\sim$25.5 kV), the silver $K\alpha$ and $K\beta$ lines appear. These lines have wavelengths $< \lambda_{\mathrm{MoK_{ab}}}$ and contribute to the excitation of the Mo K spectrum.

1.8.7.4. Comparison of Primary and Secondary Excitation

It should be obvious that excitation potentials (which, as we have seen, are numerically equal to electron energies in electron volts or kilo-electron-volts) and absorption edges are analogous quantities. Both are measures of the energy required to expel an electron from a specified level in an atom of a specified element, one applying to electrons in primary excitation, the other to photons in secondary excitation. One might say that the excitation potential is the potential or kilovolt equivalent of the absorption edge, and that the absorption edge is the wavelength or angstrom equivalent of the excitation potential. The two are related by equations having the form of Equations (1.23) and (1.17):

$$V_K = \frac{12,396}{\lambda_{K_{ab}}}; \qquad \lambda_{K_{ab}} = \frac{12,396}{V_K} \qquad (1.24)$$

$$V_{LIII} = \frac{12,396}{\lambda_{LIII_{ab}}}; \qquad \lambda_{LIII_{ab}} = \frac{12,396}{V_{LIII}} \qquad (1.25)$$

Thus, to excite, say, the molybdenum K spectrum, an x-ray tube must be operated at or above V_{MoK}. This is true regardless of whether the molybdenum is made the tube target (primary excitation) or a tube having any arbitrarily chosen target is used to irradiate the molybdenum placed external to the tube (secondary excitation). Below V_{MoK}, in the former case, the electrons will not be accelerated sufficiently to expel K electrons from the molybdenum target, and in the latter case, there will be in the primary beam no x-rays of wavelength shorter than $\lambda_{MoK_{ab}}$.

FIGURE 1.17. Primary continuous x-ray spectra from the x-ray tube in Figure 1.16. At each potential, only the shaded portion of the continuum excites the Mo K spectrum from the external specimen.

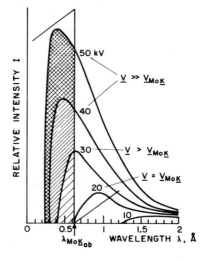

Moreover, if the molybdenum spectrum is to be excited efficiently, the tube must be operated at $V \gg V_{MoK}$—for either mode of excitation, as shown in Figures 1.16 and 1.17. In primary excitation, high potential has the effect shown in Figure 1.13 and expressed in Equation (1.22). In secondary excitation, high potential enriches the primary x-rays in the effective wavelength region below $\lambda_{MoK_{ab}}$.

Primary excitation gives a high-intensity line spectrum on a high-intensity continuous spectrum. For a given x-ray tube power (kV, mA), the line spectrum is ~ 100 times as intense with primary excitation as with secondary. However, continuum is *excited* only by electrons. The continuum that appears in secondary spectra is mostly primary continuum scattered by the specimen. Consequently, secondary excitation gives lower background and much higher line-to-background ratios.

X-rays penetrate much deeper than electrons of similar energy, so the effective specimen layer—the depth from which the measured line spectrum originates—is very different for the two modes. In primary excitation, the effective layer is determined by the depth *to* which electrons penetrate and is a few micrometers at most. In secondary excitation, the effective layer is determined by the depth *from* which the measured lines can emerge and is of the order 100 μm.

The advantages of electron excitation include: (1) better sensitivity for light elements because of the greater excitation efficiency; (2) reduced absorption-enhancement effects because the electrons penetrate only a thin surface layer; (3) better sensitivity for thin films because the electrons are substantially absorbed in the film, whereas x-rays penetrate, causing little photoionization; and (4) ease of focusing the electron beam to small diameter for use with curved-crystal spectrometers and for analysis of small selected areas. The disadvantages include: (1) higher background because of the continuum; (2) severe surface effects because of low penetration of the specimen surface; (3) generally reduced sensitivity because of the lower line-to-background ratio, except for the low-Z elements; (4) the requirement for high vacuum; (5) possible alteration of specimen composition by the electron beam; (6) inapplicability to liquid specimens, except with Lenard-window instruments, because they cannot be held in vacuum (a Lenard window is a thin metal-foil window through which electrons can emerge from a vacuum enclosure); and (7) requirement that the specimen surface be an electrical conductor; this feature may be especially troublesome with powders, which may disperse on charging [however, a thin (50-Å) film of carbon or aluminum can be evaporated on bulk solids, and powders can be admixed with colloidal graphite or aluminum powder, then pelletized].

1.8.8. Other Excitation Modes

Characteristic x-ray line spectra arise in matter in many ways, including the following, all of which have been applied in x-ray spectrochemistry: (1) bombardment by electrons; (2) bombardment by protons, deuterons, α-particles, and heavier ions from particle accelerators; (3) irradiation by α-, β-, γ-, and/or x-rays from radioisotopes; (4) irradiation by x-rays from high- and low-power x-ray tubes; (5) indirect excitation by x-rays from secondary emitters ("fluorescers"), themselves excited by x-ray tubes or radioisotopes; (6) "autoexcitation" or "self-excitation;" (7) certain spontaneous radioactive decay processes; (8) plasma excitation; (9) Pauli or Fano–Lichten excitation; and (10) atomic "pseudofusion."

Electron and x-ray excitation (modes 1 and 4) are considered in Sections 1.8.6 and 1.8.7, respectively, and are illustrated in Figures 1.18A and B, respectively. Ion excitation (mode 2) may be essentially similar to electron excitation, but different processes may also be involved (modes 9 and 10; see below). Modes 3 and 5 also constitute excitation by charged particles and/or x-rays.

It is explained in Sections 1.8.7.3 and 1.8.7.4 that an x-ray spectral-line series is excited most efficiently by x-rays having wavelength just shorter than the absorption edge (photon energy just greater than the excitation potential) associated with that series. In indirect excitation (mode 5) an x-ray tube or radioisotope source excites a secondary emitter ("fluorescer") having one or more strong spectral lines at wavelength just shorter than the absorption edge to be excited. This secondary radiation then excites the specimen.

In autoexcitation, no external excitation is applied to the specimen. In one technique, the excitation radiation is generated in the specimen itself. A small amount of the stable isotope $^{10}_{5}B$ is admixed with the specimen, which is then irradiated with neutrons to transmute the $^{10}_{5}B$ to radioisotope $^{7}_{3}Li$, which emits γ-rays, which excite the x-ray spectrum of the specimen. In another technique, the untreated specimen is irradiated with neutrons, transmuting some of the atoms of each element into radioisotopes; some of these may undergo β or γ internal conversion or K- or L-electron capture (see below), with resultant emission of characteristic x-rays.

Three spontaneous radioactive decay processes result in x-ray line emission: internal γ-ray conversion, internal β-ray conversion, and orbital–electron capture; these are shown in Figures 1.18C, D, and E, respectively.

Radioactive isotopes that decay by γ-emission may undergo internal conversion (Figure 1.18C). The γ photon is absorbed within the atom of its origin, giving up its energy in expelling and imparting kinetic energy to an orbital electron, which then becomes an internal-conversion electron. Internal γ-conversion may be regarded as secondary excitation in which the ionizing photon originates within the atom. The γ-ray may also escape

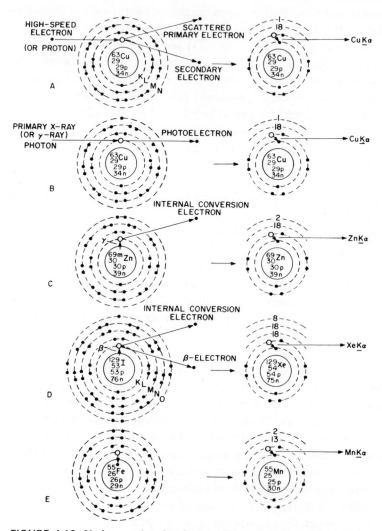

FIGURE 1.18. Various modes of excitation of characteristic x-ray line spectra. (A) Primary (electron) excitation. (B) Secondary (x-ray or γ-ray) excitation ("fluorescence"). (C) Internal γ conversion. (D) Internal β conversion. (E) Orbital–electron capture. A may also represent ion-induced excitation; C, D, and E are radioactive decay processes.

its own atom and cause secondary excitation in another atom (Figure 1.18B).

Radioactive isotopes that decay by β-emission may also undergo internal conversion (Figure 1.18D) or cause primary excitation (Figure 1.18A) of other atoms. Internal β-conversion may be regarded as primary excitation in which the ionizing electron originates within the atom. However, the emission of a β-particle (electron, e^-) results from the conversion of a neutron to a proton ($n \rightarrow p^+ + e^-$), and thereby increases the atomic number by 1. Then the internal conversion results in the emission of x-rays characteristic of this new element. In the example in Figure 1.18D,

$$^{129}_{53}\text{I} \rightarrow {}^{129}_{54}\text{Xe} + \beta^-$$

Thus, an ^{129}I source emits the xenon x-ray spectrum.

In radioactive isotopes that decay by orbital-electron capture (Figure 1.18E), a K or L electron actually falls into the nucleus, thereby "neutralizing" a proton ($p^+ + e^- \rightarrow n$) and decreasing the atomic number by 1. In the example given,

$$^{55}_{26}\text{Fe} \xrightarrow{K \text{ capture}} {}^{55}_{25}\text{Mn}$$

Thus, an ^{55}Fe source emits the manganese x-ray spectrum.

X-ray line spectra are also excited by radioisotope-induced indirect excitation (see above).

A *plasma* is a highly ionized gas containing equal numbers of positive and negative charges. The temperature in a plasma may be 10^6–10^7 °C, and the atoms may be stripped of many of their orbital electrons and therefore emit characteristic x-ray spectra. However, unlike the relatively simple x-ray spectra arising from singly ionized K-, L-, and M-state atoms, plasma spectra are very complex because they contain lines from all the multiply ionized atom states. Plasmas are formed, among other ways, by high-power pulsed laser irradiation and by "exploding" fine wires by passing high-current electric pulses through them. Plasma spectra give information about nonequilibrium atom states.

Sometimes ion-induced x-ray spectral-line intensities are 10^3–10^5 those predicted for a simple orbital-electron expulsion mechanism (Figure 1.18A). This is attributed to interpenetration of the electron-orbital systems of the bombarding ion and target atom, forming a transient pseudomolecular species having its own set of electron-energy levels. In this state, electron transitions occur from lower to higher levels, leaving inner-shell vacancies. When the two atoms later separate, they are in excited states and emit x-ray

spectra. This phenomenon is termed *Pauli excitation* because it is based on
the Pauli exclusion principle (Section 1.8.1) operating in the transient
pseudomolecular levels, and *Fano–Lichten excitation* after the scientists
who first explained it.

Pauli excitation has led some physicists to speculate that the x-ray
spectra of superheavy elements may be observable before the elements are
"discovered," that is, produced, in the usual sense. As the orbital-electron
systems of the bombarding ion and target atom interpenetrate, they undergo various transformations and may instantaneously assume the con-
figuration of the atom having atomic number equal to the sum of the atomic
numbers of the two colliding atoms. For example, during bombardment
of argon (Z 18) with argon ions, x-ray lines of krypton (Z 36) have been
observed! By the same process, it may be possible that, say, bombardment
of uranium (Z 92) with argon ions may result in x-ray lines of element 110!

Incidentally, all this transmutation of one element into another does
not violate the definition of chemical elements given in paragraph 1 of
Section 1.8.1. Transmutation and radioactive decay are subatomic processes
and certainly do not constitute "ordinary chemical means." Moreover, in
Figure 1.18E, for example, $^{55}_{26}$Fe is not "made of" $^{55}_{25}$Mn, but its atoms are
actually changed into manganese atoms by the decay process indicated.

SUGGESTED READING

Adler, I., *X-Ray Emission Spectrography in Geology*, Elsevier, Amsterdam (1966);
  Chap. 2, pp. 3–28.

Bertin, E. P., *Principles and Practice of X-Ray Spectrometric Analysis*, 2nd ed., Plenum,
  New York (1975); Chaps. 1 and 2, pp. 3–85.

Jenkins, R., *Introduction to X-Ray Spectrometry*, Heyden, London (1974); Chaps. 2
  and 3, pp. 8–51.

Jenkins, R., and J. L. DeVries, *Practical X-Ray Spectrometry*, 2nd ed., Springer-Verlag
  New York, New York (1967); Chap. 1, pp. 1–25.

Liebhafsky, H. A., H. G. Pfeiffer, E. H. Winslow, and P. D. Zemany, *X-Rays, Electrons,
  and Analytical Chemistry*, Wiley–Interscience, New York (1972); Chap. 1, pp. 1–57.

Chapter 2

Properties of X-Rays

In this chapter, we consider four x-ray phenomena of particular significance in x-ray spectrochemical analysis: absorption, scatter, diffraction, and the Auger effect and fluorescent yield.

2.1. X-RAY ABSORPTION

2.1.1. Mass-Absorption Coefficients

In Figure 2.1, an absorber of uniform thickness t cm and density ϱ (rho) g/cm^3 is sandwiched between two x-ray-opaque masks (shaded) having pinhole tunnels of the same diameter in register. When monochromatic x-radiation is directed on the left mask, its tunnel permits only a collinear beam of intensity I_0 to arrive at the absorber. These incident x-rays may undergo absorption, transmission, or scatter, as shown in the figure by the top, middle, and bottom rays, respectively. The emergent collinear beam consists of the transmitted rays and has intensity I given by the Lambert law,

$$I = I_0 \exp(-\mu t) \tag{2.1}$$

FIGURE 2.1. Arrangement for measurement of x-ray absorption. The top, middle, and bottom incident rays undergo true photoelectric absorption, transmission, and scatter, respectively.

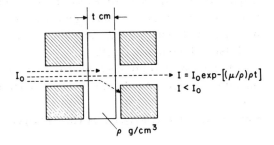

59

where μ (mu) is the *linear-absorption coefficient* of the absorber and has the unit reciprocal centimeters (cm^{-1}). The negative sign indicates that the intensity always decreases; that is, x-rays always undergo *absorption* or *attenuation* on passing through matter. A more useful form of this equation is

$$I = I_0 \exp[-(\mu/\varrho)\varrho t] \qquad (2.2)$$

where (μ/ϱ) ("mu over rho") is the *mass-absorption coefficient* (cm^2/g) of the absorber and ϱt is the *area density* (g/cm^2).

The mass-absorption coefficient μ/ϱ is an atomic property of chemical elements and is a measure of their x-ray opacity or "stopping power." For the same element, μ/ϱ is different at every wavelength, and at the same wavelength, it is different for every element. However, μ/ϱ is a function only of wavelength and atomic number and is independent of state of chemical combination or physical aggregation. Values of μ/ϱ for a compound, alloy, solution, or mixture are readily calculated from the concentrations and coefficients of their constituents. Values of different substances are directly comparable. None of these advantages applies to linear absorption coefficients, as is strikingly illustrated by a hypothetical experiment described by Sproull*: "A beam of x-rays passing from the ceiling to the floor of a chamber filled with hydrogen and oxygen may be [say] 10% absorbed, or 90% of it will reach the floor. If a spark explodes the hydrogen and oxygen, filling the chamber with steam, 90% of the x-rays will still reach the floor. Then if the chamber is chilled so that the steam condenses to a thin layer of water or ice on the floor, 90% of the x-rays will still reach the floor. This is not true for light or ultraviolet or infrared radiation, and it explains why the mass absorption coefficient of x-rays is commonly used, whereas the linear absorption coefficient is ordinarily used in optics."

For a specified element Z, $(\mu/\varrho)_Z$ is different at every wavelength and, disregarding absorption edges (see below), $(\mu/\varrho)_Z$ increases as wavelength increases; that is, the x-ray opacity of an element increases as the x-rays become less energetic and therefore less penetrating. At a specified wavelength, $(\mu/\varrho)_Z$ is different for every element and, again disregarding absorption edges, increases as atomic number increases; that is, x-ray penetration decreases as atomic number increases. These relationships are illustrated in Table 1.4, columns 17–19.

The mass-absorption coefficient of a compound, alloy, solution, or mixture of elements A, B, C, ... in concentrations (weight fraction)

---

* W. T. Sproull, *X-Rays in Practice*, McGraw-Hill Book Company, New York (1946); p. 37.

$C_A, C_B, C_C, \ldots$ is given by

$$(\mu/\varrho)_{ABC\ldots,\lambda} = \sum C_i (\mu/\varrho)_{i,\lambda} = C_A(\mu/\varrho)_{A,\lambda} + C_B(\mu/\varrho)_{B,\lambda} + \cdots \quad (2.3)$$

Sometimes a more useful form is

$$(\mu/\varrho)_{X,\lambda} = C_A(\mu/\varrho)_{A,\lambda} + (1 - C_A)(\mu/\varrho)_{(X-A),\lambda} \quad (2.4)$$

where X is the analytical specimen, A is the analyte (element to be determined), and X–A is the matrix.

A concept sometimes used in mathematical methods of x-ray fluorescence spectrometry is the *total* mass-absorption coefficient for both primary λ_{pri} and analyte-line λ_A x-rays:

$$(\mu/\varrho)' = (\mu/\varrho)_{\lambda_{pri}} + (\mu/\varrho)_{\lambda_A} \quad (2.5)$$

$$\overline{(\mu/\varrho)} = (\mu/\varrho)_{\lambda_{pri}} \csc\phi + (\mu/\varrho)_{\lambda_A} \csc\psi \quad (2.6)$$

$$= [(\mu/\varrho)_{\lambda_{pri}}/\sin\phi] + [(\mu/\varrho)_{\lambda_A}/\sin\psi] \quad (2.7)$$

where $\overline{(\mu/\varrho)}$ is substantially $(\mu/\varrho)'$ corrected for geometry, that is, for the path lengths of the primary ($\csc\phi$) and secondary ($\csc\psi$) x-rays in the specimen. Figure 4.2 shows the significance of the incident ϕ (phi) and takeoff ψ (psi) angles.

In this book, μ/ϱ indicates mass-absorption coefficient in centimeters squared per gram (cm^2/g), and $(\mu/\varrho)_{Z,\lambda}$ indicates the mass-absorption coefficient of element Z at wavelength λ. However, because it is the mass-absorption coefficient that is almost exclusively used in x-ray spectrochemistry, most writers now use μ or μ_m for simplicity and indicate the linear coefficient by μ_l.

2.1.2. X-Ray Absorption Phenomena

The linear- and mass-absorption coefficients just discussed are measures of that portion of the incident collinear x-ray beam in Figure 2.1 that does not appear in the emergent collinear beam—regardless of reason. They are total absorption coefficients and are the result of three phenomena, each having its own linear- and mass-absorption coefficients:

$$\mu = \tau + \sigma + \pi \quad (2.8)$$

$$\mu/\varrho = (\tau/\varrho) + (\sigma/\varrho) + (\pi/\varrho) \quad (2.9)$$

where τ (tau), σ (sigma), and π (pi) represent losses by, respectively, *true* or *photoelectric absorption*, *scatter*, and *pair production*.

In photoelectric absorption, x-ray photons are truly absorbed, expending all their energy in expelling and imparting kinetic energy to orbital electrons in absorber atoms. This is the process that results in emission of x-ray spectral lines. Note that, unlike electrons, photons cannot lose energy decrementally—only all at once or not at all. The difference between the absorbed photon energy and the energy required to expel an orbital electron is conserved as the kinetic energy (velocity) of the photoelectron. In the spectral region of x-ray spectrochemistry, photoelectric absorption predominates over scatter except for the lightest elements at short wavelengths, and $\mu/\varrho \approx \tau/\varrho$.

In scattering, x-ray photons are not really absorbed, but only deflected from their incident paths in the absorber, in effect disappearing from the emergent beam (Section 2.2).

In pair production, x-ray photons interact with atom nuclei, expending all their energy in creating and imparting kinetic energy to electron–positron pairs $(e^+–e^-)$. This phenomenon occurs only at photon energies $\gtrsim 1.02$ MeV and is of no importance in x-ray spectrochemistry.

2.1.3. Absorption Edges

The relationship of absorption and wavelength for a specified element is best expressed in the form of a "log–log" plot of mass-absorption coefficient *versus* wavelength. Typical absorption curves are shown in Figures 1.14, 1.15, and 2.2. As already explained, in these curves, for a specified element, $(\mu/\varrho)_Z$ should increase with increasing wavelength, and at a specified wavelength, $(\mu/\varrho)_Z$ should increase with increasing atomic number. It is evident that these predictions are essentially realized and would be wholly valid were it not for the absorption edges. These edges are discussed in Sections 1.8.7.1 and 1.8.7.2, and the reader might benefit from a review of these sections. Here we need only add a few details.

Many examples of departures from the regularity predicted above appear in Figure 2.2. The μ/ϱ of tin increases regularly with wavelength up to $\lambda_{\mathrm{Sn}K_{ab}}$, then falls abruptly and increases regularly again. Below $\lambda_{\mathrm{Pb}K_{ab}}$, μ/ϱ increases with Z, as it should; however, just above $\lambda_{\mathrm{Pb}LIII_{ab}}$, copper and zinc ($Z = 29$ and 30) have higher absorption than tin and lead ($Z = 50$ and 82), and tin and lead ($\Delta Z = 32$) differ in absorption no more than copper and zinc ($\Delta Z = 1$). The μ/ϱ of lead is greater than that of tin below $\lambda_{\mathrm{Pb}K_{ab}}$, less between $\lambda_{\mathrm{Pb}K_{ab}}$ and $\lambda_{\mathrm{Sn}K_{ab}}$, greater between $\lambda_{\mathrm{Sn}K_{ab}}$ and $\lambda_{\mathrm{Pb}LIII_{ab}}$, then slightly less at longer wavelengths. It is evident that absorption cannot be estimated simply on the basis of atomic number and wavelength.

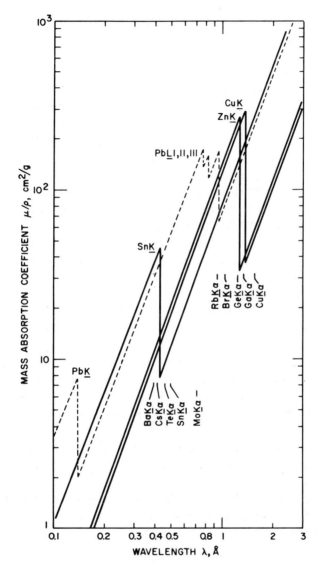

FIGURE 2.2. X-ray absorption curves of copper, zinc, tin, and lead (atomic numbers 29, 30, 50, and 82, respectively).

Referring to the absorption curves in Figure 2.2, consider a monochromatic primary x-ray beam of wavelength λ incident on a secondary target of atomic number Z. Consider what happens as λ is made progressively shorter (more energetic).

At $\lambda > \lambda_{ZK_{ab}}$, the photons do not have enough energy to expel Z K electrons; consequently, no Z K lines appear. As λ decreases, the photons become more energetic, and the absorption coefficient decreases; that is, the secondary target becomes more transparent. At $\lambda_{ZK_{ab}}$, the photons have exactly the energy required to expel Z K electrons, and the absorption increases abruptly. The absorbed photons expel K electrons, and Z K lines are emitted. This process is *photoelectric absorption*, and the expelled electrons are *photoelectrons*. The Z K electron expulsion and the line emission are most efficient at λ just less than $\lambda_{ZK_{ab}}$.

At $\lambda \ll \lambda_{ZK_{ab}}$, the photons have much more than enough energy to expel Z K electrons. However, they are so energetic that they may not be absorbed, or else, before being absorbed, they may penetrate the target to a depth from which Z K radiation cannot emerge. Thus, as λ becomes progressively shorter, line emission decreases. Note the difference in this respect between primary and secondary excitation. In primary excitation, emitted line intensity *increases* as x-ray tube potential V is made greater than the target K excitation potential V_K : $I_K \propto i(V - V_K)^{\sim 1.7}$ [Equation (1.22)]. In secondary excitation, the emitted line intensity decreases as primary wavelength is decreased below $\lambda_{ZK_{ab}}$. This difference is evident in Figure 1.13.

Even though at $\lambda < \lambda_{ZK_{ab}}$ photons have more than enough energy to expel Z K electrons, each photon that is absorbed by an electron is wholly absorbed. The surplus energy imparts kinetic energy to the photoelectron, so that the shorter the wavelength, the higher is the velocity of the photoelectrons.

The preceding discussion regarding wavelengths bracketing the K absorption edge applies as well to wavelengths bracketing the several L and M edges.

The magnitude of the change in mass-absorption coefficient at the absorption edge is expressed as the absorption-edge *jump ratio r* and *jump difference* δ (delta):

$$r = (\mu/\varrho)_s/(\mu/\varrho)_l \tag{2.10}$$

$$\delta = (\mu/\varrho)_s - (\mu/\varrho)_l \tag{2.11}$$

where s and l refer, respectively, to the short- and long-wavelength sides of the edge, that is, the "top" and "bottom" of the edge or maximum and minimum values of μ/ϱ at the edge. A pair of equations of this form can be written for each absorption edge (K, LI, LII, etc.) of each element. The values of r and δ are measures of that portion of the total absorbed x-radiation that is absorbed by the atomic energy level associated with a specific edge. The actual fraction of the total number of photoionizations that occurs in the specified shell is

$$[(\mu/\varrho)_s - (\mu/\varrho)_i]/(\mu/\varrho)_s = 1 - (1/r) = (r - 1)/r \qquad (2.12)$$

Between absorption edges, log-log plots of (μ/ϱ) *versus* λ are linear and mutually parallel, as shown in Figure 2.2 and as predicted by the Bragg–Pierce law,

$$\mu/\varrho = kZ^4\lambda_{cm}^3 \qquad (2.13)$$

2.1.4. X-Ray Cross Section

Another way to express x-ray absorption uses the concept of *cross section*. If Equation (2.1) is written in terms of numbers of x-ray photons (rather than intensities) incident upon n_0 and transmitted by n an absorber of thickness t cm and linear absorption coefficient μ cm^{-1}, one gets

$$n = n_0 \exp(-\mu t) \qquad (2.14)$$

If the volume of absorber traversed by the x-ray beam contains n_{at} atoms/cm^3, each of which presents an imaginary target area or *cross section* σ (sigma) cm^2 to the photons,

$$\mu = n_{at}\sigma \qquad \text{or} \qquad \sigma = \mu/n_{at} \qquad (2.15)$$
$$\text{cm}^2/\text{atom} \; [=] \; \text{cm}^{-1}/(\text{atoms/cm}^3)$$

Cross section and mass-absorption coefficient are related as follows:

$$\mu/\varrho = \sigma(N/A) \qquad (2.16)$$

$$\frac{\text{cm}^2}{\text{g}} \; [=] \; \frac{\text{cm}^2}{\text{atom}} \left(\frac{\text{atoms/mol}}{\text{g/mol}} \right)$$

where N is the Avogadro number (6.02×10^{23} atoms/mol) and A is atomic weight. Actually, σ is usually expressed in barns (1 barn $\equiv 10^{-24}$ cm^2).

2.1.5. Comparison of X-Ray and Optical Absorption

X-ray and optical (visible, near-infrared, and ultraviolet) absorption differ essentially only in that they occur in inner and outer orbitals of atoms, respectively. Nevertheless, their absorption spectra are strikingly dissimilar.

Optical absorption and emission spectra are complementary, the former consisting of dark lines (or bands), the latter bright lines (or bands) at the same wavelengths. X-ray absorption increases gradually with wavelength, falls abruptly at a wavelength shorter than that of the emission lines, then gradually rises again. Figure 2.3 shows this difference. X-ray photoelectric absorption is explained in Sections 1.8.7.1 and 2.1.3. In optical absorption, outer orbital electrons are elevated to higher energy levels—not clear out of the atom, as in x-ray absorption. Such *resonance* excitation requires absorption of photons having exactly the same energy as those emitted when the excited electrons return to their original levels. Thus, optical absorption and emission spectra both consist of the same discrete wavelengths.

One might wonder why x-ray absorption *lines*, analogous to optical absorption lines, do not occur. For example, why can there not be a Cu *Kα* absorption line at the same wavelength as the Cu *Kα* emission line, caused by elevation of a Cu *K* electron to the Cu *L*III shell—the reverse of the transition that causes the Cu *Kα* emission line? The reason is that ordinarily the Cu *L*III shell is full and cannot receive the Cu *K* electron.

With certain exceptions, x-ray absorption, unlike optical absorption, is substantially independent of chemical state. A striking example is provided by diamond and graphite, which are, respectively, transparent and opaque to light, but have the same x-ray mass absorption coefficient. Sproull's hypothetical experiment (Section 2.1.1) further illustrates this point. Equa-

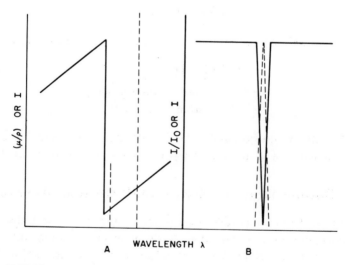

FIGURE 2.3. Comparison of x-ray (A) and optical (B) absorption.

tion (2.3) shows the simple additive nature of the mass-absorption coefficient of an element in a compound, solution, or mixture. No such equation is possible for optical absorption.

For another comparison of x-ray and optical spectra, see Sections 3.3.2 and 3.5.2.

2.1.6. Significance of X-Ray Absorption

Secondary excitation of characteristic x-ray spectra is itself based on photoelectric absorption, and in almost every component of the instrument, absorption is of great significance, sometimes beneficially, sometimes detrimentally.

Absorption of primary x-rays in the x-ray tube window becomes significant at the long wavelengths required to excite elements of low atomic number. In the specimen, absorption of primary and analyte-line radiation and absorption-enhancement effects determine sensitivity and dictate the analytical strategy. Absorption in liquid and powder cell windows becomes significant in determination of low-Z elements. Photoelectric absorption of secondary x-rays in some crystals results in line emission from the crystal. Absorption of secondary x-rays in the radiation path may require use of helium or vacuum, and for the lighter elements, even these may have significant absorption. Absorption of analyte-line radiation in the detector tube window may necessitate use of windows of thin Mylar or polypropylene or ultrathin Formvar. The gas-ionization process in gas-filled detectors and the scintillation process in scintillation counters are absorption phenomena. All filters are based on absorption: primary-beam filters to remove target lines or improve peak-to-background ratio, and secondary-beam filters to reduce spectral-line interference or permit "non-dispersive" analysis. Enhancement radiators are based on photoelectric absorption by the radiator and by the analyte.

Absorption is also the basis of several analytical methods—polychromatic and monochromatic absorption, absorption-edge spectrometry, and emission-absorption methods.

2.1.7. The Inverse-Square Law

Another form of x-ray attenuation is the reduction in intensity per unit area resulting from divergence and distance. X-rays emanating spherically from a point—or very small—source follow the inverse-square law,

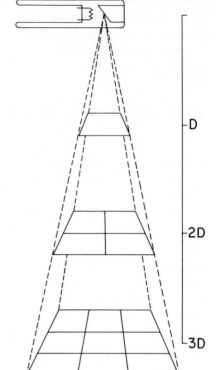

FIGURE 2.4. X-ray attenuation by divergence, the inverse-square law.

as shown in Figure 2.4 and the following equation:

$$I_2 = I_1(D_1^2/D_2^2) \tag{2.17}$$

where I_1 and I_2 are intensities per unit area at, respectively, distances D_1 and D_2 from the x-ray source. The inverse-square law does not apply to collinear beams. Soller collimators are an intermediate case, in that divergence occurs parallel to, but not perpendicular to, the foils.

2.2. X-RAY SCATTER

Figure 2.1 shows that some incident x-rays do not emerge from the absorber simply because they deviate or scatter from their collinear path. Four types of x-ray scatter must be distinguished. *Unmodified* (*elastic, Rayleigh*) scatter involves no change in wavelength. *Modified* (*inelastic, Compton*) scatter involves an increase in wavelength, that is, a decrease in photon energy. *Coherent* scatter occurs in such a way that there is a phase relationship between the incident and scattered x-rays. *Incoherent* scatter occurs without any such phase relationship. It is almost universal practice in the literature of x-ray spectrometric analysis to regard modified and incoherent scatter as synonymous. For practical purposes, this leads to no difficulty because modified scatter is necessarily incoherent. However, all incoherent scatter is not necessarily modified.

The mechanism of modified scatter is shown in Figure 2.5. The incident x-ray photon collides with a loosely bound electron in an outer orbit of an atom. The electron recoils under the impact, leaving the atom and carrying away some of the energy of the photon, which is deflected with corresponding loss of energy or increase in wavelength. The sum of the energies of the scattered photon and recoil electron equals the incident photon energy. The encounter is described by the laws of conservation of energy and momentum. The *recoil* or *Compton* electrons have the predicted direction and velocity, and the x-rays the predicted change in wavelength:

$$\lambda'_{cm} - \lambda_{cm} = \Delta\lambda_{cm} = (h/m_e c)(1 - \cos \phi) \tag{2.18}$$

where λ and λ' are wavelengths (cm) of the incident and modified scattered x-rays, respectively; h is Planck's constant (6.6×10^{-27} erg s); m_e is the rest mass of the electron (9.11×10^{-28} g); c is the velocity of light (3×10^{10} cm/s); and ϕ is the angle between unscattered and scattered x-rays. Substitu-

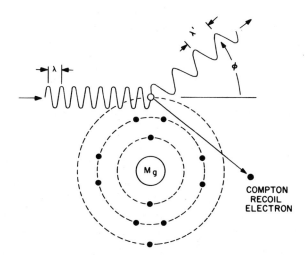

FIGURE 2.5. Modified (Compton) scatter of an x-ray photon by a magnesium atom.

tion of these values and insertion of the conversion factor 10^8 Å/cm gives

$$\Delta\lambda = 0.0243(1 - \cos\phi) \tag{2.19}$$

where $\Delta\lambda$ is in angstroms; 0.0243 Å is referred to as the *Compton wavelength*.

In Equation (2.19), $\Delta\lambda$ is independent of both x-ray wavelength and atomic number of the scatterer, and varies only with ϕ. In most commercial x-ray fluorescence spectrometers, the axis of the primary beam is perpendicular to the axis of the collimator–crystal–detector system, which then "sees" primary x-rays scattered from the specimen at 90°. Then $\phi = 90°$, $\cos\phi = 0$, and from Equation (2.19), $\Delta\lambda \approx 0.024$ Å.

The modified/unmodified scattered intensity ratio increases as: (1) x-ray photon energy increases (wavelength decreases); (2) binding of atomic orbital electrons in the scatterer decreases, that is, atomic number decreases; and (3) ϕ increases. Modified scatter predominates when x-ray photon energy greatly exceeds the orbital electron binding energy. When the binding energy approaches or exceeds the x-ray energy, unmodified scatter predominates.

In x-ray fluorescence spectrometry, scatter is responsible for the presence of primary continuum and x-ray tube target lines in the secondary spectrum. The scattered intensity increases as the mean atomic number of the specimen decreases. Modified scatter appears as a broader peak on the long-wavelength side of each unmodified scattered x-ray tube target line and separated from it by $\Delta\lambda$ [Equation (2.19)]. The lower the mean atomic number and/or the shorter the wavelength (higher the photon energy) of the target line, the more intense is the modified peak with respect to the unmodified.

Formerly, scatter was regarded only as a nuisance. Background is mostly scattered primary radiation, and both unmodified and modified scattered target lines complicate the spectrum and increase the possibility of spectral-line interference. However, scattered x-rays may be beneficial in many ways. Sometimes, there is a question as to whether a low-intensity peak is a trace analyte line *emitted* from the specimen or a target line *scattered* from the specimen. The presence of an accompanying modified scattered peak conclusively indicates the latter. Diffraction, a form of coherent scatter, is the basis of dispersion by crystals. More recently, several techniques have been developed for using scattered x-rays to correct absorption–enhancement and surface-texture effects and other difficulties. Finally, several analytical methods are based on x-ray scatter rather than emission, including a method for determining carbon in hydrocarbons.

2.3. X-RAY DIFFRACTION BY CRYSTALS

2.3.1. The Bragg Law

In the wavelength-dispersive x-ray spectrometer, an analyzer crystal disperses the characteristic x-ray spectrum so that each spectral line can be measured individually. An understanding of the function of the crystal requires an elementary knowledge of the diffraction of x-rays by crystals.

Diffraction arises from constructive interference of unmodified coherently scattered x-rays. Figure 2.6 shows interference of two rays of the same wavelength for three phase relationships. In Figure 2.6A, the rays are one half-wavelength (180°) out of phase ("peak to valley and valley to peak") and undergo complete destructive interference or cancellation. In Figure 2.6B, the rays are in phase ("peak to peak and valley to valley") and undergo complete constructive interference or reinforcement. The resultant ray has the same wavelength as the individual rays and the sum of their peak amplitudes. Figure 2.6C represents all intermediate phase relationships. The resultant ray has the same wavelength as the individual rays and the vector sum of their amplitudes—greater than zero, but less than the sum of the peak amplitudes. The more nearly in or out of phase the two interfering rays are, the more nearly they approach the extremes of Figures 2.6B and A, respectively.

A *crystal* may be defined as a solid in which the atoms or molecules are closely packed in layers in a pattern having three-dimensional periodicity. In Figure 2.7, a collimated monochromatic x-ray beam of wavelength λ is directed at angle θ (theta) on a set of crystal planes (hkl) (Section 2.3.2) having interplanar spacing d. A portion of the x-radiation arriving at each plane is scattered in all directions. In most directions, the scattered rays are out of phase and undergo destructive interference. However, in certain directions, they may be in phase and mutually reinforce. This phenomenon is *diffraction*, and a group of such reinforced rays in one direction is a diffracted x-ray beam. The conditions for diffraction are: (1) the angles made by the diffracting planes (hkl) with the incident and diffracted beams are equal; (2) the directions of the incident and diffracted beams and the normal to the diffracting plane are coplanar; and (3) reflected rays from successive planes differ in path length by an integral number of wavelengths.

In Figure 2.7, rays 1 and 2 are directed on two successive planes and scatter in all directions. Of the scattered rays, only 1′ and 2′ fill the conditions for diffraction. Rays 1 and 2 travel equal distances to AC, and rays 1′ and 2′ travel equal distances beyond AD. The path difference between

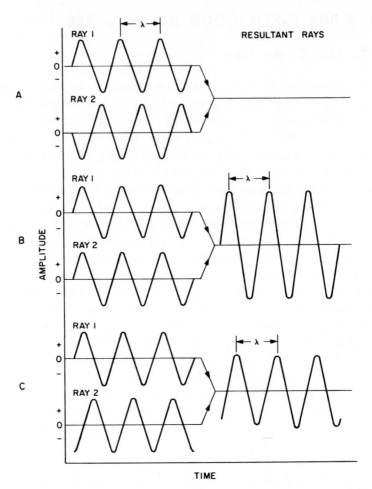

FIGURE 2.6. Destructive (A), constructive (B), and intermediate (C) wave interference.

rays $1A1'$ and $2B2'$ is then

$$CBD = CB + BD = 2AB \sin \theta = 2d \sin \theta \qquad (2.20)$$

If these rays are to be in phase, this path difference must be an integral number of wavelengths, that is,

$$n\lambda = 2d \sin \theta \qquad (2.21)$$

This is the Bragg law, where n is the *order* of the diffracted beam and is numerically equal to the path difference, in wavelengths, for successive planes; d is the interplanar spacing of the diffracting planes (Å); and θ, the Bragg angle, is the angle between the incident x-rays and the diffracting planes. X-ray spectrometers are usually calibrated in terms of 2θ, the angle between the diffracted beam and the undeflected incident beam (Figure 3.2).

We have seen that complete destructive interference occurs only for two rays out of phase by exactly $\frac{1}{2}\lambda$. Two rays scattered from successive planes at an angle $\theta + \Delta\theta$ such that they differ in path length by only, say, $\frac{1}{4}\lambda$, do not cancel, but combine to a ray of smaller amplitude than if they were in phase (Figure 2.6). However, the ray scattered at $\theta + \Delta\theta$ from the third plane down in the crystal is $\frac{1}{4}\lambda$ out of phase with that from the second plane and $\frac{1}{2}\lambda$ out of phase with that from the first plane. Consequently, the $\theta + \Delta\theta$ rays from the first and third, second and fourth, fifth and seventh, etc., planes throughout the crystal cancel one another. Consider now rays scattered from successive planes at an angle $\theta + \Delta\theta$ such that they differ in path length by only, say, 0.005λ. The ray to cancel this ray from the first plane originates deep within the crystal, $0.5\lambda/0.005\lambda = 100$ planes down, that is, the 101st plane. Thus, the $\theta + \Delta\theta$ rays from planes 101 to 200, respectively, cancel those from planes 1 to 100. Thus, diffraction occurs only at the Bragg angle. Incidentally, if the crystal is so thin that it does not have a 101st plane, the ray from the first plane is not canceled, and the diffracted beam is broadened accordingly. This phenomenon provides the basis for measurement of single-crystal film thickness and polycrystal particle size by x-ray diffraction-line broadening.

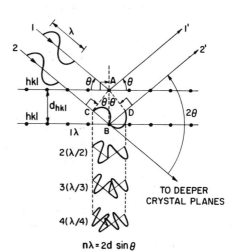

FIGURE 2.7. Diffraction by crystals, the Bragg law, showing origin of orders 2, 3, and 4.

The analogy of the Bragg version of diffraction with optical reflection is evident, but the two phenomena are quite different. Optical reflection is a wholly surface effect and occurs at all incident angles greater than a certain small critical angle.

Figure 2.7 also illustrates the phenomenon of diffraction orders. Note that at the same θ angle at which wavelength λ is diffracted in first order ($n = 1$, one wavelength path difference between successive planes), wavelength $\lambda/2$ is diffracted in second order ($n = 2$, two wavelengths path difference), $\lambda/3$ in third order, etc. Thus, the method of crystal diffraction cannot separate two wavelengths λ_1 and λ_2 related by the equation

$$\lambda_2 = \lambda_1/n \qquad (2.22)$$

If the crystal is rotated so that θ is increased progressively, the wavelength required to satisfy the Bragg law [Equation (2.21)] also increases progressively, and a polychromatic spectrum can be diffracted in sequence. At any θ, λ, $\lambda/2$, $\lambda/3$, etc., may satisfy the equation in orders 1, 2, 3, etc.

2.3.2. Miller Indexes (*hkl*)

The purpose of this section is to explain the significance of the Miller indexes (*hkl*) used to identify specific sets of planes in analyzer crystals.

A *crystal* is a solid in which the atoms or molecules are closely packed in layers in a pattern having three-dimensional periodicity. It is convenient to regard the atoms or molecules as occupying points in an imaginary framework or *lattice* in space. Figure 2.8 shows a small portion of a typical crystal lattice.

The *unit cell* is the fundamental unit of the crystal lattice, that is, the smallest portion of the lattice that can exhibit—at least in principle—the crystallographic properties of the bulk crystalline substance. The bulk crystal is built up by repetition of the unit cell in all directions. Figure 2.8 shows the unit cell (accented) and how its repetition builds up the crystal lattice. Incidentally, the unit cell is analogous to the atom and molecule, which are the fundamental units of elements and compounds, respectively; that is, the atom and molecule are the smallest particles of an element and compound, respectively, that can—in principle—exhibit the *chemical* properties of that element or compound.

The unit cell is a parallelepiped, that is, a polyhedron bounded by six parallelogram faces, characterized by the lengths of its sides a, b, c, and their included angles α, β, γ. These six features are known as the *lattice*

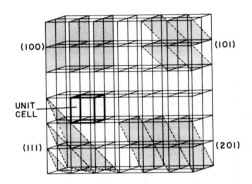

(100)

(101)

UNIT
CELL

(111)

(201)

FIGURE 2.8. Portion of a crystal lattice showing: (1) the unit cell (accented); (2) generation of the lattice by repetition of the unit cell in all directions; and (3) four of the many possible sets of parallel planes (hkl) that can be passed through the lattice, (100), (101), (111), and (201). This portion of lattice consists of $6 \times 2 \times 6 = 72$ unit cells.

constants or *lattice parameters* and are shown in Figure 2.9. On the basis of the lattice constants, seven types of unit cell are possible, permitting classification of all crystals in seven *crystal systems*. These are shown in Figure 2.10 and defined in Table 2.1. Different substances in the same crystal system may have different values of the lattice constants, but within the limits specified in Table 2.1.

Many sets of parallel planes can be passed through a crystal lattice; four such sets are shown in Figure 2.8. Crystallographers must be able to refer to these planes and so require a system of notation that permits identification of all sets of crystal planes. Miller indexes (hkl) constitute such a system of notation. Each set of planes is indexed by that portion of it that passes through any individual unit cell. The assignment of Miller indexes is illustrated for some typical planes in Figure 2.11 and Table 2.2 and involves the following procedure.

1. Note the intercepts of the plane on the three unit cell axes, that is, the distances from a unit cell corner designated as the origin, at which the plane intersects the axes; express these intercepts as fractions of the lattice constants a, b, c. If the plane lies parallel to an axis, the intercept on that axis is infinity.

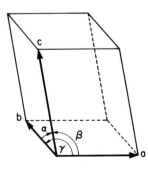

c

b

α

β

γ

a

FIGURE 2.9. Unit cell, showing lattice parameters: axial lengths a, b, c, and included angles α, β, γ; $\alpha = b \wedge c$, $\beta = a \wedge c$, $\gamma = a \wedge b$.

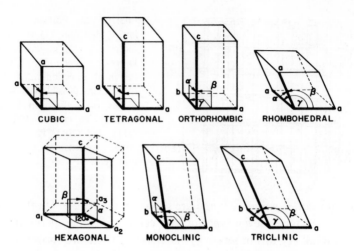

FIGURE 2.10. The seven fundamental unit cell geometries based on the relative lengths of a, b, and c and the angles α, β, and γ; these basic unit cells define the seven crystal systems (Table 2.1).

2. Take the reciprocals of the three intercepts; the reciprocal of infinity is zero.

3. If one or more of the reciprocals is fractional, that is, not an integer or zero, reduce all reciprocals to fractions having a common denominator and clear in the usual way, for example, plane I in Figure 2.11 and Table 2.2.

TABLE 2.1. Basic Unit Cell Shapes; Crystal Systems
(Refer to Figures 2.9 and 2.10)

Crystal system	Axial lengths	Angles
Cubic	$a = b = c$	$\alpha = \beta = \gamma = 90°$
Tetragonal	$a = b \neq c$	$\alpha = \beta = \gamma = 90°$
Orthorhombic	$a \neq b \neq c$	$\alpha = \beta = \gamma = 90°$
Rhombohedral (trigonal)	$a = b = c$	$\alpha = \beta = \gamma \neq 90°$
Hexagonal	$a_1 = a_2 \neq c$	$\alpha = \beta = 90°; \quad \gamma = 120°$
Monoclinic	$a \neq b \neq c$	$\alpha = \gamma = 90° \neq \beta$
Triclinic	$a \neq b \neq c$	$\alpha \neq \beta \neq \gamma \neq 90°$

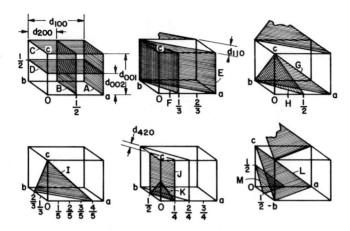

FIGURE 2.11. Designation of Miller indexes for some typical crystallographic planes (Table 2.2).

TABLE 2.2. Designation of Miller Indexes
(Refer to Figure 2.11)

Plane (Figure 2.11)	Axial intercepts			Reciprocals			Miller indexes (hkl)
	a	b	c	$\dfrac{1}{a}$	$\dfrac{1}{b}$	$\dfrac{1}{c}$	
A	1	∞	∞	1	0	0	(100)
B	$\frac{1}{2}$	∞	∞	2	0	0	(200)
C	∞	∞	1	0	0	1	(001)
D	∞	∞	$\frac{1}{2}$	0	0	2	(002)
E	1	1	∞	1	1	0	(110)
F	$\frac{1}{3}$	1	∞	3	1	0	(310)
G	1	1	1	1	1	1	(111)
H	$\frac{1}{2}$	1	1	2	1	1	(211)
I	$\frac{4}{5}$	$\frac{2}{3}$	1	$\frac{5}{4}$	$\frac{3}{2}$	1	
				$\frac{5}{4}$	$\frac{6}{4}$	$\frac{4}{4}$	(564)
J	$\frac{1}{4}$	$\frac{1}{2}$	∞	4	2	0	(420)
K	$\frac{1}{4}$	$\frac{1}{2}$	$\frac{1}{2}$	4	2	2	(422)
L	1	-1	1	1	-1	1	($1\bar{1}1$)
M	$\frac{1}{2}$	$-\frac{1}{2}$	$\frac{1}{2}$	2	-2	2	($2\bar{2}2$)

Step 2 or 3 results in a three-number code known as the *Miller indexes* (*hkl*), which identify the particular set of parallel planes. Sometimes hexagonal crystals are indexed in terms of the three basal planes $a_1 = a_2 = a_3$ and the vertical axis c (Figure 2.10), giving Miller indexes (*hkil*). However, the i index may be omitted because $i = -(h + k)$, giving (*hkl*) or (*hk·l*).

Negative indexes indicate planes that intersect unit cell axes on the other side of the origin from the side regarded as positive; for example, compare planes (111) and ($1\bar{1}1$) (G and L in Figure 2.11).

Each set of planes has its own *interplanar spacing* or *d spacing* d_{hkl}, the perpendicular distance between any two adjacent planes in the set. Some of these d spacings are shown in Figure 2.11.

Planes having indexes $n(hkl)$ are parallel to (*hkl*) planes, and $d_{n(hkl)} = d_{hkl}/n$. For example, in Figure 2.11, (200) is parallel to (100) and $d_{200} = d_{100}/2$ (planes B and A); ($2\bar{2}2$) is parallel to ($1\bar{1}1$) and $d_{2\bar{2}2} = d_{1\bar{1}1}/2$ (planes M and L). Similarly, (440) ‖ (330) ‖ (220) ‖ (110) with relative d-spacings $\frac{1}{4} : \frac{1}{3} : \frac{1}{2} : 1$; and (936) ‖ (312) with relative d spacings $\frac{1}{3} : 1$. Although $n(hkl)$ planes are distinguishable crystallographically, x-ray spectrometric analyzer crystals designated (*hkl*) and $n(hkl)$ are identical; for example, LiF(200) = LiF(100) and LiF(220) = LiF(110).

If one or more Miller indexes, h, k, or l, consist of two digits, the Miller indexes are separated by commas, for example, (12,10,0).

Indexes *hkl* (without parentheses) represent an *individual* specific plane singled out for discussion; for example, the surface plane of a silicon crystal may be a 111 plane. Indexes (*hkl*) represent the entire set of parallel planes in the crystal.

2.4. AUGER EFFECT; FLUORESCENT YIELD

In Figure 2.12, consider a magnesium atom in which a K-shell vacancy has just been created. If the K electron was expelled by an electron, the incident electron becomes a scattered electron and the K electron a secondary electron. If the electron was expelled by an x-ray photon, it becomes a photoelectron. The K-shell vacancy is filled by an electron from, say, the L shell. Ordinarily this transition would result in the emission of a Mg $K\alpha$ photon (long dashed arrow). However, suppose that this atom undergoes the phenomenon known variously as *radiationless transition*, *internal conversion*, or the *Auger effect* (French pronunciation oh-jay'). In this case, the atom emits an electron instead of a Mg $K\alpha$ photon. The phenomenon may be explained in either of two ways: It may be assumed that the atom

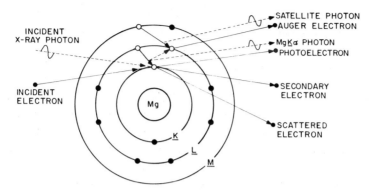

FIGURE 2.12. The Auger effect, internal photoelectric absorption. If the initial *K*-shell vacancy is produced by an electron, the incident electron becomes a scattered electron after ejecting the *K* electron, which becomes a secondary electron. If the initial *K*-shell vacancy is produced by a photon, the ejected *K* electron becomes a photoelectron. Filling of this *K*-shell vacancy by an *L* electron produces an *L*-shell vacancy and a Mg *Kα* photon. This photon may be emitted (long dashed arrow). However, internal absorption of the Mg *Kα* photon (Auger effect) produces, say, a second *L*-shell vacancy (short dashed arrow), resulting in an *LL*-state atom. Filling of one of the *L* vacancies by an *M* electron produces a satellite-line photon and results in an *LM*-state atom.

releases the energy of the *L*-to-*K* electron transition by emission of, say, an *L* or *M* electron instead of a photon. Alternatively, it may be assumed that the *L*-to-*K* electron transition results in generation of a Mg *Kα* photon. However, in this case, the photon does not leave the atom of its origin, but is absorbed within the atom (short dashed arrow) with consequent expulsion of, say, an *L* electron. This process may be regarded as internal photoelectric absorption and is analogous to the internal *γ* conversion described in Section 1.8.8 and shown in Figure 1.18C.

Either version results in emission of an Auger electron rather than an x-ray photon, and leaves the atom doubly ionized, that is, with two orbital electron vacancies—the one created by filling of the initial vacancy and the one created by the Auger process. Such atoms are said to be in the *LL*, *LM*, *MM*, etc., state. Incidentally, multiply ionized atoms are also produced by multiple excitation.

The Auger effect is more common in elements of low atomic number because their atomic electrons are more loosely bound and their characteristic x-ray photons more readily absorbed. The effect is more common for *L* series than for *K* series for the same reason.

   The Auger effect is the basis for satellite lines (Section 1.8.2), fluorescent yield (see below), and the analytical method of Auger-electron spectrometry.

   In Figure 2.12, the Auger process leaves the atom in the LL state. If now an M electron falls into one of these L vacancies, an x-ray photon is emitted having wavelength slightly different than would have been produced by the same electron transition in a singly ionized atom. X-ray spectral lines arising in this way are known as *satellite* or *nondiagram* lines.

   One consequence of the Auger effect is that the lines in a given series are not as intense as would be predicted from the number of vacancies created in the associated orbital. The *K fluorescent yield* or *K characteristic photon yield* ω_K (omega) is the number of photons of all lines in the K series emitted in unit time divided by the number of K-shell vacancies formed during the same time:

$$\omega_K = \frac{\sum (n_K)_i}{N_K} = \frac{n_{K\alpha_1} + n_{K\alpha_2} + n_{K\beta_1} + \cdots}{N_K} \tag{2.23}$$

where ω_K is K fluorescent yield; N_K is the rate at which K-shell vacancies are produced; and $(n_K)_i$ is the rate at which photons of spectral line i are emitted. The L and M fluorescent yields ω_L and ω_M are defined similarly. *Auger yield*, or *Auger-electron yield* is the ratio of the numbers of Auger electrons and orbital electron vacancies produced in the same time and is equal to $1 - \omega$. Were it not for the Auger effect, ω would always be 1. Actually, ω varies with atomic number and line series, as shown in Figure 2.13 and Table 1.4, columns 14–16. Fluorescent yield is a major limitation of sensitivity for elements of low atomic number.

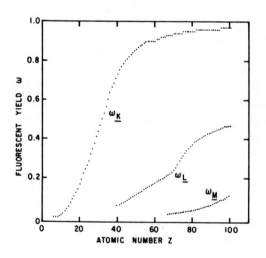

FIGURE 2.13. *K*, *L*, and *M* fluorescent yields as functions of atomic number.

Note that the *fluorescent* yield is the probability that the filling of a vacancy in a specified shell will result in emission of a characteristic x-ray photon—regardless of whether the vacancy arose from primary or secondary excitation. The term was originally defined on the basis of secondary x-ray intensity produced on irradiation of a target with x-rays, but now has the more general significance.

Chapters 1 and 2 in general and Table 1.4 in particular show that many parameters of importance in x-ray spectrometry are functions of atomic number: wavelengths of spectral lines and absorption edges, photon energies of spectral lines, excitation potentials, mass-absorption coefficients, fluorescent yields, x-ray scattering power, ratio of Compton and Rayleigh scatter, etc. Thus, the x-ray spectrochemist should strive to think of chemical elements in terms of their atomic numbers to aid him in making quick mental comparisons of x-ray properties of specified elements of interest. Throughout this book, wherever a series of elements is listed in text or tables, they are listed in order of increasing or decreasing atomic number.

SUGGESTED READING

See the list given at the end of Chapter 1.

Chapter 3

X-Ray Emission Spectrometric Analysis: General Introduction

Having considered the excitation, nature, and properties of x-rays and x-ray emission spectra, we now begin our study of the application of x-ray spectra to chemical analysis. However, first we define some terms used throughout the remainder of the book.

3.1. NOMENCLATURE

The term *x-ray spectrochemical analysis* signifies all the various analytical chemical methods based on x-ray emission and absorption spectra, including electron-probe microanalysis. *X-ray emission spectrometry* and *x-ray absorption spectrometry* signify the several analytical methods based on x-ray emission and absorption spectra, respectively. In Section 3.2, it is explained that x-ray emission spectrometry may be classified further on the basis of dispersion mode as *wavelength-dispersive* and *energy-dispersive*, or on the basis of excitation mode as *x-ray-, electron-, ion-, radioisotope-excited*, etc. The term *x-ray fluorescence* indicates the phenomenon (*not* the method) of x-ray excitation of x-ray emission spectra. Consequently, x-ray-excited x-ray emission spectrometry has come to be known as *x-ray fluorescence spectrometry*. Thus, there is *wavelength-dispersive x-ray fluorescence spectrometry, energy-dispersive x-ray fluorescence spectrometry, radioisotope-excited energy-dispersive x-ray spectrometry*, etc.

There has long been the need for a word to replace those cumbersome terms inevitable in all papers and books on chemical analysis—"element sought," "element to be determined," "element of interest," "analytical element," etc. In this book, the term *analyte* is used. The analyte may be a *major*, *minor*, or *trace* constituent of the material analyzed, depending on whether its concentration is of the order 10, 1, or ≤ 0.1 wt%.

Qualitative, *semiquantitative*, and *quantitative* analyses consist in *detection*, *estimation*, and *determination* of the analyte, respectively. Terms such as "quantitative determination," although widely used, are redundant because determinations are necessarily quantitative.

The term *specimen* indicates the object placed in the instrument specimen compartment and on which x-ray intensity measurements are made. The specimen may be: (1) a *sample*—a material of unknown composition to be analyzed or a preparation derived from it; (2) a *standard*—a similar material or preparation having known composition for calibration purposes; (3) an *intensity-reference standard*—a specimen having high chemical and physical uniformity and stability on which intensity is measured to permit adjustment of the excitation conditions to duplicate previous results; or (4) a *test specimen* for some special purpose, for example, a piece of paraffin to permit recording of the scattered primary x-ray beam.

In this chapter, we consider the instrumentation and methods of x-ray emission spectrometric analysis in general terms. Excitation and wavelength (crystal) dispersion are considered in Chapter 4; detection, readout, and x-ray intensity measurement are considered in Chapter 5; and energy dispersion is considered in Chapter 6.

3.2. INTRODUCTION

X-ray emission spectrometry is a group of nondestructive instrumental methods of qualitative and quantitative analysis for chemical elements based on measurement of the wavelengths and intensities of their characteristic x-ray spectral lines.

An x-ray spectrochemical analysis involves six essential phases, as follows: (1) selection of the analytical method (Chapter 9) and specimen preparation and presentation (Chapter 10); (2) excitation of the characteristic x-ray line spectra of the chemical elements in the specimen (Section 4.1); (3) dispersion (separation) of these lines so that a line of any specified element can be measured individually (see below, Section 4.2, and Chapter 6); (4) detection, or conversion of x-ray photons into pulses of electric current (Section 5.1); in nearly all x-ray detectors used in x-ray spectrometers, the amplitude of each pulse is proportional to the photon energy of the x-ray photon producing it; (5) readout, display, and measurement of the detector output data (Sections 5.2 and 5.3); and (6) identification of the elements present in the specimen (qualitative analysis, Chapter 7) and/or reduction of x-ray intensity data to analytical concentrations (quantitative analysis, Chapter 9). Phase 1 is effected by the analyst. Phases 2–5 are functions of the x-ray spectrometer. The qualitative aspect of phase 6 is done by the analyst (Chapter 7), although in energy-dispersive spectrometers, it may be effected electronically (Sections 6.1.3 and 7.1.2.2). The quantitative aspect of phase 6 may be done manually or by computer (Chapter 9).

In any method of chemical analysis based on measurement of x-ray spectral lines, some means must be provided to permit the lines of each element in the specimen to be measured individually. There are two basic approaches to this objective: The elements may be *excited* individually, or they may be excited all at once and some means provided to isolate the lines of each element so they can be *measured* individually. Most x-ray spectrometers by far use the latter approach, and it is convenient to classify all such spectrometers in two categories on the basis of the two means most commonly used to isolate the individual lines: *wavelength dispersion* and *energy dispersion*.

In wavelength-dispersive spectrometers, the several x-ray lines emitted by the specimen are dispersed spacially by crystal diffraction, prior to detection, on the basis of their wavelengths (Section 2.3.1). Thus, in principle at least, the detector receives only one wavelength at a time.

In energy-dispersive spectrometers, the detector receives all excited lines of all the specimen elements at once. For each incident x-ray photon, the detector generates a pulse of electric current having height proportional to its photon energy. The detector output is amplified and subjected to electronic pulse-height analysis to separate the pulses arising from the several detected wavelengths on the basis of their heights and thereby on the basis of the photon energies of the incident x-ray lines. By appropriate setting of the operating parameters, the pulses of each line of interest can be measured individually.

The wavelength- and energy-dispersive modes are illustrated simply in Figure 3.1. The polychromatic primary x-ray beam emitted by the x-ray tube is represented by a single short wavelength. The primary beam irradiates the specimen and excites three spectral lines having short, intermediate, and long wavelength, respectively. The separation of these three lines by wavelength and energy dispersion is represented to the right and left of the specimen, respectively.

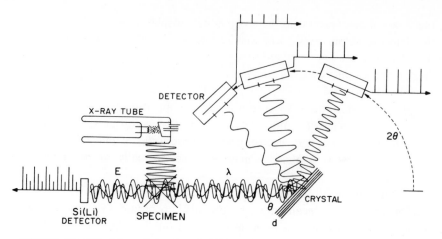

FIGURE 3.1. Principles of wavelength-dispersive (λ) and energy-dispersive (E) x-ray spectrometry.

In a wavelength-dispersive spectrometer (λ), the three wavelengths fall upon a crystal. The crystal and detector are made to rotate synchronously through successive angles θ and 2θ, respectively. As the crystal rotates, it passes the correct angle θ for its interplanar spacing d to diffract each wavelength λ. The detector receives the diffracted x-rays, converting each incident photon into a pulse of electric current having height proportional to the photon energy. If diffraction orders (Section 2.3.1) are disregarded, at any θ setting, the detector output consists of pulses having the same average height.

In an energy-dispersive spectrometer (E), all three wavelengths enter the detector at once so that all three pulse heights are present in the detector output at once. These pulses are separated electronically by height.

It is evident that only in wavelength (crystal) dispersion are the several wavelengths emitted by the specimen actually dispersed *spacially*, and only in this mode is it necessary to move the detector for each wavelength measured. Hence, the wavelength-dispersive and energy-dispersive modes are sometimes referred to as *dispersive* and *nondispersive*, respectively. However, pulse-height analysis is dispersive in the very real sense that the pulses of the several lines are dispersed on the basis of their heights and thereby their original photon energies; energy- and wavelength-dispersed spectra are shown in Figures 7.1 and 7.2, respectively. Thus, one should refer to the two modes as *wavelength-dispersive* and *energy-dispersive* rather than *dispersive* and *nondispersive*. I prefer to reserve the term *nondispersive* to all modes in which the emitted spectral lines are not separated on the basis of either wavelength or photon energy, such as selective excitation, selective filtration, and selective detection. These modes are defined in appropriate sections later.

3.3. WAVELENGTH-DISPERSIVE X-RAY SPECTROMETRY

The wavelength-dispersive x-ray spectrometer is represented by the standard commercial flat-crystal x-ray fluorescence spectrometer, which until recently has been by far the most widely used type of x-ray spectrometer.

3.3.1. Principle

Wavelength-dispersive x-ray secondary-emission spectrometry, or x-ray fluorescence spectrometry (XRFS), is a nondestructive instrumental method of qualitative and quantitative analysis for chemical elements based on measurement of the wavelengths and intensities of their x-ray spectral lines emitted by secondary excitation. The primary beam from an x-ray tube irradiates the specimen (sample or standard), exciting each chemical element to emit secondary spectral lines having wavelengths characteristic of that element (basis of qualitative analysis) and intensities related to its concentration (basis of quantitative analysis). The spectral lines are dispersed spacially by crystal diffraction prior to detection.

3.3.2. Basic Instrument and Method

The essential features of an x-ray fluorescence crystal spectrometer are shown in Figure 3.2. The instrument consists of three principal sections which effect excitation, dispersion, and detection and readout, respectively. The primary x-ray beam irradiates the specimen A held in some type of drawer or holder B in a specimen compartment C. A leaded shutter mechanism (not shown) is provided to block the primary x-ray beam at the x-ray tube window when the specimen is inserted and removed from its compartment. The secondary radiation thus excited contains many wavelengths and is emitted in all directions, but only those rays directed substantially parallel to the foils of the source Soller collimator D emerge from the specimen compartment and reach the analyzer crystal E. The crystal is mounted on a stage and may be rotated, the axis of rotation of the crystal

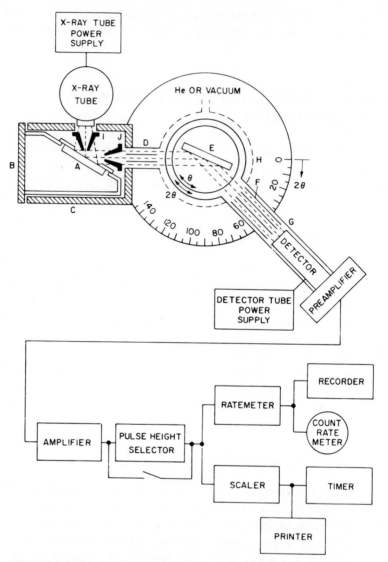

FIGURE 3.2. Wavelength-dispersive flat-crystal x-ray fluorescence spectrometer.

being also the goniometer axis. As the crystal rotates, the angle θ it presents to the secondary beam varies, and each wavelength in the beam, in turn, satisfies the Bragg law and is diffracted. In this way, the crystal disperses the secondary spectrum, in effect causing each wavelength to go in a different direction.

Note the difference in this respect between wavelength-dispersive x-ray spectrometry and optical emission spectrography. In x-ray spectrometry, the spectral lines are diffracted individually in sequence. In optical spectrography, the entire spectrum is produced at once, either by a diffraction grating or refraction prism. However, in *energy*-dispersive x-ray spectrometry (Section 3.4), the entire spectrum *is* produced simultaneously.

A second detector collimator *F* and the detector are mounted on a common arm *G* that rotates on the same axis as the crystal and is geared to it with a "2-to-1" ratio, that is, as the crystal turns through an angle θ, the detector arm turns through twice that angle, 2θ. Thus, the detector and incident secondary beam always present equal angles to the crystal, and the detector is always in the correct position to receive any diffracted x-rays (Figure 2.7). Incidentally, x-ray spectrogoniometers are graduated in degrees 2θ, rather than degrees θ, because the detector is always at angle 2θ to the direction of the secondary beam incident on the crystal. The detector converts each x-ray photon it absorbs into a pulse of electric current having amplitude proportional to the photon energy. These pulses are amplified by the preamplifier, which is mounted close to the detector on the detector arm.

When wavelengths greater than $\sim$2 Å are measured, it is necessary to reduce the absorption of the radiation path (x-ray tube window to detector window) with a helium or vacuum enclosure *H*. When very small specimens or small selected areas on large specimens are measured, apertures *I* and *J* may be placed in the primary or secondary x-ray beam (Section 10.9.1.)

The instrument just described—consisting of the specimen chamber, crystal, source and detector collimators, helium or vacuum enclosure, detector, preamplifier, and the drive mechanism for the crystal and detector arm—is known as the *goniometer* (angle-measuring device). When used for spectrometry in this way, the goniometer is referred to as a *spectrogoniometer* or *spectrometer*. (The alternative mode of operation, described briefly below, is as a *diffractometer*.) The crystal and detector arm may rotate in a horizontal or vertical plane—that is, the goniometer axis may be vertical or horizontal. In general, different sets of components—x-ray tube, collimator(s), crystal, and detector—are required for optimum performance in the long-, intermediate-, and short-wavelength x-ray spectral regions (Chapters 4 and 5).

The preamplifier is connected by cable to the amplifier, which further amplifies the detector output pulses. As we have seen, the crystal can diffract more than one wavelength at a given Bragg angle, and when this happens, each wavelength gives detector output pulses of different amplitude. The pulse-height selector sorts these pulses, passing only those having a certain preselected height. The pulse-height selector may be by-passed, as shown schematically in Figure 3.2. The amplified pulses—with or without selection—may be registered by the ratemeter, which integrates them and displays on a meter or strip-chart recorder the rate at which pulses enter the ratemeter, and thus the rate at which x-ray photons are absorbed in the detector. Alternatively, the pulses may be individually counted by the scaler-timer, the data from which may be recorded manually or by a printer. All these components and the detector-tube power supply are mounted in the electronic-circuit panel or console.

The x-ray tube and power supply, spectrogoniometer, and electronic-circuit panel collectively constitute the x-ray fluorescence spectrometer. The particular instrument just described is a single-channel manual spec-trometer—that is, it can measure only one wavelength at a time, and it is operated manually. Dual and multichannel and automatic instruments are described in Section 3.3.3.

In qualitative analysis, the spectrometer is made to scan through the entire 2θ region. The goniometer motor drive rotates the crystal and detector arm at rates proportional to θ and 2θ, respectively, per unit time, as detected x-ray intensity is recorded by the ratemeter-recorder. The result is a chart of peaks on a grid of intensity *versus* 2θ or wavelength (Figure 7.2). The elements present are identified by the 2θ angles—or wavelengths—at which peaks occur on the chart, and their concentrations may be estimated from the peak heights—or intensities. Analyte peaks may be recorded from standards for comparison. Qualitative and semiquantitative analyses are described in Chapter 7.

In quantitative analysis, the intensities of analyte lines are measured from samples and standards. The measurements are almost always made with the scaler and timer, although the ratemeter may be used. Calibration curves or mathematical relationships are established from the standard data. Quantitative analysis is described in Chapter 9.

Incidentally, the x-ray spectrogoniometer and diffractometer are es-sentially similar instruments. The spectrogoniometer permits qualitative and quantitative analysis for chemical elements in any kind of specimen by x-ray spectrometry, that is, by measurement of the wavelengths and in-tensities of their characteristic x-ray emission spectral lines. The diffractom-eter permits qualitative and quantitative analysis for chemical *species* (elements and/or compounds) in *crystalline* specimens by measurement of interplanar spacings (Section 2.3.2) of and diffracted intensities from the crystallographic planes. However, although both the spectrometer and

TABLE 3.1. Comparison of the X-Ray Spectrometer and Diffractometer

Bragg law	$n\lambda$	$=$	$2d$	$\times$	$\sin \theta$
Spectrometer	Calculated		Known		Measured
Diffractometer	Known		Calculated		Measured

diffractometer are based on the Bragg law, their modes of operation are entirely different. In a diffractometer, the specimen drawer B (Figure 3.2) and compartment C are omitted, and the x-ray tube is rotated about its axis so that the primary beam is directed through the collimator D. The single crystal E is replaced with the polycrystalline sample to be investigated. Thus, a known wavelength—the filtered $K\alpha$ line of the x-ray tube target—is directed on the sample, and as the sample is rotated through successive θ values, each set of crystal planes (hkl) having its own interplanar spacing d_{hkl} in turn satisfies the Bragg law and diffracts the incident wavelength. The functions of the spectrometer and diffractometer may be contrasted in terms of the Bragg law as shown in Table 3.1. Both instruments produce charts consisting of peaks on an intensity *versus* 2θ ("two theta") scale. On the spectrometer chart, each peak represents a *different* wavelength diffracted by the *same* set of planes in the analyzer crystal. On the diffractometer chart, each peak represents the *same* wavelength— the target $K\alpha$ line—but diffracted by a *different* set of planes in the crystalline sample.

3.3.3. Manual, Semiautomatic, and Automatic Spectrometers

3.3.3.1. Introduction

Commercial x-ray diffractometers were available long before commercial x-ray spectrometers. Figure 3.2 is typical of the earliest commercial spectrometers, which were actually converted diffractometers. These older goniometers with specimen "drawers" are still available and have many advantages. In small laboratories and schools, they perform both spectrometric and diffractometric functions economically and with relative ease of interchangeability. They are excellent for teaching purposes because the components and operation are easily observable. They also provide a versatility in choice of modes of operation, components, accessories, and specimen presentation not practical with the newer instruments (see below).

In the spectrogoniometer in Figure 3.2, it is evident that the primary x-ray beam from the x-ray tube is directed *downward* onto the *upper* surface of the specimen. The first basic modification made in this goniometer was *inverted geometry*, in which the primary beam is directed *upward* onto the *lower* surface of the specimen. This arrangement accommodates much larger specimens, has certain advantages for liquid and loose-powder specimens, and is more compatible with modern automatic spectrometers (see below). With the original geometry (Figure 3.2), the specimen size is limited by the space between the x-ray tube window and goniometer base or instrument tabletop. With inverted geometry, there is no such limitation. If appropriate provision is made for x-ray safety, very large specimens can be accommodated, including engine parts and paintings, antiquities, and other art objects.

The x-ray spectrometer shown in Figure 3.2 is a *manual, single-channel, wavelength-dispersive* instrument; that is, it has only one crystal–collimator–detector "channel," and. although qualitative 2θ scans are motor-driven, all settings and measurements for quantitative analysis are made manually, and components are changed manually.

Suppose that such a manual single-channel spectrometer is to be used for production control. It may be required to analyze large numbers of similar samples submitted continuously throughout the day, and in such cases, it is common for results to be required very soon after the samples are submitted. Even instruments used for incoming inspection and lot approval are required to make large numbers of multielement analyses from time to time, perhaps with somewhat less urgency. It has been mentioned already that for optimum performance, different x-ray tube targets, crystals, collimators, and detectors are required for different x-ray spectral regions. If the elements in a multielement specimen are relatively close in atomic number, the same set of components is likely to apply to all. However, if the analytes have a substantial range of atomic number, no one set of components is likely to give optimal—or even acceptable—results for all of them. In such cases the same components may be used for the entire multielement analysis with a certain amount of sacrifice of sensitivity and statistical precision; otherwise, one or more components must be changed during the course of the analysis, with substantial loss of time. In multielement analyses, regardless of the spread in atomic number, many individual settings may be required for each analyte and perhaps for one or more internal standards—kV, mA, 2θ, preset count or time, pulse-height selector, etc.—and the accumulated count or lapsed time must be read and recorded. For large numbers of multielement samples, the number of settings and readings becomes very large, and, aside from the tedium, the opportunity for error increases. If the specimen compartment can accommodate only one specimen and there is no provision for continuous loading, the instrument is completely idle while one sample is removed and the next one loaded. For large numbers of samples, the cumulative loss of time can be substantial.

Consequently, ever since the first basic commercial x-ray spectrometers were introduced, modifications and additions have been made to facilitate changing of instrument components and settings in multielement analyses. The first innovations were the continuous specimen loader, which permits loading the next specimen while one is being measured, and the multiple specimen holder, which permits indexing a series of preloaded specimens into the measuring position. The dual-target x-ray tube, multiple crystal changer, coarse–fine collimator changer, and dual detector permit convenient interchange of these components without serious interruption of a series of multielement analyses. Provision was made to permit selection of one of several preset x-ray tube operating potentials (kV) or currents (mA) by means of pushbuttons or selector switches. Coupling of the pulse-height selector baseline and window settings to the goniometer 2θ drive eliminated the need for manual setting of these parameters. A typewriter printed out the scaler-timer data.

All these features represent accessories on, or relatively minor modifications of the basic spectrometer shown in Figure 3.2. Today's instruments are designed specifically for x-ray spectrometry, and, although they have the same components and operate in the same way, the goniometer is no longer visible nor accessible outside the instrument housing. These modern instruments are designed to make intensity measurements from large numbers of multielement samples and to compute and print out the analytical concentration data, all quickly and conveniently and with a minimum of operator attention. These modern computerized automated instruments are of two types—sequential and simultaneous.

3.3.3.2. Sequential Automatic Spectrometers

A sequential automatic x-ray spectrometer is basically an automated single-channel instrument programmed to index automatically to a preselected series of analyte lines, stopping at each to accumulate a preset count from the sample and from a standard. The instrument is programmed to automatically select components and conditions optimal for each analyte line. This type of instrument greatly expedites routine analyses. The operator is relieved of the tedium of making the many instrument settings and readings required for multielement analyses of large numbers of routine samples, and the opportunity for error is minimized.

All these instruments provide for automatic selection of a number of the following components and conditions for each programmed analyte, but none provides for all. Of course, any component or parameter for which selection is not provided is the same for all analytes.

1. *Specimen*: Various instruments accommodate 2–30 specimens, which are automatically indexed into position in turn by means of a carousel arrangement.

2. *Specimen mask area*: Choice of two or three window sizes.

3. *Specimen motion*: Fixed or rotated.

4. *Element*: The goniometer automatically indexes to up to 15–24 2θ angles, depending on the instrument. In one instrument, the optimal 2θ angle for, say, Al $K\alpha$ can be programmed for both the element and aluminum oxide to allow for wavelength shift due to chemical effects.

5. *X-ray tube target*: Tungsten or chromium, platinum or chromium, or molybdenum or chromium (only with a dual-target tube).

6. *Primary-beam filter*: In or out; if in, choice of up to five to remove target lines and/or to reduce analyte-line intensity and thereby preclude coincidence loss without changing x-ray tube operating conditions.

7. *X-ray tube potential*: A selection of preset values up to 50, 60, 75, or 100 kV, depending on the instrument.

8. *X-ray tube current*: A selection of preset values up to 50, 75, or 100 mA, depending on the instrument.

9. *Collimator*: Coarse, medium, or fine.

10. *Crystal*: Choice of up to six.

11. *Diffraction order*: The instrument can be instructed to measure the first- or second-order line.

12. *Path*: Air, helium, or vacuum.

13. *Detector*: Most instruments have a dual scintillation and gas-flow proportional counter for short- and long-wavelength lines, respectively, and the output can be taken from either or both.

14. *Delay time*: The instrument may be set to delay a preset time before making a measurement each time a new specimen is indexed into position to ensure adequate helium flushing or vacuum and/or to allow primary-beam equilibration after changing x-ray tube kV and mA.

15. *Amplifier gain*: A selection of several preset values.

16. *Automatic pulse-height selection*: In or out, and window width.

17. *Preset count or preset time*: Which mode, and a selection of values. For each analyte, that count (or time) is programmed that gives the required statistical precision.

18. *Repetition*: The instrument can be set to repeat a measurement a specified number of times.

Some instruments also provide for selection of analytical measurement modes as follows:

1. *Specimen or element sequence*: The instrument can be programmed to measure all analytes on the same specimen, then advance to the next specimen, or to measure the same analyte on all specimens, then advance to the next analyte.

2. *Peak or profile mode*: The instrument can be programed to "slew" to the peak 2θ, scale and print out or store the peak intensity, then slew to the next peak, etc.; alternatively, it can be made to slew to a specified low-2θ edge of the peak, step-scan or continuously scan through the peak to a specified high-2θ edge of the peak, print out or store the profile or integrated count, then slew to the low-2θ side of the next peak, etc. ("Slewing" is very rapid scanning—up to $600°/\text{min}$—along the 2θ scale.)

These spectrometers can be operated in any of three modes as follows.

1. *Manual*. In this mode, the instrument becomes substantially a standard single-channel scanning spectrometer useful for qualitative analysis (2θ scans) and for quantitative analysis of individual samples.

2. *Semiautomatic*. For each sample, the analytes are selected in turn, and for each analyte, the components and conditions are selected—all by manual operation of pushbuttons and multiposition selector switches on the control panel. As the operator makes each setting, the instrument indicates that it has executed its "instructions" correctly. This mode is used for analysis of large numbers of nonroutine samples for which no program has been established.

3. *Fully Automatic*. The samples are indexed into position and, on each sample, the specified analytes are measured, each with its specified components and conditions—all automatically. This mode is used for analysis of large numbers of routine samples that warrant establishment of a program. The program may be set manually, or by means of a patchboard, switchboard, prewired plug-in casette, punched card, or punched tape specific for a certain series of analytes in a certain type of sample. Alternatively, the program can be stored in the computer memory (see below) and selected by addressing it on the computer input–output typewriter.

The spectrometer output may be printed out directly or "interfaced" with a computer, which may be "dedicated," that is, a permanent complement to the spectrometer, or a central "time-shared" computer. The computer converts intensity data measured from the samples directly to concentration by means of calibration functions automatically established from intensity data measured from the standards. The calibration may be based on a single standard, a linear function derived from two or more standards, or even a nonlinear function divided into regions, each considered to be linear. Alternatively, the computer can be programmed to calculate concentrations by application of mathematical intensity corrections (Section 9.9).

The equipment is usually designed so that the basic manual instrument can be purchased first, then later expanded to a semiautomatic, then to a fully automatic, instrument by addition of controller and programming units, respectively. Finally, the computer may be added.

The output may be read out on a printer, a digital display, or the computer typewriter. A strip-chart recorder may be provided for ratemeter readout with manual operation. The output data for each analyte in each sample may be read out in any of four forms: (1) absolute accumulated count (or counting time); (2) ratio of accumulated count (or time) for the sample and standard; (3) ratio of accumulated count (or time) for the specimen (sample or standard) and monitor; and (4) concentration.

The specimens may be solids, briquets, powders, slurries, or liquids. Provision is usually made for continuous loading—through an air lock if helium or vacuum path is used. Most sequential spectrometers are useful for all elements down to atomic number 9 (fluorine).

Sequential x-ray spectrometers, compared with simultaneous spectrometers (Section 3.3.3.3), have the advantages that: (1) they are much more compact because of the absence of a multiplicity of spectrometers arrayed about the specimen; (2) they are economical of collimators, crystals, detectors, and electronic readout components; (3) x-ray tube target, primary-beam filter, kV, and mA can be programmed parameters; (4) the measured intensity for all analytes leaves the specimen in the same direction, so that homogeneity and surface texture are not so critical; and finally (5) because there are no fixed preset spectrometers, they are readily indexed to any specified 2θ angle for measurement of any analyte line, scattered target line, or background. The principal disadvantages are that the indexing and control mechanisms are very complex, use of curved crystals is not convenient, and the advantages of simultaneous measurement of analyte and internal-standard lines, analyte line and background, and all analyte lines cannot be realized. Moreover, sequential spectrometers are not necessarily faster than manual ones: They must perform all the same operations as a human operator. Finally, automatic spectrometers are not necessarily more accurate than manual ones: Accuracy still depends largely on the standards and calibration function. Rather, the benefits of the automatic sequential instruments lie in their reduction of operational and instrumental error, and in freeing the operator for other work while the analysis is in progress.

3.3.3.3. Simultaneous Automatic Spectrometers

A simultaneous automatic x-ray spectrometer is basically a number— up to 22—of single-channel instruments, each havings its own crystal, collimator(s), and detector, arrayed radially about a common x-ray tube and specimen. Each detector has its own amplifier, pulse-height selector, and count scaler or integrator. Most of the channels are *fixed* or *preset*, that is, set at the 2θ angle for a specified analyte line and fitted with components optimal for that line, and thereafter left undisturbed; these channels have no 2θ motor drive. However, one or more of the channels may be *scanning* channels. These have motor drives and may be used to make 2θ scans for qualitative analysis. If three scanning channels are provided, they may be fitted with components optimal for the short-, intermediate-, and long-wavelength x-ray spectral regions, respectively, and their outputs read out on a three-channel recorder. Then a qualitative analysis can be made in all three spectral regions simultaneously. These scanning channels are also conveniently set to measure lines of elements determined less frequently than those for which the fixed channels are set.

In operation, all channels accumulate counts for their respective analyte lines for the same preset time or until a standard channel accumulates a preset count. Any one of the channels may be designated as the standard channel, and its crystal may be set to diffract an internal standard line, a scattered x-ray tube target line, or a selected wavelength in the scattered primary continuum. Alternatively, the standard channel may be a nondispersive monitor channel receiving scattered primary x-rays. (All these techniques are considered in appropriate subsequent sections.) On completion of the counting period, the accumulated count in each channel is read out in rapid sequence and either printed out or entered into a computer, as in a sequential spectrometer (see above).

Simultaneous instruments usually have a modular construction. The basic unit can accommodate up to six channel positions arranged radially about the specimen compartment. Each position receives a metal plate on which can be placed one or two curved-crystal channels. Thus, the basic unit can accommodate up to 12 analytical channels. The single unit can be expanded to a dual unit by addition of a semicircular annex, not unlike a bay window, that can accommodate five more positions—10 more channels—up to a total of 22.

The simultaneous technique has some of the advantages of the sequential and some additional ones. Simultaneous instruments are extremely rugged because of the fixed preset spectrometers and the absence of moving parts. The preset spectrometers can be of the full-focusing curved-crystal type. The components (crystal, detector, etc.) of each channel can be optimized for its specific element. The simultaneous measurement of all analytes permits greater economy of analysis time and cost. Analysis time is limited by the counting or integrating time required to give acceptable statistical precision for the analyte having lowest intensity, plus short loading and readout times. If no element requires more than 1 min of counting time, an analysis for 22 elements can require as little as 2 min for both counting and readout. A sequential instrument requires the sum of the individual counting times plus the operation time. Another advantage is that many instrument and specimen variables are compensated by simultaneous measurement of all intensities. Also, it is very convenient to be able to measure analyte and internal-standard lines, and even line and background, simultaneously. Finally, in trace analysis, higher sensitivity and statistical precision can be realized in relatively short time by setting two channels at the same trace analyte line and combining their outputs.

The principal disadvantage of simultaneous spectrometers are that: (1) they are very bulky because of the multiplicity of spectrometers; they become especially cumbersome when provision must be made for large numbers of elements; (2) they require large numbers of expensive components (collimators, crystals, and detectors); (3) the electronic readout system is elaborate and complex; (4) x-ray tube target, primary-beam filter, kV, and mA cannot be programmed parameters; (5) because each spectrometer views the specimen from a different direction, homogeneity and surface topography are critical, and specimen rotation is usually required; and (6) it is inconvenient to change a preset channel to a different spectral line. Because of this last limitation, simultaneous instruments are relatively inflexible and best suited to analysis of very large numbers of similar samples, especially in production control. For example, these instruments are ideally suited for analysis of iron-base alloys in iron foundries and steel mills, copper-base alloys in brass and bronze foundries, briqueted cements in cement factories, and hydrocarbon liquids in petroleum refineries. The flexibility of simultaneous instruments is substantially increased by inclusion of one or more "scanning"—as distinguished from preset fixed—spectrometers that can be used to record 2θ scans and can readily be set to any specified analyte line.

3.3.3.4. Special Spectrometers

A portable x-ray fluorescence spectrometer is available commercially consisting of two units connected by a 9-ft cable. The power unit weighs 100 lb and houses a 30-kV, 1–4-mA constant-potential supply that operates from the 110-V ac line and uses 220 W at full power. The x-ray unit weighs 23 lb and houses the x-ray tube, specimen compartment, goniometer, detector, ratemeter, and detector-tube power supply. The x-ray tube has a tungsten target and beryllium end window. The goniometer consists of source and detector collimators, LiF crystal, and an argon-filled detector; the 2θ angle is varied by means of an external knob in the range 13–89°, thereby permitting measurement of all elements down to atomic number 22 (titanium). The ratemeter output is indicated on a panel meter. A goniometer drive motor and strip-chart recorder are available to permit making 2θ scans. A scaler-timer unit is also available. A specimen holder is available for small solids, powders, and liquids, another for bar, rod, and sheet stock. Sensitivity down to 0.01 wt% is claimed for favorable cases. The instrument can be used in the plant and warehouse for identification and sorting of metal stock, chemicals, leaded fuels, etc. With a mobile generator, it can be used in the field for geological and mining applications. The instrument has also been used to indicate the lead content of painted surfaces for investigation of sources of lead poisoning.

Standard x-ray tubes are relatively inefficient for excitation of long-wavelength x-ray spectra, so electron excitation becomes attractive in this region. However, some of the advantage is offset by the increased continuum intensity and resultant higher background. Of course, the electron source and specimen must be placed in a vacuum enclosure. Electron penetration is much shallower than x-ray penetration, so that the effective specimen layer is only a few micrometers at most. This reduces absorption–enhancement effects, but greatly increases surface-texture effects. Also, it is necessary that the surface be representative of the bulk specimen. At the point of impact, the electron beam heats the specimen, possibly expelling volatile constituents, and also causes deposition of carbon from residual vacuum-pump oil vapors. These conditions are avoided by use of a large-area electron beam, or by scanning a focused beam over a large area, $\sim$1 cm². A much more sophisticated electron-excited x-ray spectrometer, the electron-probe microanalyzer, is described in detail in Chapter 11.

In x-ray spectrometry, the ultralong-wavelength (ultrasoft) x-ray region extends from 15 to 150 Å, or about 1–0.1 keV. Its short-wavelength limit lies at the long-wavelength limit of standard commercial x-ray spectrometers. The region contains the $K\alpha$ lines of elements 9 (fluorine, 18.3 Å) to 4 (beryllium, 114 Å) and the $L\alpha_1$ lines of elements 26 (iron, 17.6 Å) to 14 (silicon, 136 Å). These lines have very low excitation potentials and are very strongly absorbed. These features dictate the design and construction of spectrometers for the ultralong-wavelength region. The specially designed, demountable, continuously pumped x-ray tubes operate at low potential but high current—up to 15 kV, 500 mA—and the power supplies are designed accordingly. KHP (potassium hydrogen phthalate), lead stearate, and lead lignocerate analyzers permit measurement of K spectra of elements down to oxygen, boron, and beryllium, respectively. The detectors are of the gas-flow proportional counter type (Section 5.1.3) using P90 gas (90 vol% methane, 10 vol% argon), methane, propane, or neon. The x-ray tube and detector have ultrathin windows.

Ultralong-wavelength x-ray spectrometry extends the applicability of x-ray spectrometric analysis from fluorine to beryllium. Because of the long wavelengths, the method is particularly applicable to relatively thin surface layers. The method is extremely valuable for investigation of chemical state and valence-electron band structure because the spectra in the ultralong region are relatively highly influenced by chemical state (Section 8.1). The chemical effects cause the spectral lines or bands to shift in wavelength, change in profile, or split into multiple bands. Such investigations are of great significance in chemistry, solid-state physics, and semiconductor electronics.

Because ultralong-wavelength x-rays penetrate only a few micrometers of matter, surface contamination of the specimen, crystal, detector window, etc. becomes a serious problem. Even thin films of back-streamed vacuum-pump oil vapor can be detrimental if the equipment is not adequately cold-trapped.

Process-control instruments are designed for continuous, on-line, largely unattended analysis of a product during some phase of its processing, or of a by-product, residual or waste material, or other material that can indicate the condition of the process. The output is read out under continuous observation, and any changes may be used to guide adjustment of the process. The ultimate objective of the process-control instrument is to use the output to automatically control the process. This has been realized, at first only for some simple processes, such as plating of metal sheet or recovery of a single element from a slurry of powdered ore. However, today instruments are in use that can analyze for up to nine elements in up to 21 "streams" in a complex process. If no material in the process has suitable form, sample preparation is required, such as finishing, grinding, drying, briqueting, dilution, fluxing, or addition of an internal standard; this preparation must also be automated. It is fair to state that the principal problem in design of process-control equipment is the design of reliable dynamic sample-presentation systems. The most common specimen forms are continuous strip, liquids and solutions, slurries, and briquets.

3.4. ENERGY-DISPERSIVE X-RAY SPECTROMETRY

3.4.1. Principle

Essentially, energy-dispersive x-ray spectrometry differs from wavelength-dispersive only in the means used to disperse (separate) the several spectral lines emitted by the specimen. However, *practically*, this distinction results in marked differences in the instrumentation. In wavelength-dispersive (crystal) spectrometers, the several wavelengths are dispersed *spacially* on the basis of their wavelengths *prior* to detection; thus, in principle at least, the detector receives only one wavelength at a time. In energy-dispersive spectrometers, the detector receives the undispersed secondary beam comprising all excited lines of all specimen elements. The detector converts each absorbed x-ray photon into a pulse of electric current having amplitude proportional to the photon energy. The amplified detector output is then subjected to electronic pulse-height selection: The pulses arising from the several detected wavelengths are separated on the basis of their average pulse heights—thereby on the basis of the photon energies of the corresponding x-ray lines. Qualitative and quantitative energy-dispersive analyses are based, respectively, on the pulse heights present and their intensities.

3.4.2. Basic Instrument and Method

The essential features of an energy-dispersive x-ray spectrometer are shown in Figure 3.3. The instrument consists of three principal sections, which effect excitation, detection, and dispersion and readout, respectively. The instrument shown in the figure uses x-ray tube excitation, although other means of excitation are feasible in energy-dispersive spectrometry,

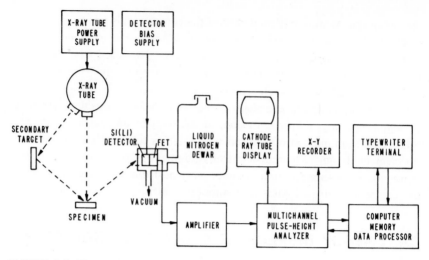

FIGURE 3.3. X-ray-tube-excited energy-dispersive x-ray fluorescence spectrometer. The x-ray tube may excite the specimen directly. Alternatively, the x-ray tube may excite a secondary target having a strong spectral line at wavelength just shorter than that of the absorption edge of an analyte; this secondary x-radiation then excites the target.

including radioisotope and ion excitation. The primary x-ray beam may irradiate the specimen directly, as in x-ray fluorescence spectrometry. Alternatively, the primary beam may irradiate a secondary target consisting of an element having a strong spectral line at wavelength just shorter than the analyte absorption edge. This secondary radiation then excites the specimen. Both these modes are shown in the figure.

The x-rays excited in the specimen consist of many discrete wavelengths and are emitted in all directions. The detector intercepts a substantial fraction of these x-rays, and the solid angle intercepted is greater the closer the detector is to the specimen. The detector consists of the solid-state, semiconductor, lithium-drifted silicon detector itself, Si(Li), and its associated field-effect transistor preamplifier, FET, housed in a vacuum cryostat cooled by liquid nitrogen (Section 5.1.5).

As stated above, the detector converts each absorbed x-ray photon into a pulse of electric current having pulse amplitude proportional to the photon energy. However, in Section 5.1.2, it is explained that even for strictly monochromatic incident x-rays, the detector output pulses are not all of *exactly* the same amplitude, but rather have a statistical spread about a *mean* amplitude proportional to the photon energy. The amplitudes of the pulses in this group have a Gaussian distribution (Section 8.4.2), and the set of pulses is known collectively as a *pulse-height distribution*.

The detector output is then amplified and subjected to electronic pulse-height analysis, as explained above. The several pulse-height distributions present are read out as peaks on a scale of intensity *versus* pulse height or photon energy, either sequentially on an *X–Y* recorder or simultaneously on a cathode-ray tube display (Figures 7.1 and 7.10). In qualitative analysis, the elements present are identified by the mean pulse heights (photon energies) at which peaks occur on the chart or display; concentrations may be estimated from the peak heights (intensities).

A qualitative x-ray spectrometric analysis by either wavelength- or energy-dispersive mode results in a series of peaks on a chart or on some other type of display, and each peak represents an x-ray spectral line of an element in the specimen. In wavelength dispersion, the display is a plot of intensity *versus* 2θ and therefore intensity *versus* wavelength (Figure 7.2). In energy dispersion, the display is a plot of intensity *versus* pulse height and therefore intensity *versus* photon energy (Figures 7.1 and 7.10).

In quantitative wavelength-dispersive analysis, the *peak* intensities of the analyte lines are measured from samples and standards. In quantitative energy-dispersive analysis, the *integrated* intensities of the analyte pulse-height distributions are measured from samples and standards.

3.5. APPRAISAL OF X-RAY SPECTROMETRIC ANALYSIS

3.5.1. Scope

The method of wavelength-dispersive x-ray fluorescence spectrometry is applicable to all chemical elements down to atomic number 4 (beryllium), but standard commercial instruments are limited to atomic number 9 (fluorine). *In favorable cases*, the method, practiced on standard commercial flat-crystal spectrometers, is applicable to: concentrations down to $\sim$0.0001 wt% in solids and $\sim$0.1 µg/ml (ppm) in liquids; isolated masses down to $\sim$1 ng (10^{-9} g); and films as "thin" as $\sim$0.01 monolayer. The specimen can have practically any form, including solid, briquet, fusion product, powder, film, liquid, slurry, and fabricated forms (such as rod and wire) and parts of any form and size. The useful specimen area for flat-crystal instruments is usually 2–5 cm², but areas as small as a few square micrometers can also be analyzed, especially with curved-crystal instruments. The precision of an individual determination may be of the order 0.1% relative, and the accuracy generally lies between $\sigma = 0.01C^{1/2}$ and $0.05C^{1/2}$, where

σ and C are standard deviation and analyte concentration, respectively. The principles and practice of the method have been summarized in several excellent reviews (see the reading list at the end of the chapter).

The method of energy-dispersive x-ray spectrometry is applicable to all chemical elements down to atomic number 6 (carbon), but standard commercial instruments are limited to atomic number 11 (sodium). The sensitivity and precision are perhaps an order poorer than those given above for fluorescence spectrometry, but the specimen may have all the same forms and sizes.

3.5.2. Advantages

No instrumental method of chemical analysis has as impressive a list of advantages as x-ray spectrometry.

1. X-Ray Spectra. Many advantages of x-ray spectrometry arise from the relative simplicity of x-ray emission spectra, compared with optical emission spectra. In general, each element has fewer x-ray lines than optical lines, and many of these x-ray lines are so weak as to be insignificant. However, weak lines of major constituents may interfere with detection and determination of minor and trace constituents. In general, each element has the same K, L, and M lines (commensurate with its atomic number), and the wavelength of each line varies regularly with atomic number, regardless of the nature of the element—metal, nonmetal, liquid, or gas. These features are illustrated in Figure 1.11 and by comparison of the optical emission spectra of iron, copper, and an Alnico alloy in Figure 3.4 with the x-ray spectra in Figure 7.2. Another advantage of x-ray spectra, compared with optical spectra, is their relative independence of the state of chemical combination of the emitting element. Finally, the simplicity

FIGURE 3.4. Optical emission spectra of iron, copper, and an Alnico alloy (53Fe–17Ni–13Co–9.5Al–6Cu–0.5Ti–0.5Zr). Compare this Alnico spectrum with the x-ray spectrum in Figure 7.2.

of x-ray spectra and their relationship to atomic number make selection of internal standard elements much easier than it is in optical spectrography.

2. *Spectral-Line Interference.* Because of the simplicity of x-ray spectra, spectral-line interference is relatively infrequent. However, when it does occur, there are many ways to deal with it: (1) choice of an alternative spectral line or a higher order; (2) prevention or reduction of excitation of the interfering line by operating the x-ray tube below the interferant excitation potential, selection of the x-ray tube target, or monochromatic excitation; (3) use of a blank to measure the interfering line intensity; (4) increased collimation; (5) selection of the crystal for greater dispersion or resolution or to eliminate even-order interfering lines; (6) filtration; (7) use of air path to eliminate long-wavelength interfering lines; (8) selection of the detector for maximum efficiency for the analyte line or minimum efficiency for the interfering line; (9) pulse-height selection; (10) dilution; and (11) mathematical correction, including unfolding (stripping) of overlapped lines.

3. *Absorption–Enhancement Effects.* The elements occurring with the analyte in the specimen may cause absorption–enhancement effects ("matrix" effects) in both x-ray and optical emission (Section 8.2). In optical emission, these effects are largely unpredictable, except by experience. In x-ray emission, they are systematic, predictable, and readily evaluated. Moreover, there are many methods of dealing with them (Chapter 9).

4. *Excitation and Absorption.* X-ray spectral-line excitation and absorption vary with atomic number in the same uniform, regular manner as x-ray spectra, and this regularity is also uninterrupted by the physical and chemical state of the element. Elements having about the same atomic number have very similar excitation and absorption properties. An interesting example of this is provided by the determination of occluded argon in argon-sputtered silicon dioxide (SiO_2) films. Argon calibration standards would be extremely difficult to prepare. However, the analysis is readily calibrated by use of films of potassium chloride (KCl). Argon (Z 18) lies between chlorine (Z 17) and potassium (Z 19) in the periodic table.

5. *Stability.* Electronic and/or magnetic line, kV, and mA stabilization result in extremely well-regulated excitation conditions. Except for very small variations due to slight instability of x-ray tube operating conditions, x-ray-excited x-ray emission intensity is constant with time. Radioisotope sources are even more stable. Optical emission intensity from an arc or spark may vary markedly with time.

6. *Nondestruction of Specimen.* X-ray secondary-emission spectrometry is nondestructive in the sense that the specimen placed in the instrument

usually remains substantially unchanged during the analysis. The method is also nondestructive in the sense that, frequently, sampling is not required; the instrument can be arranged to accommodate very large objects, which then need not be damaged by cutting samples from them. These features are advantageous in many ways. Specimen composition does not change during analysis. Since the instrument can accommodate the specimen in nearly any form, little or no specimen preparation is required in many cases. Other analyses and tests can be made, and questionable analyses can be repeated on the *same* sample. Samples can be subjected to additional treatment or processing, then reanalyzed. Standards, test specimens, expensive or precious samples, criminal evidence, etc., are not consumed. Fabricated parts, art objects, antiquities, etc., are not damaged.

However, certain materials may undergo some deterioration under prolonged intense x-irradiation. Minerals, ceramics, glasses, and other inorganic materials may become colored, usually temporarily. Biological, organic, and liquid specimens may undergo radiolysis (radiation-induced chemical change). Plastics and electronically active materials may sustain permanent radiation damage. Finally, on repeated use, standards may suffer damage due partly to repeated prolonged exposure to the intense primary x-rays and partly to oxidation by the ozone (O_3) generated in the specimen chamber by the primary x-rays. These effects are minimized in energy-dispersive spectrometers using low-power x-ray-tube, secondary-emissive, or radioisotope excitation.

7. Specimen Versatility. No analytical method can deal with as wide a variety of specimen forms as the x-ray spectrometer. The physical form may be solid, briquet, powder, slurry, liquid, or even gas. The material may be metal, mineral, ceramic, plastic, rubber, textile, paper, or practically anything else. The size may be anything from a barely visible speck or a film having less material than a monolayer, to a very large, massive object. The shape may be plane, cylindrical, filamentary, or irregular, and the method is applicable to small, fabricated parts of all shapes. The analyte may be distributed on filter paper, Mylar film, cellulose tape, ion-exchange resin, etc. The method is applicable to special conditions such as high and low temperatures, special atmospheres, remote sensing, and cases where preparation or other contact with the specimen is not permissible or is difficult or impossible. The method is applicable to on-line, closed-loop, continuous automatic plating thickness gaging and chemical process control.

8. Operational Versatility. The x-ray spectrometer can be fitted with a variety of components and accessories, and operated in a variety of modes under a variety of conditions, all selected to give optimum results in a given analysis. The selectable components include: (1) the x-ray tube target—about 10 are available commercially, including dual-target types (chromium and tungsten, platinum, or molybdenum); (2) collimators of various lengths and foil spacings; (3) crystals, selected for "reflectivity," dispersion, resolution, or elimination of orders; and (4) detectors—sealed and flow proportional counters having various gas fillings, scintillation counters, or Geiger counters. The selectable accessories include rotating specimen stages, inverted optics, bulk-specimen holders, selected-area apertures, curved-crystal accessories, and helium or vacuum enclosures. It is fair to state than an x-ray spectrometer can usually be fitted with accessories optimal for a specified analysis more conveniently than any other analytical chemical instrument. The modes of operation, in addition to x-ray spectrometry, include polychromatic and monochromatic absorption; absorption-edge spectrometry; methods based on scattered, rather than emitted, x-rays; and even polychromatic and monochromatic contact microradiography.

9. Versatility of Analytical Strategy. X-ray spectrometry lends itself to a wide variety of techniques of specimen preparation and presentation (Chapter 10) and analytical methods (Chapter 9).

10. Speed and Convenience. Because wavelength and intensity are measured directly on the instrument and because of the nondestructive feature, x-ray spectrometry is extremely rapid and convenient and readily applicable to automation. The method is applicable to 100% inspection of large numbers of multielement samples. Typical x-ray measurement times are of the order 20–60 s, and a multichannel instrument can measure up to 20 or 30 elements in a minute or two. In fact, it is rarely the spectrometer that limits the specimen throughput rate. In general, no analytical method can deal as quickly and conveniently with very large numbers of samples when they can be analyzed in the as-received form. Even when specimen preparation is required, the method is often more rapid and convenient than other methods. X-ray spectrometry permits extremely rapid confirmation, identification, or selection of known alternatives on the basis of the presence or absence, or high or low concentration, of some element or the intensity ratios of two or more elements. For example, Kovar and stainless steel parts can be sorted on the basis of presence or absence of cobalt; 52-alloy and Invar parts (48–52 and 65–35 iron–nickel alloys, respectively) can be distinguished on the basis of relative intensities of Fe $K\alpha$ and Ni $K\alpha$ lines. With energy-dispersive spectrometers, such identifications can be made almost immediately even without any prior knowledge of the specimen whatever. Errors, anomalous results, out-of-specification results, etc., can often be noted while data is being taken, and checked immediately.

11. Selected-Area Analysis. Small selected areas can usually be analyzed in place so that contamination and loss by chemical separation are avoided. Uniformity and homogeneity of large-area samples and uniformity of thickness and composition of films and platings can be evaluated.

12. Semiquantitative Analysis (Estimation). Semiquantitative analyses can be made from data taken without standards other than the pure analyte to within a factor of 2 or 3 in x-ray spectrometry, to within an order in optical spectrography. Simple calculations based on absorption coefficients can improve the x-ray estimates.

13. Concentration Range. X-ray spectrometry is applicable over an extremely wide concentration range, from 100% for any element above fluorine to 0.0001% for sensitive elements in favorable matrixes. Small differences in concentration of major constituents are readily measured. For most elements, the method is applicable to micro and trace analysis. In favorable cases, traces can be determined without separation, so that contamination or loss by chemical or physical separation is avoided. Quantities < 0.1 μg and films "thinner" than one monolayer can be detected for most elements above magnesium on modern commercial flat-crystal spectrometers. In general, the method is applicable to detection, estimation, and determination of one, several, or all elements in a multi-element system.

14. Sensitivity. The sensitivity, although in general not as good as that of optical emission spectrography, is still very high. Instrument components, accessories, and conditions may be chosen for optimal sensitivity for a given analysis. With a given x-ray tube target, crystal, collimator system, and detector, sensitivity of the pure analyte is a relatively simple function of atomic number. Elements having about the same atomic number are likely to have about the same sensitivity in a given system. No such simplicity exists in optical emission spectrography. In Table 3.2, the sensitivities of x-ray fluorescence spectrometry and other analytical chemical methods applicable to trace and micro analysis are compared in terms of minimum detectable mass of element under *most favorable conditions.*

15. Precision and Accuracy. These features are generally very good. In modern x-ray spectrometers, instrumental and operational errors are minimized, and the relatively large specimen area (~ 2 cm^2) and specimen rotation reduce heterogeneity effects. Thus, the precision may well be limited by the counting error and may be made as good as desired by appropriate selection of the accumulated count. The accuracy is usually

TABLE 3.2. Ultimate Sensitivities of Microanalytical Methods for the Most Sensitive Elements under the Most Favorable Conditions

Method	Minimum detectable amount (g)
Microchemistry	10^{-9}
Polarography (volt-ammetry)	10^{-9}
X-ray fluorescence spectrometry (XRFS)	10^{-9}
Optical absorption spectrometry	10^{-10}
Optical emission spectrography	10^{-10}
Flame ("atomic") absorption spectrometry ("AA")	10^{-11}
Flame emission spectrometry	10^{-12}
Ion-induced x-ray spectrometry	10^{-12}
Ion-scattering spectrometry (ISS)	10^{-13}
Mass spectrometry (MS)	10^{-13}
Neutron-activation analysis (NA)	10^{-13}
Optical fluorescence microscopy	10^{-14}
X-ray-induced photoelectron spectrometry (XPES)	10^{-14}
Auger-electron spectrometry (AES)	10^{-15}
Electron-probe microanalysis (EPMA)	10^{-15}
Radioactive tracer analysis	10^{-18}
Secondary-ion mass spectrometry (SIMS)	10^{-18}

limited by the accuracy with which the compositions of the standards are known, and compares favorably with that of other instrumental analytical methods.

16. Automation and Computer Control. Modern automated computer-controlled sequential and simultaneous x-ray spectrometers index up to 20 specimens into measurement position, measure lines of up to 30 elements on each, and print out the analytical compositions (Sections 3.3.3.2 and 3.3.3.3). Sequential instruments automatically select optimum kV, mA, 2θ, crystal, collimator, detector, pulse-height selector settings, and the number of counts to be scaled for each element. The analysis time is the sum of the counting and indexing times. In simultaneous instruments, each channel is preset for a certain analyte and fitted with components optimum for that element. The total analysis time is a very few minutes. On-line process-control instruments control plating thickness and various chemical and metallurgical processes.

17. Use with Other Methods. Because of its nondestructive feature, x-ray spectrometry can be used to complement other analytical methods, such as x-ray diffraction and absorption, infrared absorption spectrometry, electrophoresis, chromatography, gas chromatography, and radiochemical tracer techniques.

18. Operating Costs. Although the initial cost of the instrumentation may be high (Section 3.5.3), and x-ray tubes must be replaced occasionally, operating costs are low because there are no consumed supplies.

3.5.3. Disadvantages and Limitations

1. Light Elements. Standard commercial x-ray spectrometers are subject to certain difficulties and inconveniences for elements having atomic number 22 (titanium) or less. These difficulties become progressively more severe as atomic number decreases and may be very serious for atomic numbers 9–14 (fluorine to silicon). The difficulties also become more severe as concentration decreases. Present commercial wavelength- and energy-dispersive spectrometers are limited to atomic numbers 9 (fluorine) and 11 (sodium), respectively. However, specially designed ultralong-wavelength x-ray fluorescence spectrometers and electron-probe microanalyzers are capable of measuring K spectra of elements down to atomic number 4 (beryllium).

2. Penetration. Only a relatively thin surface layer, ≤ 0.1 mm for solids, contributes to the measured x-ray line intensities. Consequently, the relationship between analyte-line intensity and concentration is affected by surface texture, corrosion, particle size, etc. It follows that x-ray spectrometric methods give accurate bulk compositions only for homogeneous specimens. These effects become more severe as analyte-line wavelength increases (photon energy decreases). Surface-texture and particle-size problems usually can be avoided by suitable specimen preparation (Chapter 10).

3. Absorption–Enhancement Effects. The relationship between analyte-line intensity and concentration may be affected by absorption–enhancement effects caused by other elements present in the specimen. In general, the slope of the intensity *versus* concentration calibration curve is inversely proportional to the matrix absorption for the analyte line. However, these effects are systematic and predictable, and can usually be avoided by suitable choice of the analytical method (Chapter 9).

4. Sensitivity. In general, x-ray spectrometry is less sensitive than optical spectrography by two or more orders, but there are exceptions to this, especially the nonmetallic elements.

5. Qualitative Analysis. By use of a wavelength-dispersive x-ray fluorescence spectrometer, recording one x-ray spectrum with the ratemeter–recorder may require 30 min or more, during which time 10 or more optical emission spectra can be photographed and developed. More rapid techniques for recording x-ray wavelength spectra have been developed but have not come into widespread use. Of course, *energy-dispersed* x-ray spectra are recorded or displayed in 30 s to several minutes.

6. Standards. For quantitative analysis, the x-ray spectrometric method is a comparative one, like all instrumental analytical methods, and suitable standards of accurately known composition are required for calibration. The standards must have as nearly as possible the same chemical composition as the samples in both x-ray and optical spectrochemical analysis. However, in x-ray spectrometry, standards must also have the same physical form as the samples: solid metal (blanks, sheet, and foil ≥ 0.1 mm thick) can be analyzed only with solid standards, powders packed in cells only with similarly packed powder standards, briquets only with briquets, etc. The same applies to analysis of analytes distributed on filter paper, Mylar film, ion-exchange resins, etc. Moreover, solid standards must have the same surface texture as the samples, powders the same particle-size distribution and packing density, etc. If the analyte line is subject to chemical effects, the analyte should be present in the standards in the same chemical state as in the samples. No such restrictions apply in optical spectrography, where, for example, metal drillings may often be analyzed with synthetic standards consisting of mixed metal powders—or even metal oxide powders.

This limitation is particularly inconvenient when the same substance must be analyzed in a number of physical forms—for example, metal blanks, drillings, shot, filings, powder, thin rod, wire, and small fabricated parts. In optical spectrography, one set of standards would suffice—and each could be in a different form, but not in x-ray spectrometry.

When standards are not available in the same physical form as the samples, and when the same substance must be analyzed in several forms, all standards and samples must be reduced to a common form—powder, fusion product, solution, etc. In such cases, much of the advantage of speed and convenience is lost.

7. Instrument Components and Settings. Among the advantages cited above for the x-ray spectrometric method are its versatility with respect to wide selection of components (x-ray tube targets, crystals, collimators, detectors) and operating conditions (x-ray tube potential and current, detector potential, amplifier gain, pulse-height selector settings). However, it must be admitted that in a laboratory equipped with a *manual* wavelength-dispersive x-ray spectrometer, if the variety of elements to be determined necessitates frequent change of components and/or settings, much time can be lost. Moreover, in a series of multielement samples of the same type, frequently all analytes cannot be measured optimally with the same components and/or settings. In such cases, if all measurements are made with the same components and/or settings, a compromise must be made in the interest of convenience and economy at the expense of intensity. Usually, some loss of intensity can be afforded for major and even minor constituents, but not for traces. The alternative is to measure one or more analytes of about the same atomic number in all specimens with one set of components and settings, then others with another set, etc. The problem is much less severe in energy-dispersive spectrometry simply because there are fewer components and conditions. The problem is substantially eliminated by use of fully automated computer-controlled sequential and simultaneous spectrometers.

8. Error. In the operation of single-channel manual instruments, there are many opportunities for error and reduction of precision when two or more analytes are to be measured in each of many specimens. The likelihood increases as the number of settings to be changed increases—2θ (line and background), kV, mA, detector-tube potential, amplifier gain, pulse-height analyzer baseline and window, and preset time or count. One must read the scaler or timer at least once for each analyte in each specimen, and usually twice (line and background). A printer eliminates this last source of error, and automatic instruments eliminate many of the others.

9. Tedium. In the operation of a single-channel manual instrument, data taking may be very tedious. A printer may record the data, but specimens must be loaded and settings made manually, so that constant attention is required throughout the analysis. Here again, automatic instruments minimize the problem.

10. Cost. The initial cost of x-ray spectrometric instrumentation is relatively high. A single-channel manual instrument, such as that shown schematically in Figure 3.2, costs about $35,000. A fully automated computer-controlled sequential spectrometer costs about $110,000. The cost of a fully automated computer-controlled simultaneous spectrometer depends on the number of channels, and is typically about $110,000 for seven channels to $200,000 for 20 channels. A fully computer-controlled energy-dispersive spectrometer having low-power x-ray tube excitation costs $50,000–$70,000; units having radioisotope excitation cost $5000–$8000 less. An energy-dispersive unit for attachment to a scanning electron microscope or electron-probe microanalyzer costs about $40,000. A fully automated electron-probe microanalyzer having four wavelength-dispersive spectrometers costs about $170,000.

Some additional equipment may be required for specimen preparation. The instrument should be housed in a separate room, especially if it is used in a factory environment. This room should be air conditioned. Certain radiation safety requirements must be complied with. Operating costs are low, but for wavelength-dispersive instruments, x-ray tubes must be replaced occasionally.

3.5.4. Wavelength and Energy Dispersion Compared

The principal advantages of the energy-dispersive mode, compared with wavelength dispersion, arise from the simplicity of the instrumentation, economy of emitted analyte-line intensity, and feasibility of simultaneous accumulation and display of the entire x-ray spectrum.

Elimination of the crystal means elimination of the goniometer with its collimators and precision θ–2θ drive mechanism; in fact, there are no moving parts. An incidental advantage of absence of the crystal is absence of higher orders in the spectrum.

Elimination of crystal diffraction and collimation eliminates the severe loss of emitted intensity from these processes. Also, the detector may now be placed very close to the emitting specimen, providing even greater economy of intensity by reduced distance and increased solid angle of interception. The resulting 100-fold or more increased conservation of intensity is advantageous when selective excitation and filtration are used. The conservation of intensity permits use of excitation sources other than high-power x-ray tubes with their large, heavy, expensive, power-consuming supplies. These alternative sources include low-power x-ray tubes, secondary monochromatic radiators, radioisotopes, and ion beams. These low-flux excitation sources are particularly advantageous for specimens subject to radiation damage, such as liquids, organics, biological and medical specimens, glasses, art objects, and antiquities.

In energy-dispersive mode, photons of all x-ray spectral lines emitted by the specimen enter the detector simultaneously. This condition provides the basis for simultaneous accumulation and display of the entire energy spectrum, including background, by means of a multichannel analyzer and cathode-ray tube display. Consequently, semiconductor-detector x-ray spectrometers permit qualitative analyses to be performed in 30 s to a few minutes, depending on the intensities of the lines of minor and trace elements. Background and spectral interferences are observable. Also, simultaneous display permits observation and measurement of elements other than those specified for analysis and reduces the possibility of oversight of unsuspected elements.

Simultaneous accumulation has several other benefits: it is advantageous to measure all analyte and internal-standard lines and backgrounds simultaneously. It is also advantageous to measure the integrated analyte line, rather than just the peak, as is customary in wavelength dispersion. The effects of instrument drift and chemical wavelength shift are reduced. Finally, the accumulated spectrum can be entered in a computer and processed in various ways, including calculation and printout or display of analytical concentrations.

The two principal limitations of the energy-dispersive spectrometer are imposed by the Si(Li) detector (Section 5.1.5) and have to do with resolution and tendency to coincidence loss. At wavelengths longer than ∼0.8 Å (15 keV), the resolution of a Si(Li) energy-dispersive spectrometer is poorer than that of a crystal spectrometer. (The resolution of an energy-dispersive instrument using a proportional or scintillation counter is poorer at any wavelength.) Thus, the crystal spectrometer may be preferable for the intermediate-wavelength region, certainly for the ultralong-wavelength region—for the K lines of elements of atomic number $\lesssim 11$ (sodium). As mentioned above, in energy dispersion, all x-rays emitted and scattered by the specimen enter the detector simultaneously—not just the discrete wavelength diffracted by a crystal. Thus coincidence losses, pulse-height shift, and detector choking occur much more readily. Higher individual intensities can be measured with crystal spectrometers because only a narrow spectral region is admitted to the detector and because gas-filled proportional and scintillation counters are capable of higher count rates than Si(Li) detectors. When the multichannel analyzer is used, it is difficult to realize high sensitivity for weak lines in the presence of strong lines; the instrument accumulates all lines simultaneously, and the strongest line determines the counting time and stops the count. Other disadvantages apply when radioisotope sources are used.

Perhaps the best comparison of wavelength- and energy-dispersive modes to date is that by Gilfrich, Burkhalter, and Birks (see the reading list at the end of the chapter). They evaluated sensitivities and detection limits for 19 elements ranging in atomic number from 13 (aluminum) to 82 (lead) in micro quantities in filter paper disks in the following spectrometer arrangements under conditions as nearly identical as possible:

1. Wavelength-dispersive spectrometer with chromium-, rhodium-, and tungsten-target x-ray tubes; LiF(200), graphite, and KHP crystals; and gas-flow proportional counter with P10 gas (90 vol% argon, 10 vol% methane).

2. Energy-dispersive spectrometer with Si(Li) detector and the following excitation sources: (1) x-ray tubes: molybdenum and tungsten targets at 50 kV(CP), 3 mA; (2) secondary radiators: manganese, copper, silver, and chromium–zirconium alloy, excited by a tungsten-target tube at 45 kV(CP), 20 mA; (3) radioisotopes: 7-mCi ^{55}Fe and 70-mCi ^{109}Cd; and (4) ions: 5-MeV protons and alpha particles.

These workers conclude that: (1) major concentrations can be determined in either mode with any of the excitation sources; (2) intermediate concentrations can be determined in either mode, but direct x-ray tube excitation must be used for energy dispersion; and (3) concentrations approaching the detection limit can be determined only by wavelength dispersion.

3.6. PERSONNEL SAFETY

The potential personnel safety hazards involved in operation of x-ray spectrometers arise from three sources: (1) *radiation* from x-ray tubes and radioisotopes; (2) *ingestion* of and/or laboratory contamination by radioactive material (only for radioisotope excitation); and (3) *electric shock* from high-potential x-ray generators.

A *careful, alert, competent* worker aware of the potential hazards of x-rays need have little apprehension for his/her personal safety in working with commercial x-ray spectrometers properly assembled. These instruments are adequately shielded to confine the x-rays and prevent irradiation of the laboratory environment. Most serious radiation accidents occurring to operators of x-ray spectrometers have involved direct irradiation by the primary beam from a high-power x-ray tube *and occurred when the operators were unaware that the tube was energized!* In some cases they simply neglected to ascertain that the x-ray generator was off, in others, they relied on warning and/or protective devices later shown to be faulty.

The principal sources of possible extraneous x-rays from an x-ray-tube-excited spectrometer are: (1) the interface between the x-ray tube head and the specimen chamber; (2) the specimen chamber itself when specimens are inserted and removed, especially if the x-ray shutter assembly is faulty;

(3) the high-potential electron-tube rectifiers, which may radiate through the relatively thin steel of the oil tank on the peak-inverse half-cycles; solid-state rectifiers do not produce x-rays; and (4) the crystal, which diffracts, scatters, and emits x-rays in directions other than that intercepted by the detector collimator. Of course, the extraneous intensity from any source increases with increasing x-ray tube operating potential and current, and is substantially greater from high-power than from low-power tubes, and from 100-kV than from 50-kV generators. Also, the hazard is greater from the older tabletop goniometers of the type shown in Figure 3.2 than from modern wholly enclosed units.

The greatest possible radiation hazard in any x-ray generator is the primary beam, particularly close to the x-ray tube window. Even in a 50-kV x-ray spectrometer, the primary intensity at the x-ray tube window may be of the order of megaroentgens per minute (see below), and even momentary exposure to it is almost certain to result in a lesion at the irradiated site. The sources of extraneous radiation listed above are likely to have very low intensity. However, if the shielding is inadequate or improperly installed, an operator taking data for several hours each day may well receive a substantial fraction of his/her maximum permissible dose (see below) from this source.

Incidentally, x-ray diffraction units are potentially more hazardous than x-ray spectrometers. The x-ray diffraction tube has three or four windows from which x-ray beams emerge. Normally, the windows are shuttered except when covered by diffraction cameras or diffractometers. However, if a window has a faulty shutter and no instrument, the direct beam radiates across the laboratory unobstructed.

Although the radioisotope sources used in energy-dispersive x-ray spectrometry emit radiation at intensities lower by several orders than that of high-power x-ray tubes, their potential hazard is actually substantially greater. An x-ray tube can be turned off when not in use and cannot contaminate the laboratory environment; also, there is nothing to ingest. None of these favorable safety features applies to radioisotope sources. Consequently, such sources must be kept shielded at all times and must be stored and handled—in the prescribed way—with great care to prevent rupture of their encapsulation and dispersion of the radioactive material. Radioisotopes emit several types of radiation, principally α-, β-, x-, and γ-rays. The penetrating power increases in the order given, but ingestion of β- and, especially, α-emitting isotopes may be extremely hazardous.

The hazard from ionizing radiation in general (α, β, γ, x) is compounded by the fact that such radiation is completely insidious. It is quite possible that one could be receiving a lethal dose of ionizing radiation and be wholly unaware of any sensation. Further, the physiological effects are delayed, and symptoms may not appear for days or even weeks after the exposure. Finally, the physiological effects are cumulative if one receives radiation at a rate faster than that at which the body can repair the physiological damage, that is, faster than the maximum permissible dose rate (see below).

For purposes of radiation health and safety, radiation is measured in terms of dose or dose rate. The unit of biological radiation exposure or dose is the *rad* (radiation *a*bsorption *d*ose), which is that quantity of ionizing radiation of any kind (α, β, γ, x) that results in absorption of 100 ergs of energy per gram of biological tissue, or equivalent material. The unit of x- or γ-ray dose is the *roentgen*, R, which is that quantity of x- or γ-radiation that produces in 1 cm^3 of dry air at 0°C and 760 torrs (1 atm), ions carrying 1 esu of charge, regardless of sign; this is equivalent to 2.08×10^9 ion pairs/cm^3, or 84 ergs. The unit of biological *radiation damage* is the *rem* (*r*oentgen *e*quivalent, *m*an), which is equivalent to 1 rad of x- or γ-radiation, giving average specific ionization of 100 ion pairs per micrometer in water. Dose rate is given in terms of dose per unit time, for example, milliroentgens per hour (mR/h), millirems per hour (mrem/h). For practical purposes, for x-radiation produced in x-ray spectrometers excited by tubes operating at $\lesssim$100 kV, the doses in roentgens, rems, and rads are substantially equivalent.

The maximum permissible whole-body dose rate is 5 rems/year or 3 rems/quarter (13 weeks); of course, if one receives 3 rems in one quarter, he/she may receive a total of not more than 2 rems in the following 3 quarters. This is substantially equivalent to 5 R/year or 100 mR/week; for a 40-h work week, this is 2.5 mR/h. The maximum permissible accumulated lifetime whole-body dose is $5(y - 18)$ R, where y is age in years and must be >18. Certain radiologically insensitive portions of the body may receive substantially greater doses; for example, 30 R/year for the skin, 75 R/year for hands, forearms, feet, ankles, neck, and head (excluding the eyes). Background radiation at sea level from natural radioactivity in the environment and from cosmic radiation is typically 0.01–0.1 mR/h, say, 0.05 mR/h; this is $0.05 \times 24 = 1.2$ mR/day, $1.2 \times 365 \approx 500$ mR/year.

All evidence indicates that if a healthy normal adult receives x-radiation at a dose rate no greater than the maximum permissible whole-body dose, his body can easily repair the minor physiological damage. However, every effort should be made to reduce the dose rate to as low a level as possible, and, pursuant to this objective, certain simple precautions are prudent.

The x-ray spectrometer should be thoroughly monitored with a radiation-survey meter when it is installed and at routine 3–6-month intervals thereafter. In x-ray tube-excited spectrometers, particular attention should be given to the regions of the sources of extraneous radiation given above and to the place at which the operator sits when taking data. The region of the x-ray tube head and specimen compartment should also be monitored each time the x-ray tube is changed or the compartment disturbed. All monitoring should be done with the x-ray tube operating at maximum potential (50 or 100 kV) and maximum permissible current, and with an emission specimen having a strong spectral line of short wavelength—molybdenum, for example. The survey meter should be a model having good response for x-rays having energy up to 50 keV (or 100 keV) and should be properly calibrated to read milliroentgens per hour.

In general, the survey meter should indicate not more than 0.5 mR/h at 5 cm from any accessible surface or part of the instrument. Actually, the 5 cm should be measured to the center of the active volume of the detector in the survey meter. Thus, in practice, the survey meter should be held as close to the spectrometer as possible without danger of damaging the meter probe. If the reading exceeds 0.5 mR/h, the source of the "leak" should be examined to ascertain that the components involved are properly assembled and that any shielding provided by the manufacturer is securely in place. If all is proper, and a radiation leak still exists, additional shielding must be provided.

Particularly hazardous conditions may exist when the x-ray tube is operated without its housing in place, that is, with its glass end exposed, or without its shutter, or without the specimen compartment in place. Laboratory-designed and built instruments and accessories and, especially, temporary arrangements should be especially carefully monitored.

All x-ray laboratory personnel should be provided with film badges and/or pocket dosimeters. The person responsible for the x-ray laboratory must be familiar with all federal, state, and municipal regulations pertaining to radiation safety and must take all necessary measures to comply with them (see the reading list).

Another hazard is that of electric shock. Contact with the high-potential supply while in operation may well result in electrocution. Fortunately, when the high-potential cable is secured at both the oil-tank and x-ray tube ends, contact with the high potential is unlikely. However, even when the power is off, if the cable is unplugged, there may be danger of shock from charge stored in capacitors in the high-potential circuit if there is no "bleeder" resistor or if the resistor is open. Also, great care must be taken

when it is necessary to work on the equipment with the cabinet door interlocks thwarted. Contact with the 220-V or 110-V line power is very
hazardous.

SUGGESTED READING

AMERICAN NATIONAL STANDARDS INSTITUTE, "American National Standard: Radiation
   Safety for X-Ray Diffraction and Fluorescence Analysis Equipment," *Nat. Bur.
   Stand. (U.S.) Handb.* **111**, 20 pp. (1972).
AMERICAN SOCIETY FOR TESTING AND MATERIALS, "Symposium on Fluorescent X-Ray
   Spectrographic Analysis," *Spec. Tech. Publ.* **157**, 68 pp. (1954); photo copies available
   from University Microfilms, Inc.
AMERICAN SOCIETY FOR TESTING AND MATERIALS, "Symposium on X-Ray and Electron-
   Probe Analysis," *Spec. Tech. Publ.* **349**, 209 pp. (1964).
BERTIN, E. P., *Principles and Practice of X-Ray Spectrometric Analysis*, 2nd ed., Plenum
   Press, New York (1975); Chaps. 3 and 9, pp. 89–112, 405–433.
CARR-BRION, K. G., and K. W. PAYNE, "X-Ray Fluorescence Analysis, a Review,"
   Analyst (London) **95**, 977–991 (1970).
GILFRICH, J. V., P. G. BURKHALTER, and L. S. BIRKS, "X-Ray Spectrometry for Particulate
   Air Pollution—Quantitative Comparison of Techniques," *Anal. Chem.* **45**, 2002–2009
   (1973).
HAKKILA, E. A., "X-Ray Spectroscopic Methods;" in *Modern Analytical Techniques for
   Metals and Alloys*, Part 1; R. F. Bunshah, ed.; Techniques of Metals Research
   Series, Vol. 3, Part 1; Wiley–Interscience, New York, pp. 275–323 (1970).
JENKINS, R., "X-Ray Fluorescence Analysis;" in *Analytical Chemistry*—Part 2, T. S.
   West, ed.; Vol. 13 of *Physical Chemistry, Series One*; MTP [Medical and Technical
   Publishing Co.] *International Review of Science*; Butterworths, London, and Uni-
   versity Park Press, Baltimore; pp. 95–126 (1973).
JENKINS, R., and D. J. HAAS, "Hazards in the Use of X-Ray Analytical Instrumentation,"
   X-Ray Spectrom. **2**, 135–141 (1973).
JENKINS, R., and J. L. DE VRIES, "X-Ray Spectrometry" [metallurgical application],
   Met. Rev. **16**, 125–141 (1971).
LAMBERT, M. C., "Some Practical Aspects of X-Ray Spectrography," *Norelco Rep.*
   6, 37–51 (1959).
PARRISH, W., "X-Ray Spectrochemical Analysis," *Philips Tech. Rev.* **17**, 269–286 (1956);
   Norelco Rep. **3**, 24–36 (1956).

Chapter 4

Excitation;
Wavelength Dispersion

Having considered the method and instrumentation of x-ray spectrometric analysis in general terms in Chapter 3, we now devote two chapters to a more detailed consideration of the instrument components and their functions.

4.1. EXCITATION

The second phase of an x-ray spectrometric analysis (Section 3.2) is excitation, the creation of vacancies in the inner orbitals of atoms of the elements in the specimen with resultant emission of their characteristic x-ray spectral lines. The principles of 10 basic modes of excitation of x-ray spectra are stated briefly in Section 1.8.8. The principles of primary (electron) and secondary (x-ray) excitation are considered in more detail in Sections 1.8.6 and 1.8.7, respectively. This section is limited to the four excitation modes most commonly applied in practical x-ray spectrochemical analysis: x-ray tubes, electron guns, radioisotopes, and ion accelerators. X-ray tubes are considered last, but in most detail.

4.1.1. Electron Excitation

Electron excitation is applied in x-ray tubes, electron-probe microanalyzers (Chapter 11), and specially designed commercial and laboratory-built instruments. The principles of electron excitation are considered in Section 1.8.6, its advantages and limitations in Section 1.8.7.4. An electron excitation equation is given in Section 11.6.

4.1.2. Ion Excitation

In ion excitation, a 10-keV to 10-MeV ion beam excites the specimen ("target"), and the emitted spectrum is measured with an energy-dispersive spectrometer. The ion beam is produced by an ion accelerator, which may be a relatively small laboratory type or a large cyclotron or synchrotron. Protons ($_1^1H^+$) are perhaps the most commonly used "projectile" ions; deuterons ($_1^2H^+$), tritons ($_1^3H^+$), and α-particles ($_2^4He^{2+}$) are also used, as well as heavier ions, such as $_6O^+$, $_{10}Ne^+$, $_{18}Ar^+$, and $_{36}Kr^+$.

Ion excitation has several very valuable features. (1) Continuum is substantially nonexistent. For example, the continuum with proton excitation is weaker than for electron excitation by a factor equal to the square of the ratio of the electron and proton masses, or $\sim(1/1800)^2$. The reduced background favors sensitivity compared with electron and x-ray excitation, where there is, respectively, emitted and scattered continuum. (2) Ion excitation is applicable to trace analysis of the top several monolayers of the specimen surface if low-energy and/or high-mass ion beams are used to reduce penetration. (3) Heavy ions, such as Ne^+, Ar^+, and Kr^+, have higher x-ray excitation efficiency than expected due to the Pauli and Fano–Lichten effects (Section 1.8.8), and also well-defined excitation threshold energies. These thresholds often permit selective excitation of a specific analyte to the exclusion of matrix elements. For example, 100-kV Kr^+ ions bombarding a silicon target having implanted antimony atoms excite Sb *M* x-rays, but not Si *K* x-rays. (4) Finally, the analyte distribution as a function of depth near the specimen surface can be mapped with high resolution.

The limitations of ion excitation include the following. In general, the x-ray yield from ion bombardment is much lower than from electron or even x-ray excitation, but the yield increases with incident ion energy. The ion accelerator becomes increasingly elaborate and costly the higher the mass, energy, and current of the ion beam. The specimen spectrum may be complicated by the spectrum of heavy projectile ions, which become implanted in the target. Heavy ions may also remove target atoms by sputtering. This phenomenon may be used to advantage to strip the target surface layer by layer for determination of the analyte depth distribution.

4.1.3. Radioisotope Excitation

Radioisotopes are defined and discussed in Section 1.8.1. A radioisotope *source* is a specified amount of a specified radioisotope fabricated in a form suitable for a specified application. The radioactive material is encapsulated to prevent its dispersion into and thereby contamination of the laboratory environment, and shielded to retain radiation, except in certain directions.

Radioisotope sources for x-ray spectrometric application are characterized by four principal properties as follows: (1) radioactive decay process and type of emitted radiation—α-, β-, or γ-emission, or K- or L-orbital–electron capture, which results in x-ray spectral-line emission (Section 1.8.8); (2) energy of the emitted radiation—1–150 keV; (3) activity of the source—1 mCi to 5 Ci (the *curie*, Ci, is that mass of radioisotope in which 3.7×10^{10} disintegrations occur each second); and (4) half-life, the time required for half the atoms of the radioisotope to disintegrate, that is, for the activity of the source to fall to half its initial value. Table 4.1 lists these properties for each of the radioisotope sources commonly used for x-ray spectrometric excitation.

In addition to these parameters, some other considerations determine the suitability of a radioisotope for x-ray spectrometric excitation. The radioactive emission should be as simple as possible, preferably a single intense γ- or x-ray line. It follows that radioisotopes with stable (that is, nonradioactive) decay ("daughter") products are preferable. Also, one must be aware of possible spectral interference from the emission lines from radioisotopes, especially in trace and micro analysis.

The three principal source–specimen–detector geometry arrangements are shown in Figure 4.1: the central or button source, the annular or ring source, and the source–target arrangement. In the central or annular geometries, the source irradiates and excites the specimen directly. In the source–target geometry, the source irradiates and excites the characteristic x-ray spectrum of the target, which is selected to have its principal line at wavelength just shorter than that of the analyte absorption edge. This target x-radiation then irradiates and excites the specimen, which must be shielded from direct irradiation by the radioisotope. The targets are interchangeable and may consist of pure metals in the form of rings or small thimbles (Figure 4.1, bottom) or oxide powders bonded with epoxy resin to ring or thimble supports. This arrangement gives extremely pure analyte spectra having high peak-to-background ratios, but requires high-activity sources because of the inefficiency of the secondary excitation process, and even then gives relatively low analyte-line intensity. For the source–target arrangement, γ-emitting isotopes are most commonly used, and such sources are also known as γ-x sources. A feature of these sources that can be both an advantage and a disadvantage is its relative specificity. The analyte having a major absorption edge at wavelength just longer than that of the secondary

TABLE 4.1. Commonly Used Radioisotope Sources for X-Ray Spectrometric Excitation[a]

Radioisotope source	Principal radioactive decay process[b]	Half-life	Useful radiation[c]		Typical practical source activity (Ci)	X-rays excited efficiently
			Type	Energy (keV)		
$^{3}_{1}$H–Ti	β^-	12.3 y	Continuum	3–10	5[d]	$_{11}$Na–$_{29}$Cu K
			Ti K x-rays	4–5		
$^{3}_{1}$H–Zr	β^-	12.3 y	Continuum	2–12	5	$\lesssim_{30}$Zn K
				2		
$^{55}_{26}$Fe	OEC	2.7 y	Mn K x-rays	5.9	0.020[d]	$_{13}$Al–$_{24}$Cr K
$^{57}_{27}$Co	OEC	270 d	Fe K x-rays	6.4	0.5	$\lesssim_{98}$Cf K
			γ	14		
			γ	122		
			γ	136		
$^{109}_{48}$Cd	OEC	1.3 y	Ag K x-rays	22	0.003[d]	$_{20}$Ca–$_{43}$Tc K
			γ	88		$_{74}$W–$_{92}$U L
$^{125}_{53}$I	OEC	60 d	Te K x-rays	27		$\lesssim_{54}$Xe K
			γ	35		
$^{147}_{61}$Pm–Al	β^-	2.6 y	Continuum	12–45	0.5	$\lesssim_{60}$Nd K
$^{153}_{64}$Gd	OEC	236 d	Eu K x-rays	42	0.010[d]	$_{42}$Mo–$_{58}$Ce K
			γ	97		$_{69}$Tm–$_{92}$U L
			γ	103		
$^{210}_{82}$Pb	β^-	22 y	Bi L x-rays	11	0.010	$\lesssim_{62}$Sm K
			γ	47		
$^{238}_{94}$Pu	α	89.6 y	U L x-rays	15–17	0.030[d]	$_{20}$Ca–$_{35}$Br K
						$_{74}$W–$_{82}$Pb L
$^{241}_{95}$Am	α	470 y	Np L x-rays	11–22	0.010[d]	$_{50}$Sn–$_{69}$Tm K
			γ	26		
			γ	59.6		

[a] L. S. Birks, *X-Ray Spectrochemical Analysis*, 2nd ed., Wiley–Interscience, New York (1969), p. 64; J. R. Rhodes and T. Furuta, *Analyst* (*London*) **91**, 683 (1966); *Advan. X-Ray Anal.* **11**, 249 (1968); *Amer. Soc. Test. Mater., Spec. Tech. Publ.* **STP-485**, 243 (1971).
[b] OEC, orbital–electron capture.
[c] For the $^{3}_{1}$H–Ti, $^{3}_{1}$H–Zr, and $^{147}_{61}$Pm–Al sources, the useful radiation is that from the overall source, rather than from only the radioisotope.
[d] The x-ray output of each of these sources is $\sim 10^8$ photons/s.

FIGURE 4.1. Radioisotope source–specimen–detector geometry arrangements: center or button source (top left), annular or ring source (top right), and two types of source–target (γ–x) arrangements (bottom), the left of which has annular form. (*A*) Radioisotope. (*B*) Specimen. (*C*) X-ray filter (if used); the filter has an open center for the source–target geometry. (*D*) Detector—lithium-drifted silicon, scintillation, or proportional. (*E*) Secondary target, radiator, or "fluorescer." (*F*) Shielding for radioactive radiation. *A* and *F* together constitute the radioisotope source.

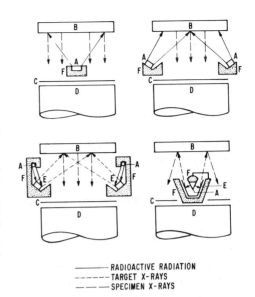

———— RADIOACTIVE RADIATION
– – – – TARGET X-RAYS
— — — SPECIMEN X-RAYS

x-rays is excited most efficiently, each successively lower atomic number less so.

The advantages of radioisotopes include the following. Radioisotope source–detector assemblies are very small, light, and inexpensive, making them applicable to specimens difficult to present to a tube-excited spectrometer, portable and field use, inaccessible places, on-line process control, and space applications. Excitation energies of 100–150 keV are attainable, so the *K* lines of the heaviest elements can be excited. Excitation conditions during analysis of a series of specimens are, for practical purposes, absolutely stable; this permits use of longer counting times to compensate the lower intensity of radioisotope sources. This low intensity, a disadvantage in some ways, can be advantageous in that it prevents choking of semiconductor detectors and substantially eliminates radiolysis of liquid, organic, and biological specimens and discoloration of glasses, ceramics, art objects, etc. The low penetration of α- and β-rays permits surface analysis and reduces absorption-enhancement effects in bulk specimens. Source–target and α sources give practically no continuous background, so that high peak-to-background ratios are realized. The high specificity of source–target sources is used to advantage in light, compact, survey-type instruments for single elements, for example, instruments for environmental surveys for lead in paint.

The principal disadvantages of radioisotope excitation are the following. There is some personnel health hazard and possibility of contamination of the laboratory environment. The output intensity from a radioisotope source is much lower than that from even low-power x-ray tubes. However, this feature can be compensated by longer counting times made feasible by the low background and high source stability. In source–target and orbital-electron capture sources, some of the source x-radiation may scatter coherently and incoherently from the specimen. The target must be chosen so that neither type of scattered target line interferes with the analyte line. Long-term radioisotope source intensity gradually decreases with time so that occasional recalibration is required.

4.1.4. X-Ray Tube Excitation (X-Ray "Fluorescence")

4.1.4.1. General

X-ray tubes can be used for excitation in any of three modes: (1) direct excitation by the unfiltered primary beam in the usual way so that both continuum and target-line x-rays irradiate the specimen; (2) direct excitation by the primary beam filtered so that substantially monochromatic target-line x-rays irradiate the specimen; and (3) excitation of a secondary target having its strongest line at wavelength just shorter than that of the analyte absorption edge; the secondary x-rays in turn irradiate the specimen. The x-ray tube may be of the high-power (2–5-kW) or low-power ($\lesssim$100-W) type.

For wavelength dispersion, direct excitation by high-power tubes is used almost exclusively to compensate the high-intensity losses in the crystal and collimator.

For energy dispersion, unfiltered direct excitation by high-power tubes has the serious disadvantages that, even at the lowest power attainable with conventional high-power generators, the tubes produce high-intensity continuum and relatively high-intensity specimen emission. The continuum causes high scattered background, reducing the sensitivity attainable, and the high total background and emission intensity may cause coincidence loss, pulse-height shift, and/or choking in the detection system. Consequently, for energy dispersion, high-power tubes are usually used in filtered or indirect modes, and often only high-power tubes give primary intensity adequate for these modes. For direct excitation, low-power tubes are suitable.

Continuous radiation permits relatively efficient simultaneous excitation of substantially all the elements in the specimen, whereas the two monochromatic modes are much more selective. Thus, excitation by continuum is advantageous for qualitative and semiquantitative analyses, comparisons, and "signature" or "fingerprint" analyses (Section 7.2.2).

The disadvantages of excitation by continuum cited above are minimized by filtered primary excitation, but the excitation is then more specific. The element having a major absorption edge at wavelength just longer than that of the strongest target line is excited most efficiently, each successively lower atomic number less so. Elements six or seven atomic numbers lower are excited very inefficiently and may not be detected if present in low concentration.

Excitation by secondary emission is also selective and permits selection of optimal excitation wavelength, gives more suitable specimen emission intensity, and gives lower background and consequently higher peak-to-background ratios.

Low-power x-ray tubes, compared with high-power tubes, have the advantages that they: (1) are much more compact; (2) are much less expensive; (3) can be energized by very compact, light, inexpensive, highly regulated, solid-state power supplies; (4) require no water cooling; and (5) have much longer life.

X-ray tubes, particularly low-power tubes, have certain advantages as excitation sources for energy-dispersive spectrometry, compared with radioisotope sources:

The effective wavelength and intensity of the excitation radiation can be varied much more widely with tubes than with isotopes by selection of x-ray tube target; operating conditions (kV, mA); unfiltered or filtered mode; and secondary target.

The effective wavelength and intensity of an x-ray tube output are readily varied over a wide range; those of a radioisotope source are largely limited by the specific isotope and its activity. Almost any element in some form or other can serve as a secondary target for use with an x-ray tube, so that substantially any wavelength is available to excite the specimen. The choice of targets for radioisotopes is limited to elements efficiently excited by the radiation emitted by available practical isotopes (Table 4.1).

As already mentioned, excitation by intense primary continuum, although producing high background and perhaps undesirably high total intensity, permits efficient simultaneous excitation of all elements in the specimen.

The intensity from low-power tubes is higher than that from practical radioisotope sources, and accordingly, analysis time is shorter. For example, a 100-mCi source—a relatively high activity source—emits 3.7×10^9 photons/s, whereas an x-ray tube operating at only 100-μA tube current emits $\sim 10^{12}$ photons/s.

With power supplies having well-regulated ($<$0.1%) kV and mA stabilization, the long-term stability of low-power tubes is excellent, far exceeding that of radioisotopes. For example, over a 4-day period, ^{55}Fe and ^{109}Cd sources (half-lives 2.7 and 1.3 y, respectively) decay by 0.29% and 0.59%, respectively, whereas low-power tubes may remain stable within 0.25% for 1–2 weeks.

Finally, x-ray tubes are safer than radioisotopes. They can be turned off when not in use and, if desirable, even when changing specimens. They cannot contaminate the laboratory environment, and there is nothing to ingest. Also, there is no high-energy ($>$100-keV) radiation to be shielded from the environment—and from the detector.

Conversely, although low-power x-ray tubes do not require water-cooling and can be energized by very small, compact, light, solid-state power supplies, an x-ray tube source is much larger than a radioisotope source and does require a source of power (ac line, battery, motor–generator, etc.) These considerations place x-ray tubes at a disadvantage and may even preclude their use in certain field and space applications, and in applications involving small inaccessible places, such as wells and bore holes.

To summarize, x-ray tube sources are much more versatile and flexible than radioisotope sources, but much more cumbersome, complex, and expensive.

4.1.4.2. Theory—Excitation Equations

In quantitative x-ray fluorescence spectrometric analysis, it is the secondary spectral-line intensity emitted by the analyte that is measured and used to determine analyte concentration. The useful emitted intensity is affected by: (1) the spectral distribution (both continuous and target-line spectra) of the primary x-ray beam; (2) the absorption of the analyte and matrix for the primary x-rays; (3) the excitation probability and fluorescent yield of the analyte line; (4) the absorption of the analyte and matrix for the analyte line; and (5) the geometry of the x-ray spectrometer. In this section, equations based on these parameters are derived for the relationship between emitted analyte-line intensity and concentration. An electron excitation equation is given in Section 11.6.

The process of secondary excitation of an analyte line is most easily understood for the case of a monochromatic primary beam. In Figure 4.2, consider an incremental layer of thickness Δt at depth t in a specimen of density ϱ. The incident angle of the primary beam is ϕ, and the takeoff

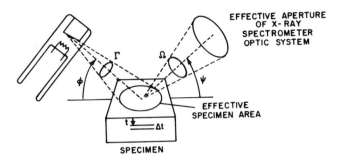

FIGURE 4.2. Geometry for secondary excitation.

angle of the secondary beam is ψ. Now consider a single wavelength λ_{pri} having intensity $I_{0,\lambda_{\text{pri}}}$ in the incident primary beam. If the specimen has mass-absorption coefficient $(\mu/\varrho)_{\text{M},\lambda_{\text{pri}}}$ for λ_{pri}, the fraction of incident photons penetrating to depth t and arriving at layer Δt is

$$\exp[-(\mu/\varrho)_{\text{M},\lambda_{\text{pri}}}\,\varrho t \csc \phi]$$

Of the photons arriving at layer Δt, the fraction absorbed in the layer is

$$(\mu/\varrho)_{\text{M},\lambda_{\text{pri}}}\,\varrho\,\Delta t \csc \phi$$

Of the photons absorbed in layer Δt, the fraction absorbed by the analyte A having concentration C_{A} (weight fraction) and mass-absorption coefficient $(\mu/\varrho)_{\text{A},\lambda_{\text{pri}}}$ for λ_{pri} is

$$\frac{C_{\text{A}}(\mu/\varrho)_{\text{A},\lambda_{\text{pri}}}}{(\mu/\varrho)_{\text{M},\lambda_{\text{pri}}}}$$

Of the photons absorbed by the analyte in layer Δt, the fraction that create electron vacancies in the atom shell corresponding to the series of the analyte line is related to the absorption-edge jump ratio of the analyte for that shell r_{A}; the fraction is

$$1 - (1/r_{\text{A}}) = (r_{\text{A}} - 1)/r_{\text{A}} \tag{4.1}$$

Of the photons causing such vacancies, the fraction that lead to actual emission of photons of the spectral lines in that series is given by the fluorescent yield of the analyte for that series ω_{A}.

Of the primary photons causing such secondary photon emission, the fraction that lead to emission of the particular analyte line L to be measured is given by g; g is the fractional value or relative intensity of the analyte line in its series, or the probability of the orbital electron transition causing the analyte line, or the fraction of, say, the total K-series x-ray photons emitted by the analyte that are $K\alpha$ photons; for example,

$$g_{K\alpha} = \frac{I_{K\alpha_1} + I_{K\alpha_2}}{\sum I_K} \tag{4.2}$$

where $\sum I_K$ is the sum of the intensities of all the analyte K lines, including $K\alpha_1$ and $K\alpha_2$.

Of the analyte-line photons emitted by layer Δt, only that fraction directed toward the source collimator can contribute to the measured line intensity; this fraction is given by $d\Omega/4\pi$ (Figure 4.2).

Of the analyte-line photons directed toward the collimator, the fraction that emerges from the overlying layer t having mass-absorption coefficient $(\mu/\varrho)_{M,\lambda_L}$ for the analyte line L is

$$\exp[-(\mu/\varrho)_{M,\lambda_L}\varrho t \csc \psi]$$

The foregoing processes may be grouped for convenience into five factors as follows:

1. Incident photon intensity: $I_{0,\lambda_{pri}}$.
2. Attenuation of the primary beam in reaching layer Δt:

$$\exp[-(\mu/\varrho)_{M,\lambda_{pri}} \varrho t \csc \phi]$$

3. Probability of excitation of analyte line L:

$$\frac{C_A(\mu/\varrho)_{A,\lambda_{pri}}}{(\mu/\varrho)_{M,\lambda_{pri}}} \frac{r_A - 1}{r_A} \omega_A g_L \left[\left(\frac{\mu}{\varrho} \right)_{M,\lambda_{pri}} \varrho \, \Delta t \csc \phi \right]$$

4. Fraction of analyte-line photons emitted toward source collimator: $d\Omega/4\pi$.

5. Attenuation of this analyte-line radiation in emerging from the specimen:

$$\exp[-(\mu/\varrho)_{M,\lambda_L} \varrho t \csc \psi]$$

Combination of all these expressions gives the analyte-line intensity ΔI_L emitted from layer Δt that actually enters the source collimator:

$$\Delta I_L = I_{0,\lambda_{pri}} \left\{ \exp\left[-\left(\frac{\mu}{\varrho} \right)_{M,\lambda_{pri}} \varrho t \csc \phi \right] \right\} \left(\frac{\mu}{\varrho} \right)_{M,\lambda_{pri}} \varrho \, \Delta t \csc \phi$$

$$\times \frac{C_A(\mu/\varrho)_{A,\lambda_{pri}}}{(\mu/\varrho)_{M,\lambda_{pri}}} \frac{r_A - 1}{r_A} \omega_A g_L \frac{d\Omega}{4\pi} \exp\left[-\left(\frac{\mu}{\varrho} \right)_{M,\lambda_L} \varrho t \csc \psi \right] \tag{4.3}$$

Simplification of Equation (4.3), followed by integration from $t = 0$ to t, gives

$$I_L = I_{0,\lambda_{\mathrm{pri}}} \omega_A g_L \frac{r_A - 1}{r_A} \frac{d\Omega}{4\pi} \frac{C_A(\mu/\varrho)_{A,\lambda_{\mathrm{pri}}}\varrho \csc \phi}{(\mu/\varrho)_{M,\lambda_{\mathrm{pri}}}\varrho \csc \phi + (\mu/\varrho)_{M,\lambda_L}\varrho \csc \psi} \quad (4.4)$$

If we define

$$P_A = \omega_A g_L \frac{r_A - 1}{r_A} \frac{d\Omega}{4\pi} \quad (4.5)$$

and

$$A = \sin \phi / \sin \psi \quad (4.6)$$

Equation (4.4) becomes

$$I_L = P_A I_{0,\lambda_{\mathrm{pri}}} C_A \frac{(\mu/\varrho)_{A,\lambda_{\mathrm{pri}}}}{(\mu/\varrho)_{M,\lambda_{\mathrm{pri}}} + A(\mu/\varrho)_{M,\lambda_L}} \quad (4.7)$$

Equations (4.7) and (4.4) may be regarded as the fundamental excitation equations (intensity formulas) for secondary x-ray excitation with monochromatic primary x-radiation. These equations are valid only for smooth homogeneous specimens, for collinear primary beams, and in the absence of multiple scatter and enhancement effects.

In Equation (4.7), P_A may be regarded as an excitation factor, A as a geometric factor, and the fractional term as an efficiency factor. P_A is constant for a given analyte A, matrix, and spectrometer, including a specified set of components. A influences the relative importance of absorption of the primary and secondary (analyte-line) x-rays. At low takeoff angles ψ, A becomes large, and secondary absorption predominates. This relationship can be useful in excitation by continuous spectra because secondary absorption effects apply to the monochromatic analyte line λ_L, so calculations are much simpler than when absorption of the primary continuum predominates.

If the analyte line is excited by a polychromatic continuous x-ray beam rather than a monochromatic beam, it is necessary to consider all incident wavelengths between the short-wavelength limit $\lambda_{\min}$ [Equation (1.17)] and the absorption edge λ_{ab} associated with the analyte line. One must consider the spectral distribution of this region and the mass-absorption coefficient of the specimen for each wavelength in this region.

The *spectral distribution* $J(\lambda_{\mathrm{pri}})$ is the plot of the intensity of each wavelength in the effective region between $\lambda_{\min}$ and λ_{ab}. This distribution may be measured, calculated, replaced with a monochromatic effective wavelength λ_{eff} (Section 1.7), or corrected for by means of influence coefficients (Section 9.9.4). When excitation is predominantly by continuum, the effective wavelength is given by

$$\lambda_{\text{eff}} \approx \tfrac{2}{3}\lambda_{\text{ab}} \tag{4.8}$$

When excitation is predominantly by target-line x-rays $\lambda_{L_{\text{tgt}}}$,

$$\lambda_{\text{eff}} \approx \lambda_{L_{\text{tgt}}} \tag{4.9}$$

Fortunately, the value of λ_{eff} is not critical, and differences of $\pm 25\%$ have relatively little effect on the calculations.

Mass-absorption coefficients must be considered for all elements in the specimen matrix for all effective primary wavelengths and for the analyte line. Expressions of the following form are obtained from Equation (2.3):

$$(\mu/\varrho)_{\text{M},\lambda_{\text{pri}}} = \sum C_i(\mu/\varrho)_{i,\lambda_{\text{pri}}} \tag{4.10}$$

$$(\mu/\varrho)_{\text{M},\lambda_L} = \sum C_i(\mu/\varrho)_{i,\lambda_L} \tag{4.11}$$

where $(\mu/\varrho)_{\text{M},\lambda_{\text{pri}}}$ and $(\mu/\varrho)_{\text{M},\lambda_L}$ are mass-absorption coefficients of the specimen for a specific primary wavelength λ_{pri} and the analyte line λ_L; $(\mu/\varrho)_{i,\lambda_{\text{pri}}}$ and $(\mu/\varrho)_{i,\lambda_L}$ are mass-absorption coefficients of a specific element i for λ_{pri} and λ_L; and C_i is concentration of element i (weight fraction). In these equations, i includes A, the analyte.

Substitution of an expression for $J(\lambda_{\text{pri}})$ and of Equations (4.10) and (4.11) in Equation (4.7) gives a general expression for I_L:

$$I_L = P_A C_A \int_{\lambda_{\text{min}}}^{\lambda_{\text{ab}}} J(\lambda_{\text{pri}}) \left(\frac{\mu}{\varrho}\right)_{A,\lambda_{\text{pri}}} \frac{1}{\sum C_i[(\mu/\varrho)_{i,\lambda_{\text{pri}}} + A(\mu/\varrho)_{i,\lambda_L}]} \, d\lambda \tag{4.12}$$

P_A and A are defined by Equations (4.5) and (4.6), respectively; their significance is given following Equation (4.7). Equation (4.12) may be regarded as the fundamental excitation equation (intensity formula) for secondary x-ray excitation with polychromatic primary x-radiation. Like Equations (4.7) and (4.4), this one is valid only for smooth homogeneous specimens, for collinear primary beams, and in the absence of multiple scatter and enhancement effects.

4.1.4.3. Practice—Operating Conditions

Figure 4.3 shows the wave forms of 50-kV ac, half-wave and full-wave rectified pulsating dc, and constant-potential dc. The K-excitation potentials of barium (37.4 kV), molybdenum (20.0 kV), and iron (7.1 kV) are in-

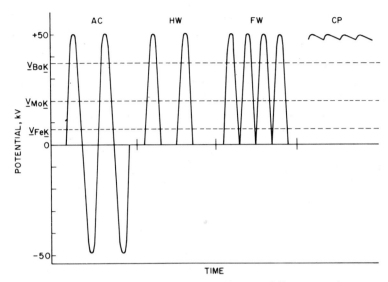

FIGURE 4.3. Secondary excitation by half-wave, full-wave, and constant-
potential wave forms.

dicated. The K lines of these elements are excited only when the x-ray tube
potential exceeds their respective excitation potentials, and they are excited
efficiently only when the excitation potentials are exceeded substantially.
It is evident that the smaller the difference between operating and excitation
potentials, the greater is the advantage of constant-potential (CP) compared
with full-wave (FW) power. At 50 kV, the relative intensities for half-wave,
full-wave, and constant-potential excitation are for Fe $K\alpha$ 43, 86, and 100,
respectively, and for Ba $K\alpha$ 13, 26, and 100. It is also evident that constant
potential is particularly advantageous for determination of minor and trace
constituents having high excitation potentials.

Commercial x-ray spectrometers are available with maximum high
potentials of 50, 75, and 100 kV peak or CP. A potential of 50 kV excites
K lines up to europium (Z 63), 75 kV up to osmium (Z 76), and 100 kV
up to radon (Z 86). Of course, each of these potentials excites K lines
efficiently only up to about five atomic numbers below those cited above.
The advantage of the K lines of the heaviest elements is questionable.
However, K lines are fewer in number and more intense than L lines, and
all K and L lines excited at 50 kV are excited more efficiently at 75 or 100 kV
[Equation (1.22)]. For example, La $K\alpha$ net intensity at 100 kV is said to
be five times La $L\alpha_1$ net intensity at 50 kV, and the advantage increases for
the heavier lanthanons. Background is also higher, but peak-to-background

ratio stays about the same, so there is some net advantage because shorter
counting times are feasible. No really good crystal having both high dispersion and high "reflectivity" is available for these short-wavelength lines;
LiF(422), LiF(420), and quartz(502) are the most applicable.

Choice of x-ray tube current (mA) is less a problem than choice of
potential (kV) because current affects both line and background intensity
linearly and therefore does not affect the peak-to-background ratio. However, potential affects the two intensities very differently.

Generally, it is preferable to select the optimum potential, then set the
current to the maximum permitted by the target power rating (kW or kVA).
Actually, it is prudent to set the current at ∼5 mA less than this value:
a few milliamperes costs very little intensity, but may substantially increase
the tube life.

For qualitative analysis, it is prudent to operate at maximum potential
so that lines having high excitation potentials are excited. This is particularly
important when the spectrum is recorded in air path, and K lines of the
heavy elements must be excited because the L lines may lie in the helium
or vacuum region.

For quantitative analysis, maximum potential always gives maximum
intensity for all lines, regardless of whether the line is excited primarily
by continuum or by target lines. The intensities of all primary wavelengths
increase with x-ray tube potential. Continuum intensity increases with the
square of the x-ray tube potential [Equation (1.18)]. Target-line intensities
increase with approximately the 1.7th power of the difference between the
x-ray tube and excitation potentials [Equation (1.22)]. However, maximum
operating potential often is not used. It may be necessary to avoid excitation
of a certain element having high excitation potential (selective excitation).
Also, it is inefficient and uneconomical to operate at a potential so high
that a substantial portion of the primary x-rays have wavelength so short
that they penetrate the specimen to a depth from which analyte-line radiation cannot emerge.

For quantitative analysis, optimum operating potential for a specified
analyte line may be defined as that potential that gives the highest peak-to-
background ratio. This potential is determined experimentally by measuring
peak and background intensities for the analyte line at a series of potentials.
The optimum potential is likely to be different for analyte lines of substantially different wavelength. For excitation of short-wavelength spectra
by continuum, optimum potential is usually that which places the continuum "hump" on the short-wavelength side of the analyte absorption
edge. For excitation of long-wavelength spectra by continuum, peak-to-

background ratio decreases as potential increases because intensity increases in the hump more rapidly than at the longer wavelengths most efficient in exciting the long-wavelength spectra. The wavelengths in the hump may be scattered by the specimen and diffracted by the crystal in higher orders, thereby increasing the background in the long-wavelength region. However, pulse-height selection discriminates this higher-order background, so if analyte-line intensity cannot be spared, maximum x-ray tube potential can be used.

The considerations in the choice of x-ray tube operating potential and target are summarized in Figures 4.4 and 4.5.

Figure 4.4 illustrates excitation by the continuum. The figure shows the continuum "hump" for an x-ray tube operating at each of several potentials in the range 35–50 kV, and the K-absorption edges of a series of elements having atomic number 46–92 (palladium to uranium). At any specified potential: (1) only those elements having K edges at wavelength longer than $\lambda_{\min}$ are excited at all; (2) only those elements having K edges

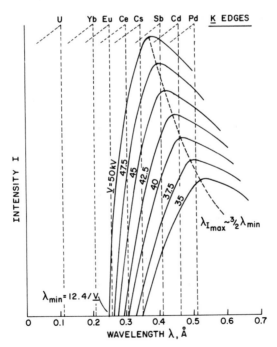

FIGURE 4.4. Secondary excitation by the primary continuum, showing the effects of x-ray tube operating potential and analyte absorption-edge wavelength.

at wavelength substantially longer than λ_{min} are excited efficiently; and (3) those elements having K edges at wavelength a little longer than $\lambda_{I_{max}}$ are excited most efficiently. Of course, similar considerations apply to excitation of the LIII edges of the elements. As x-ray tube potential is increased: (1) λ_{min} and $\lambda_{I_{max}}$ move to shorter wavelengths; (2) elements of successively higher atomic number are excited; and (3) all elements excited at lower potential are excited more efficiently.

Figure 4.5 illustrates excitation by both continuum and target-line radiation. The figure shows the continuum and line spectrum from tungsten and chromium targets operating at 50 kV, and the K-absorption edges of a series of elements having atomic number 17–92 (chlorine to uranium). For excitation by the continuum, the tungsten target is preferable because intensity is proportional to atomic number [Equation (1.18)]. However, when a strong target line lies near the short-wavelength side of an analyte absorption edge, the target line is likely to make the predominant contribution to the excitation of that element, and the nearer the line lies to the edge, the more efficient is the excitation. Thus, Figure 4.5 shows that W $L\gamma_1$ excites the K lines of germanium and lighter elements, W $L\beta_1$ excites those of zinc and lighter elements, and W $L\alpha_1$ excites those of nickel and lighter elements. It is evident that tungsten is the optimal target for nickel because

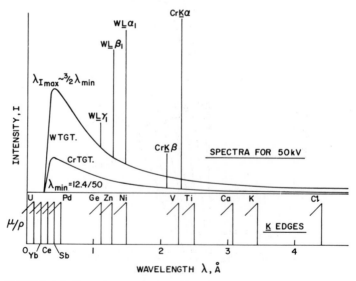

FIGURE 4.5. Secondary excitation by the continuum and spectral lines of tungsten (atomic number 74) and chromium (atomic number 24) targets.

W $L\alpha_1$, $L\beta_1$, and $L\gamma_1$ all lie close to the short-wavelength side of the Ni K edge, and W $L\alpha_1$, the strongest line, lies extremely close to the edge. It is also evident why chromium is preferable to tungsten for excitation of vanadium and lighter elements: The continuum for both targets is very weak in this region, so the relative proximity of the Cr K lines to the low-Z absorption edges assumes the predominant role.

On prolonged use, the x-ray tube target and window may become contaminated with elements from various structural components in the tube: tungsten from the filament, copper from the anode block, chromium, iron, and nickel from stainless steel parts, silver from brazes, etc. Spectral lines of these contaminants occur in the primary beam with increasing intensity as the tube ages. The target spectrum can be observed as follows. A specimen containing no elements of atomic number $\gtrsim 9$ (fluorine) and having average atomic number as low as possible is placed in the specimen compartment to scatter the primary beam into the spectrometer. Suitable scatter targets include a piece of paraffin or spectroscopically pure carbon, a briquet of starch, or a cell of distilled water covered with 3- or 6-μm (0.00012- or 0.00025-inch) Mylar. A 2θ scan is made of the scattered x-rays and recorded on the ratemeter–recorder. Such spectra should be recorded whenever a new tube is first used to identify the extraneous lines, and occasionally thereafter to observe increase of spectral contamination with life.

The specimen area irradiated by the target is observed by placing a piece of unexposed x-ray film or a glass microscope slide in the specimen plane without the specimen mask and irradiating it at full power for 3–5 min. The (undeveloped) film or glass is colored in the irradiated area. This technique aids in aligning the x-ray tube for optimal irradiation of the specimen.

X-ray tube output flux at constant kV and mA decreases ~3% for each 1000 h of life as a result of target pitting and sublimation of tungsten on the inside of the tube window. This condition is evaluated by periodic measurements on a sturdy, inert test specimen.

4.2. WAVELENGTH DISPERSION

The third phase of an x-ray spectrometric analysis (Section 3.2) is dispersion—the separation of the x-ray spectral lines emitted by the elements in the specimen so that a line of a specified analyte can be measured individually. Wavelength dispersion is considered here, energy dispersion in Chapter 6.

4.2.1. Radiation Path

The radiation path is the path of the primary and secondary x-rays from the outside of the x-ray tube window to the outside of the detector window. As the wavelength of the analyte line and of the primary x-rays most efficient in exciting it increases (photon energy decreases), the absorption of the path medium progressively increases.

Early spectrometers had air paths. Later, rubber bags were provided to enclose the path as shown in Figure 3.2, and helium was passed through the bag at $\sim$10 ft^3/h while measurements were made. In modern instruments, the path is enclosed and can be operated with air or helium at atmospheric pressure, or vacuum at $\leq$0.5 torr (mm Hg) from a rotary pump.

Measured intensities in air and helium paths are substantially equal at wavelengths shorter than $\sim$1 Å ($_{35}$Br $K\alpha$, $_{88}$Ra $L\alpha_1$). Measured intensities in helium and vacuum are substantially the same at wavelengths shorter than $\sim$6 Å ($_{15}$P $K\alpha$, $_{40}$Zr $L\alpha_1$, $_{78}$Pt $M\alpha$). In general, air path is entirely satisfactory for wavelengths shorter than $\sim$2.5 Å ($_{22}$Ti $K\alpha$, $_{59}$Pr $L\alpha_1$) and for major constituents in full-size specimens down to $\sim$4 Å ($_{19}$K $K\alpha$, $_{48}$Cd $L\alpha_1$, $_{92}$U $M\alpha$). However, vacuum or helium is beneficial at wavelengths greater than $\sim$2.5 Å and must be used beyond $\sim$4 Å. Helium path is satisfactory for wavelengths shorter than $\sim$6 Å and for major constituents in full-size specimens down to $\sim$18 Å ($_9$F $K\alpha$). However, vacuum is beneficial at wavelengths greater than $\sim$6 Å, should be used beyond $\sim$12 Å ($_{11}$Na $K\alpha$), and is required beyond $\sim$18 Å. Vacuum path should be used for all L and M lines in the region beyond $\sim$6 Å.

Liquid specimens may evaporate and loose powders outgas and spatter in vacuum path, so helium may be more convenient for these specimens. However, these problems are surmountable, and vacuum can be used.

When using helium or vacuum path, a certain time must be allowed after changing samples to permit air to be flushed out or evacuated. About 10 s is usually sufficient for vacuum or helium in modern instruments, but the older helium bags required 30–60 s.

4.2.2. Collimators

Flat-crystal x-ray fluorescence spectrometers use *Soller* collimators, which consist of spaced parallel metal foils. Collimators intercept divergent x-rays so that substantially parallel beams arrive at the crystal and detector. In Figure 3.2, the excited specimen emits x-rays in all directions, but only those rays directed substantially parallel to the collimator foils pass through to the crystal. Incidentally, the projection of this collimator on the specimen plane defines the maximum useful specimen area.

Soller collimators are placed at either or both of two positions in the goniometer (Figure 3.2). The *primary, source,* or *divergence* collimator is placed between the specimen and crystal, and is fixed. The *secondary, detector,* or *receiving* collimator is placed between the crystal and detector, and moves with the detector. In principle, the two positions are equally effective for collimation of the specimen emission. However, the collimator between crystal and detector is helpful in excluding from the detector any secondary emission and spurious reflections from the crystal itself. Alternatively, the collimator between specimen and crystal makes the 2-to-1 alignment of the spectrogoniometer less critical. When collimators are used in both positions, one is usually coarse (short and/or wide-spaced foils), the other fine (long and/or close-spaced foils).

In some instruments, a very short, coarse Soller collimator is placed between the front flow-proportional and rear sealed-proportional or scintillation counters of a combination (tandem) detector.

The x-ray beam transmitted by a Soller collimator has a triangular intensity *versus* 2θ profile. For a collimator of length L and foil spacing S, the width of this profile at half-maximum in degrees 2θ is

$$W_{1/2} = \tan^{-1}(S/L) \tag{4.13}$$

The spacing S is measured *between* the foils, and is not the width of the collimator aperture divided by the number of foils. Collimators are specified either in terms of length and foil spacing (for example, 100×0.250 mm) or by the angular divergence in degrees as given by Equation (4.13).

The effects of increased collimation, that is, longer and/or closer-spaced collimators are: (1) narrower peaks; (2) narrower "tails" (or "skirts" or "wings") on the peaks; (3) increased resolution, as a consequence of (1) and (2); (4) decreased background; and (5) decreased measured analyte-line intensity. In short, the higher the collimation, the higher is the resolution and the lower is the intensity. As a rule, the shortest, most widely spaced collimator is used that gives resolution adequate for the measurements to be made. At wavelengths greater than 3 or 4 Å, the separation of the K lines of neighboring elements increases as atomic number decreases (Figure 1.11). Consequently, low collimation is adequate unless M or higher-order K or L lines of high-Z elements are present. This applicability of low collimation is fortunate because it partially compensates the low fluorescent yield and high absorption losses in this spectral region.

When not mounted on the goniometer, collimators must be handled carefully and stored in dustproof boxes to prevent crimping the ends of the foils and accumulation of dirt between the foils.

4.2.3. Analyzer Crystals

4.2.3.1. Introduction

The *crystal, analyzer crystal, analyzer,* or *monochromator* performs the same function in an x-ray emission spectrometer as the prism, or, more analogously, the grating, in an optical emission spectrograph: It spacially disperses the secondary x-ray beam into a wavelength spectrum. As the crystal is rotated from $0°\ 2\theta$, where it is parallel to the collimated secondary x-ray beam, through progressively larger angles, it passes successively through the Bragg angle for each wavelength present, up to a certain limit (see below), and diffracts each wavelength in turn. X-ray diffraction, the basic phenomenon involved in crystal dispersion, is discussed in Section 2.3. Wavelength dispersion is also known as *crystal dispersion* or *diffractive dispersion*.

An analyzer crystal consists of a thin bar of a specified single-crystalline material cut or cleaved parallel to a specified set of crystallographic planes. The crystal may be used flat, or curved cylindrically, logarithmically, spherically, or torically.

Many crystals are used in more than one "cut," each having a different crystallographic plane with different Miller indexes (*hkl*) and *d* spacing parallel to its "reflecting" surface and therefore useful in a different wavelength region. Thus, an analyzer crystal is specified by its name (or chemical formula or letter symbol) *and* the Miller indexes or *d* spacing (actually $2d$ spacing, see below) of its diffracting planes. Incidentally, in the Miller index notation, (*hkl*) and *n*(*hkl*) cuts are identical; for example, LiF(100) is the same as LiF(200), and LiF(110) is the same as LiF(220). Also, it is common practice to omit the *i* in the Miller indexes of hexagonal and orthorhombic crystals (*hkil*), since *i* is always $-(h + k)$; for example, quartz(101) is the same as quartz(10$\bar{1}$1).

4.2.3.2. Features

A crystal is chosen on the basis of the following features, the first three of which are by far the most important: (1) wavelength region appropriate for the analyte lines to be measured; (2) high resolution—high angular dispersion and narrow diffracted peak breadth; (3) high diffracted intensity; (4) absence of interfering elements, that is, elements that may emit their own characteristic lines ("crystal emission"); (5) high stability in air and vacuum, and on repeated prolonged exposure to x-rays; (6) low thermal coefficient of expansion; and (7) good mechanical strength. Table 4.2 lists some of the properties of the commonly used analyzer crystals (and of multilayer films, which are discussed in Section 4.2.3.3).

The useful region of a standard commercial flat-crystal x-ray fluorescence spectrometer is $\sim$0.2–20 Å. In principle, the longest wavelength that a crystal can diffract is equal to twice the interplanar spacing ($2d$) of the crystal planes parallel to its surface. From the Bragg law,

$$\sin \theta = n\lambda/2d \qquad (4.14)$$

If $n\lambda > 2d$, $\sin \theta > 1$, which is not possible. It follows that for dispersion of long wavelengths, long d spacing is required. Incidentally, interplanar spacings for analyzer crystals are usually given in terms of $2d$, rather than d, because of this particular significance of $2d$. However, in practice, mechanical features limit the useful 2θ region of the crystal. At small 2θ angles, the crystal approaches parallelism with the Soller collimator foils and an increasing portion of the secondary beam strikes the edge of the crystal, passes the crystal, and even enters the detector directly. At high 2θ angles, the detector arm strikes the x-ray tube head, specimen chamber, or some other structure. These mechanical conditions limit the useful crystal region to 10–15° 2θ to 120–150° 2θ.

At $2\theta = 150°$, $\sin \theta = 0.966$, and from Equation (4.14), for $n\lambda = 20$ Å, $2d \approx 21$ Å. Thus, in principle, all that is required is one crystal having $2d$ 21 Å. However, such a crystal would have prohibitively poor angular dispersion (see below) in the short-wavelength region. So in practice, for this and other reasons, any one crystal is limited to a certain spectral region, and a series of crystals having different $2d$ spacings is required to cover the entire spectral region.

High resolution is desirable to minimize spectral interference. The resolution of a crystal is its ability to distinguish or recognize as separate two spectral lines of nearly the same wavelength. Resolution is the combined effect of the angular dispersion—the separation of the two lines in 2θ—and the divergence—the breadth of the diffracted lines in 2θ. Differentiation of the Bragg equation shows that angular dispersion increases as $2d$ decreases and as order n and diffraction angle θ increase.

$$\frac{d\theta}{d\lambda} = \frac{n}{2d \cos \theta} \qquad (4.15)$$

TABLE 4.2. Most Commonly Used X-Ray Spectrometer Analyzer Crystals and Multilayer Films

Letter symbol and diffracting plane (hkl) / Common name / Chemical name	Chemical formula	$2d$ (Å)	Practical useful wavelength region (10–140° 2θ), λ (Å) $K\alpha$ lines $L\alpha_1$ lines	Optimal application Remarks
LiF(420) Lithium fluoride	LiF	1.80	λ: 0.157–1.72 $K\alpha$: $_{84}$Pb–$_{28}$Ni $L\alpha$: $\gtrsim_{70}$Yb	High-Z K lines excited by 100-kV generators
Topaz(303)	$Al_2(F,OH)_2SiO_4$	2.712	λ: 0.236–2.59 $K\alpha$: $_{70}$Yb–$_{23}$V $L\alpha$: $\gtrsim_{58}$Ce	Improves dispersion in the region V–Ni K, where Z $K\beta$ overlaps $(Z+1)$ $K\alpha$ Improves dispersion of lanthanon (Z 57–71) L lines
LiF(220) Lithium fluoride	LiF	2.848	λ: 0.248–2.72 $K\alpha$: $_{68}$Er–$_{22}$Ti $L\alpha$: $\gtrsim_{57}$La	Better than topaz for the same application Diffracted intensity $\sim$2–4× topaz, $\sim$0.4–0.8× LiF(200)
LiF(200) Lithium fluoride	LiF	4.028	λ: 0.351–3.84 $K\alpha$: $_{58}$Ce–$_{19}$K $L\alpha$: $\geq_{49}$In	Best general crystal for K K to Lr L lines Combines high intensity and high dispersion
NaCl Rock salt Sodium chloride	NaCl	5.641	λ: 0.492–5.38 $K\alpha$: $_{50}$Sn–$_{16}$S $L\alpha$: $\gtrsim_{44}$Ru	S and Cl $K\alpha$ in matrixes lighter than Mg Like LiF(200), good general crystal for S K to Lr L lines
Si(111) Silicon	Si	6.271	λ: 0.547–5.98 $K\alpha$: $_{47}$Ag–$_{16}$S $L\alpha$: $\gtrsim_{41}$Nb	Intermediate- and high-Z elements, where Si $K\alpha$ crystal emission is absorbed by the air path No second order

Crystal	Formula	$2d$ (Å)	Wavelength range	Applications and remarks
Ge(111) Germanium	Ge	6.532	λ: 0.570–6.23 $K\alpha$: $_{46}$Pd–$_{15}$P $L\alpha$: $\geq_{40}$Zr	Intermediate- and low-Z elements, where Ge $K\alpha$ crystal emission is discriminated by pulse-height selection. No second order
Graph(002) Graphite	C	6.708	λ: 0.585–6.40 $K\alpha$: $_{46}$Pd–$_{15}$P $L\alpha$: $\geq_{40}$Zr	P, S, Cl K lines. P $K\alpha$ intensity $\sim 5\times$ EDDT, but resolution poor
Quartz(100) α-Quartz Silicon dioxide	SiO$_2$	8.52	λ: 0.742–8.12 $K\alpha$: $_{41}$Nb–$_{14}$Si $L\alpha$: $_{100}$Fm–$_{36}$Kr	Same applications as PET and EDDT with higher resolution but lower intensity
PET(002) Pentaerythritol	C(CH$_2$OH)$_4$	8.742	λ: 0.762–8.34 $K\alpha$: $_{40}$Zr–$_{13}$Al $L\alpha$: $_{99}$Es–$_{36}$Kr	Al–Cl K, especially P and S K. Good general crystal for Al–Sc K. Al–S K intensities ~ 1.5–$2\times$ EDDT and KHP. Very low background
EDDT(020), EDT Ethylenediamine-d-tartrate (d for $dextro$-, not dl)	NH$_2$—CH$_2$—CH$_2$—NH$_2$ COOH—(CHOH)$_2$—COOH	8.808	λ: 0.768–8.40 $K\alpha$: $_{41}$Nb–$_{13}$Al $L\alpha$: $_{99}$Es–$_{35}$Br	Same applications as PET, but lower intensity. Very low background
ADP(101) Ammonium dihydrogen phosphate	NH$_4$H$_2$PO$_4$	10.640	λ: 0.928–10.15 $K\alpha$: $_{37}$Rb–$_{12}$Mg $L\alpha$: $_{91}$Pa–$_{33}$As	Mg $K\alpha$. Can be used for Al and Si $K\alpha$ when Mg $K\alpha$ is also to be measured, but lower intensity than PET or EDDT
Gyp(020) Gypsum Calcium sulfate dihydrate	CaSO$_4\cdot$2H$_2$O	15.185	λ: 1.32–14.49 $K\alpha$: $_{81}$Ga–$_{11}$Na $L\alpha$: $_{77}$Ir–$_{29}$Cu	Na $K\alpha$, but not as good as KHP. Can be used for Mg–Cl $K\alpha$ when Na $K\alpha$ is also to be measured

continued overleaf

TABLE 4.2 *(continued)*

Letter symbol and diffracting plane (hkl) Common name Chemical name	Chemical formula	$2d$ (Å)	Practical useful wavelength region (10–140° 2θ), λ (Å) $K\alpha$ lines $L\alpha_1$ lines	Optimal application Remarks
RHP(100), RAP^a Rubidium hydrogen phthalate	RbOOC—C$_6$H$_4$—COOH (COORb / COOH ring)	26.12	λ: 2.28–24.92 $K\alpha$: $_{24}$Cr–$_8$O $L\alpha$: $_{61}$Pm–$_{23}$V	Good general crystal for all low-Z elements down to O. Diffracted intensity $\sim 3\times$ KHP for Na, Mg, Al $K\alpha$, and Cu $L\alpha$; $\sim 4\times$ KHP for F $K\alpha$, $\sim 8\times$ KHP for O $K\alpha$. Peak/background ratio higher than for KHP
KHP(100), KAP^a Potassium hydrogen phthalate	KOOC—C$_6$H$_4$—COOH (COOK / COOH ring)	26.632	λ: 2.32–25.41 $K\alpha$: $_{23}$V–$_8$O $L\alpha$: $_{60}$Nd–$_{23}$V	Same application as RHP, but not as good
LST, LOD Lead stearate Lead octadecanoate	[CH$_3$(CH$_2$)$_{16}$COO]$_2$Pb	100.4	λ: 8.75–95.75 $K\alpha$: $_{12}$Mg–$_5$B $L\alpha$: $\lesssim {}_{34}$Se	Ultralong-wavelength region down to B $K\alpha$
LTE Lead lignocerate, lead carnaubate Lead tetracosanoate	[CH$_3$(CH$_2$)$_{22}$COO]$_2$Pb	126	λ: 11.35–124 $K\alpha$: $_{11}$Na–$_4$Be $L\alpha$: $\lesssim {}_{31}$Ga	Ultralong-wavelength region down to Be $K\alpha$

^{a} For rubidium *acid* phthalate and potassium *acid* phthalate, which are not preferred nomenclature.

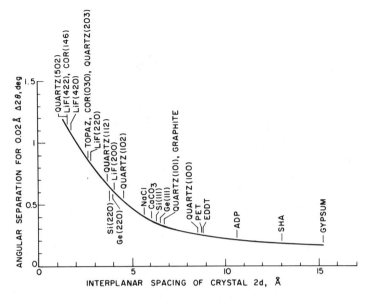

FIGURE 4.6. Angular dispersion of common analyzer crystals as a function of interplanar spacing, 2d.

For a specified wavelength, best dispersion occurs at a large θ angle with a crystal having $2d$ only slightly greater than the wavelength. Resolution is poorest at small θ angles. Dispersion is shown as a function of $2d$ in Figure 4.6 and in Table 4.3. Figure 4.7 shows the effect of $2d$ on resolution in an even more striking way. The $K\alpha$ lines of all the lanthanon elements are

TABLE 4.3. Crystal Dispersion as a Function of $2d$: Dispersion of Mn $K\alpha$ (2.103 Å) and Cr $K\beta$ (2.085 Å), $\Delta\lambda$ 0.018 Å

Crystal	$2d$ (Å)	$\Delta 2\theta$ (deg)
Topaz	2.712	1.21
LiF	4.028	0.61
NaCl	5.641	0.40
EDDT	8.808	0.24
ADP	10.640	0.21
KHP	26.632	0.08
Clinochlore	28.392	0.07

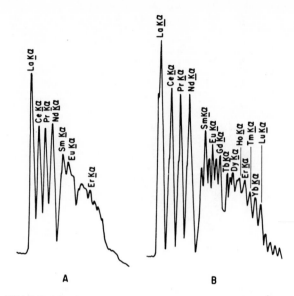

FIGURE 4.7. Comparison of resolution of the $K\alpha$ lines of the lanthanon elements in an aluminum oxide (Al_2O_3) matrix by LiF(200) ($2d$ 4.028 Å) (A) and LiF(220) ($2d$ 2.848 Å) (B) analyzer crystals. [K. W. Payne, *Proceedings of the 5th Conference on X-Ray Analytical Methods*, Swansea, Wales, U.K. (1966); published by N. V. Philips Gloeilampenfabrieken, Scientific and Analytical Equipment Dept., Eindhoven, Netherlands; courtesy of the author and publisher; British Crown copyright, reproduced by permission.]

resolved much better by LiF(220) ($2d$ 2.848 Å) than by LiF(200) ($2d$ 4.028 Å). The combined effect of collimation and crystal $2d$ spacing are shown in Figure 4.8.

An increase in $\Delta\theta/\Delta\lambda$ does not necessarily result in good resolution if the diffracted peaks are too broad and have broad tailings. The line breadth is the combined effect of the natural breadth of the line and the rocking curve of the crystal. The greater the line breadth, the lower is the intensity at the peak and the more difficult is the resolution of the line from adjacent lines. The rocking curve is a measure of the angular range over which the diffracted monochromatic wavelength is spread due to crystal imperfection. It is expressed numerically as the width in degrees 2θ of the diffracted peak at half-maximum—full width at half-maximum (FWHM), or half-width ($W_{1/2}$). The rocking curve is approximately Gaussian and is indicative of the degree of crystal imperfection. It is the combined effect of intrinsic

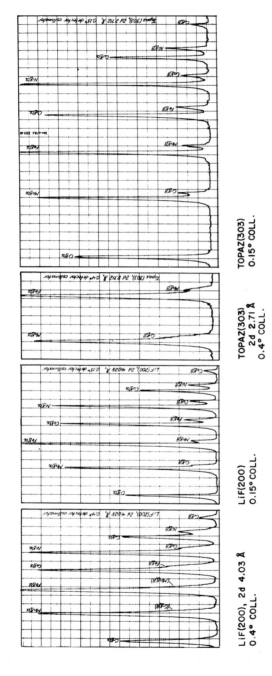

LiF(200), 2d 4.03 Å
0.4° COLL.

LiF(200)
0.15° COLL.

TOPAZ(303)
2 d 2.71 Å
0.4° COLL.

TOPAZ(303)
0.15° COLL.

FIGURE 4.8. Effect of 2d spacing of the analyzer crystal, and of the collimator on dispersion and resolution. The specimen was a mixture of oxides of Cr, Mn, Fe, Co, Ni, and Cu.

(natural) line width, mosaic misalignment, imperfections, stress, surface texture, etc.

In practice, diffracted intensity is determined by the crystal and plane (hkl), but varies also with wavelength and surface finish. In general, polished surfaces are preferable for long wavelengths, matte surfaces for short wavelengths. The second order has much lower intensity than the first and is absent in some crystals.

The crystal scatters incident secondary x-rays, and the elements of which it is composed are excited by the incident x-rays to emit their own characteristic spectra. This scattered and emitted radiation emanates in all directions, contributes to the continuous background, reduces the peak-to-background ratios of the measured analyte lines, and may cause interference if the pulse-height selector is used. When the spectrometer is used in air path, K lines of all crystal elements up to chlorine are absorbed to insignificance, and K lines of potassium and calcium are reduced to very low intensity. However, when helium or vacuum path is used, this radiation reaches the detector, higher background results, and extraneous pulse-height distributions occur which may interfere directly with those of the analyte(s) if pulse-height analysis is used. Organic crystals, such as EDDT and PET, are free from such emission.

Crystals should be stable in air and mechanically strong. Most of them are, but there are exceptions. For example, gypsum may effloresce in low humidity and in vacuum, and PET is very soft and fragile—about like paraffin. Some crystals, such as PET, tend to deteriorate with time and on prolonged exposure to x-rays. All crystals should be handled carefully and only by the edges. They should be stored in individual dust-free boxes when not in place on the instrument. Great care must be taken not to mar the diffracting surface and to keep it clean.

An increase in temperature increases the $2d$ spacings in the crystal, causing 2θ to decrease with consequent shift in the diffracted peak and error in measured wavelength and/or intensity. Figure 4.9 shows $-\Delta 2\theta/°C$ as a function of 2θ for several common crystals. Some modern instruments provide constant-temperature chambers for the crystal maintained at $\pm0.5°C$. These accessories are most useful for precise, high-resolution work, such as measurement of line profiles, absorption-edge fine structure, and chemical effects. However, air-conditioning to maintain laboratory temperature within ±0.5–$1°C$ of a mean is to be recommended as standard practice.

Crystals are available in several forms. Many instruments use un-mounted flat crystals in the form of $7.5\times2.5\times0.3$-cm ($3\times1\times\frac{1}{8}$-inch) bars.

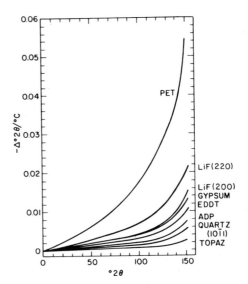

FIGURE 4.9. Curves of temperature coefficient of linear expansion for some common analyzer crystals. [R. Jenkins and J. L. de Vries, *Practical X-Ray Spectrometry*, 2nd ed., Springer-Verlag New York (1969); courtesy of the authors, publishers, and N. V. Philips Gloeilampenfabrieken.]

Other instruments, particularly those with crystal changers, use mounted crystals of about the same length and width, but ~1 mm thick cemented in accurately machined metal holders ("boats"), which are different for different instruments.

Crystals usually have one finished side that is clearly intended to be the diffracting surface. The other side is usually rougher, labeled, or cemented to a mount. When a crystal is first acquired, spectral lines should be recorded in the low-, medium-, and high-2θ regions with the crystal in both its end-for-end orientations to compare performance with respect to intensity and peak breadth. If either orientation appears to be superior, the good end should be indicated on the back. However, crystals are now of excellent quality, and it is unlikely that any significant difference will be found. The crystal surface usually lies within 5 arcmin of the nominal crystallographic plane.

4.2.3.3. Other Wavelength-Dispersion Devices

Ruled gratings and specular reflectors are used for wavelength dispersion, but are not considered here. Two other analyzers that do warrant consideration are polycrystalline graphite and multilayer lead-soap films.

Natural-mineral single-crystal graphite is not available in dimensions and quality suitable for use as x-ray analyzers. However, polycrystalline graphite analyzers have been prepared by hot-pressing graphite flakes.

These analyzers have reflectivity >5 times that of PET for P $K\alpha$ and are especially good for the K lines of phosphorus, sulfur, and chlorine. These graphite "crystals" are very rugged and can be prepared in any form, including cylindrically or logarithmically curved. In fact, the graphite layers are formed directly on their flat or curved support. As a result of their large mosaic spread, these graphite crystals also have large rocking angles.

Suitable true crystals are not available having $2d$ spacings long enough for use in the ultralong-wavelength region beyond ∼24 Å (O $K\alpha$). In this region, pseudocrystals consisting of multilayers of lead salts of long, straight-chain aliphatic organic acids are used having the general formula $[CH_3(CH_2)_nCOO]_2Pb$. These "crystals" are prepared as follows.

A solution of stearic acid [octadecanoic acid, $C_{17}H_{35}COOH$, or $CH_3(CH_2)_{16}COOH$] in benzene is added dropwise to the surface of a dilute water solution of a lead salt. The benzene spreads over the water surface and evaporates, leaving a monolayer of lead stearate molecules (Figure 4.10) with their hydrophilic (lead) ends toward the water and their hydrophobic (hydrocarbon) ends upward. By means of a floating straight-edged barrier, a two-dimensional pressure is applied to the lead stearate film

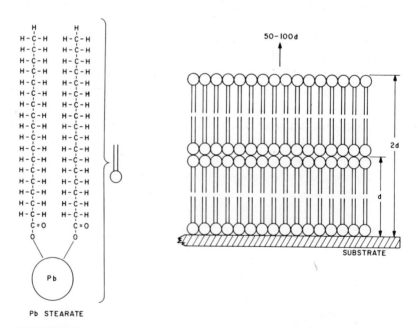

Pb STEARATE

FIGURE 4.10. Lead stearate molecule and the first four layers (first two repetition intervals d,. first $2d$ interval) of a lead stearate multilayer-film pseudocrystal.

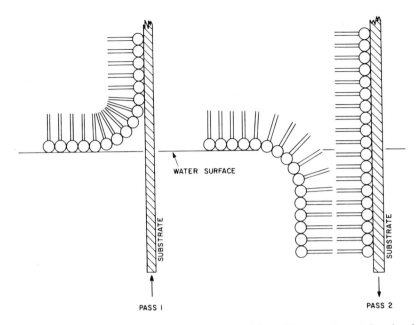

FIGURE 4.11. Preparation of lead stearate multilayer-film pseudocrystals, showing the laying down of the first two layers.

parallel to the liquid surface, compressing the film and close-packing the molecules. The method of preparation of the multilayers is shown in Figure 4.11. A clean hydrophilic plate (usually glass or mica) is immersed in the solution at near-normal incidence. The lead stearate does not adhere to the clean plate on immersion, but on withdrawal (pass 1), a monolayer of lead stearate adheres with the lead ends toward the plate, as shown. On reimmersion (pass 2), a second layer coheres to the first, but with opposite orientation. By successive immersion and withdrawal of the plate, a series of alternately oriented monolayers is built up, the first four of which are shown in Figure 4.10. After each pass, the floating barrier is readjusted to maintain the close packing of the monomolecular film. If the substrate is first made hydrophobic, by application of a thin coat of stearic acid, for example, the film does adhere on the first downward pass, with the hydrocarbon ends toward the substrate and the lead atoms outward. On withdrawal, a second layer coheres with lead to lead, etc. The resulting multilayers are identical. The stearate chain is $\sim$25 Å long, so the d spacing (repeat interval, Pb to Pb) is $\sim$50 Å, giving $2d \approx 100$ Å. By use of longer-chain acids, multilayer films having $2d$ up to 156 Å can be prepared, permitting measurement of Be $K\alpha$ (113 Å).

4.2.3.4. Curved-Crystal Spectrometers

Crystal-dispersion arrangements may be classified in two categories: flat-crystal or *nonfocusing*, or curved-crystal or *focusing*. Actually, curved crystals do not focus x-rays in the true optical sense, but diffract them in such a way that they converge to a line or point. The flat-crystal arrangement is represented by Figure 3.2 and is still the most widely applied. Curved crystals may be of the transmission or "reflection" type and may be curved to conform to small sections of cylinders, spheres, toruses, or logarithmic curves. In general, nonfocusing geometry is best suited to large-area specimens and 2θ-scanning spectrometers; focusing geometry is best suited to small-area specimens, small selected areas on large specimens, excitation by small-diameter electron or ion beams, and fixed-2θ spectrometers in multichannel instruments (Section 3.3.3.3).

Consider a spectrometer of the type shown in Figure 3.2, having a flat crystal and Soller collimators and set at the 2θ angle for a specified wavelength λ emitted by the specimen. Each point on the specimen emits this wavelength in all directions, but only a V-shaped wedge of rays arrive at the crystal and are diffracted toward the detector. The point of the V lies at the point on the specimen, its end extends from top to bottom of the crystal, and its plane is parallel to the Soller foils and crystal axis. All other rays of wavelength λ emitted from that point are wasted. These same considerations apply to all points on the specimen, so it is evident that at any specified wavelength, *most* of the emitted radiation is lost. In a flat-crystal spectrometer, this loss can be afforded because a relatively large specimen area (2–5 cm^2) is used.

However, suppose that the effective specimen area is small, for example, a single particle or fiber, or a small selected area on a large specimen. Or suppose that the specimen is excited by an electron or ion beam of small diameter, for example, in an electron-probe microanalyzer. For such applications, economy of emitted x-radiation assumes prime importance. This economy is realized in wavelength-dispersive mode by use of curved crystals to collect a relatively large solid angle of emitted radiation and "focus" it on the detector.

In focusing geometry, it is necessary that the source of emitted x-rays, crystal, and detector all lie on a common *focusing circle* having diameter R in such a way that the distances L from the center of the crystal to the source and detector are equal. Then,

$$L = n\lambda R/2d = R \sin \theta \qquad (4.16)$$

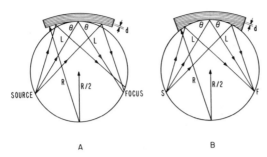

A B

FIGURE 4.12. Johann (A) and Johansson (B) curved-crystal arrangements. The Johann crystal is curved to radius R and gives imperfect focusing. The Johansson crystal is curved to radius R and ground to radius $R/2$ to bring the inside surface into congruence with the focal circle to correct the focusing defect. L is the distance from the center of the crystal to the source and to the focus. A slit perpendicular to the page is placed at the focus, and the detector is placed behind this slit. R is typically 5–20 cm; the crystals are typically 2.5–7.5 cm long by 1–2.5 cm wide. All rays make angle θ with the crystal planes at the point of contact.

where n, λ, and d are diffraction order, wavelength, and crystal interplanar spacing, respectively. The focusing circle is analogous to the Rowland circle of light optics. X-ray focusing geometry is shown in Figure 4.12.

In Figure 4.12, the x-ray source may be: (1) a small particle or fiber (perpendicular to the page) on the focusing circle excited by an x-ray tube off the circle; (2) a small selected area on a large-area specimen on the circle irradiated through a primary-beam aperture (I in Figure 3.2) by an x-ray tube off the circle; (3) a secondary-beam pinhole or slit aperture (J in Figure 3.2) on the circle over a small selected area on a large-area specimen excited by an x-ray tube, both off the circle; or (4) a near-point source of x-rays on a large-area specimen on the circle excited by a small-diameter electron or ion beam.

The curved crystal "focuses" a point or line x-ray source of any of the preceding types to a line focus at the detector window. Clearly, Soller collimators would defeat the purpose of collecting divergent rays and converging them to a line focus, and so they are omitted. Simple pinhole or slit apertures are used at the source where required, and a slit at the detector window. This replacement of relatively large, expensive Soller collimators with small, simple, inexpensive slits and pinholes makes focusing geometry very attractive for multichannel spectrometers (Section 3.3.3.3).

Only two types of curved crystal merit consideration in this introductory text, the Johann and Johansson reflecting arrangements shown in Figure 4.12.

In the Johann geometry (Figure 4.12A), the crystal is simply curved cylindrically to radius R equal to the diameter of the focusing circle. In *semifocusing* spectrometers, L and R [Equation (4.16)] are constant at all values of θ. This arrangement is optimal at that wavelength that satisfies the equation for these values of L and R, but deteriorates as wavelength increases or decreases from this wavelength. For optimal performance, either R is kept constant and L varied with θ, or L is kept constant and R varied with θ. Since L increases as θ increases, it may become inconveniently large at large angles for a crystal of given radius. If L is to have convenient length at all angles, two or three cyrstals having different radii may be required. A Johann crystal gives optimal focusing only at its vertical axis, where it is tangent to the focusing circle, so the focusing is not sharp, as shown in exaggerated scale at the focus in Figure 4.12A. The defect is most serious at very small 2θ angles, but decreases as 2θ increases and becomes insignificant at $2\theta > 90°$. The Johann arrangement is useful for crystals that are not readily ground for the Johansson arrangement (see below) and for spectrometers in which the crystal radius is varied continuously with θ.

"Perfect" x-ray focusing free of the defect noted above requires that two conditions be met which, on first consideration, appear to be mutually exclusive. (1) The *Rowland condition* requires that the x-ray point or line source, crystal, and focused image all lie on a circle of radius $R/2$; to meet this requirement, the crystal must have radius R. (2) The *Bragg condition* requires that the crystal must lie tangent to the above Rowland circle at all points; to meet this requirement, the crystal must have radius $R/2$. These conditions are indeed mutually exclusive for optical gratings, which are ruled on the surface. However, with three-dimensional gratings, which, in effect, is what crystals are, the conditions are readily met, at least for physically suitable crystals.

The Johansson arrangement is substantially the same as the Johann, except that the focusing defect is corrected by bringing the inner surface of the crystal into contact with the focal circle. This is done by cylindrically curving the crystal to radius R, then grinding it to radius $R/2$, as shown in Figure 4.12B. In effect, the crystal *planes* are curved to radius R, and the crystal *surface* is ground to radius $R/2$. The method is not applicable to crystals that cannot be ground, or to spectrometers in which the crystal radius is varied continuously with θ. Equation (4.16) applies to the Johansson arrangement, as does the requirement for two or three crystals of different radii to keep L of convenient length at all angles.

4.2.4. Spectrogoniometer Alignment

In manual single-channel instruments, whenever the crystal or collimator is changed, it is usually necessary to adjust the crystal so that the
spectral lines occur at or near their nominal 2θ angles. An adjustment is
provided for this purpose on the crystal stage. In manual instruments fitted
with crystal changers, and in semiautomatic and automatic instruments,
the crystals are of the mounted type and are aligned when installed on the
changer; thereafter, the alignment requires only occasional checking.

It may not be possible to achieve correct alignment over the full 2θ
range. If alignment is correct at, say, the low-2θ end, it may progressively
diverge as 2θ increases, or if it is correct at midscale, it may diverge toward
both ends. In such cases, when multielement specimens are to be measured,
the 2θ angles at which all divergent lines fall must be established experimentally. Alternatively, the deviation of observed from tabulated 2θ values
may be determined for several lines distributed over the goniometer range
and a curve plotted of divergence *versus* 2θ. Sometimes this can be done
with several orders of a single strong line. Such curves may have lasting
value with crystal changers; otherwise they must be replotted each time a
crystal is changed.

The procedure for precise alignment of the spectrogoniometer mechanism is given in the instruction manual. However, for routine "peaking"
after changing a crystal or collimator, or when installing crystals on a
changer, the following simple procedure may be used. It is assumed that
the detector is receiving the line at which alignment is to be made at a
relatively high intensity—several thousand counts per second. The crystal
rocking screw is set near the center of its range. The goniometer is set at
the correct 2θ angle for the line. While intensity is observed on the panel
ratemeter or recorder, the crystal is slowly "rocked" until maximum response is obtained. The line is then accurately located in 2θ by one of the
methods given below, and the magnitude and direction of the discrepancy
are noted. The crystal rocking screw is then readjusted in the direction to
correct the error and the line position checked again. This procedure is
repeated until the line falls satisfactorily near its correct 2θ value.

Some goniometers have a -2θ region 10–20° beyond zero. A line in
this region can be peaked on both sides of zero to give very accurate alignment.

After the goniometer has been aligned, any of the following techniques
may be used to determine the exact 2θ angle at which each of the analyte
lines is to be measured, or to plot the divergence *versus* 2θ curves mentioned

above. In all the following methods, it is assumed that the detector is receiving the line at a relatively high intensity—several thousand counts per second.

1. While intensity is observed on the panel ratemeter or recorder, the goniometer (not the crystal alone) is manually slowly "rocked" repeatedly through the peak until maximum response is observed.

2. The excitation is adjusted to give full-scale, or perhaps slightly off-scale, deflection at the peak 2θ angle established in method 1 above. While the panel ratemeter or recorder is observed, 2θ is increased until the intensity falls to some arbitrary value—say, 8 on a 0–10 meter or recorder scale—on the high-2θ side of the peak, then decreased to 8 on the low-2θ side. The two angles are noted, and the angle midway between them may be taken as the peak. Recorded peaks are usually asymmetrical, so the intensity should not be decreased too much on the two sides of the peak; 7–9 on a 0–10 scale is usually satisfactory.

3. The goniometer is set at, say, 0.5° below the 2θ angle found in method 1, then advanced manually in 0.1° or 0.2° increments through the peak. The recorder is allowed to trace the intensity for 15–20 s at each position. The point of maximum intensity is taken as the peak.

4. Method 3 is more conveniently and accurately done by scaling, rather than recording, the intensity at each position. A step-scanner accessory does this automatically.

5. The goniometer may be made to slow-scan the top of the peak starting, say, 0.2° below the peak at, say, 0.05° 2θ/min. The position of the peak is read from the broad profile.

Regardless of which procedure is used, several precautions should be taken:

1. In general, a specimen of pure element should always be used to preclude the possibility of misalignment on a nearby line of some other element. An oxide or other compound having no interfering element may do as well. However, for the K lines of the lightest elements and for certain L lines of intermediate elements, the wavelength may be slightly different from pure element and from specimens having that element in various chemical states.

2. Linear, rather than logarithmic, ratemeter response should be used, especially for resolving an unresolved doublet (such as $K\alpha_{1,2}$) or one of the lines in a partially resolved doublet. Logarithmic ratemeters reduce the difference in response between the two individual members of the pair, and thereby in the peak profile.

3. The line should have intensity high enough to minimize ratemeter or recorder fluctuations.

4. Lag due to ratemeter time constant and recorder response must be allowed for in adjusting the crystal or goniometer. If the adjusting is done too rapidly, the panel meter or recorder response cannot follow.

5. If the goniometer may be badly out of alignment, care must be taken not to use the wrong line. Before locating the $K\alpha$ line, one should verify that it really is the $K\alpha$ line by verifying that the weaker $K\beta$ line occurs at a lower 2θ. Before locating the $L\beta_1$ line, one should verify the presence of the (usually) stronger $L\alpha_1$ line at higher 2θ and the weaker $L\gamma_1$ line at lower 2θ.

6. The final settings of the crystal rocking screw and goniometer should always be approached from the same direction.

7. It may be wise to use a single line, such as $K\beta$, rather than an unresolved multiple line, such as $K\alpha_{1,2}$, because the latter may have an asymmetrical peak profile. This is especially important in higher orders where the α_1–α_2 separation is greater. The $L\alpha_1$ line is satisfactory because $L\alpha_2$ is relatively very weak.

8. When peaking a long-wavelength line using vacuum or, especially, helium path, it may be convenient to make preliminary adjustments in air with a higher-order, shorter-wavelength line that falls at about the same 2θ. In this way, possible difficulties with the vacuum enclosure or helium jacket (Section 4.2.1) are precluded. Once preliminary adjustment is effected, final peaking is done with the analyte line in vacuum or helium path. For example, if an EDDT crystal is to be peaked for Al $K\alpha$ (142.57° 2θ), preliminary adjustment may be done in air with fourth-order Mn $K\alpha$ (145.73° 2θ).

SUGGESTED READING

ADLER, I., *X-Ray Emission Spectrography in Geology*, Elsevier, Amsterdam (1966); Chaps. 3 and 5, pp. 29–42, 61–78.

BERTIN, E. P., *Principles and Practice of X-Ray Spectrometric Analysis*, 2nd ed., Plenum Press, New York (1975); Chaps. 4 and 5, pp. 113–217.

BIRKS, L. S., *X-Ray Spectrochemical Analysis*, 2nd ed., Wiley–Interscience, New York (1969); Chaps. 3 and 4, pp. 15–42.

JENKINS, R., *Introduction to X-Ray Spectrometry*, Heyden, London (1974); Chap. 4, pp. 52–98.

JENKINS, R., and J. L. DE VRIES, *Practical X-Ray Spectrometry*, 2nd ed., Springer-Verlag New York, New York (1969); Chap. 2, pp. 26–46.

Chapter 5

Detection and Readout; X-Ray Intensity Measurement

5.1. X-RAY DETECTION

5.1.1. Principles

The fourth phase of an x-ray spectrometric analysis (Section 3.2) is detection of the characteristic x-rays emitted by the specimen.

Three types of detector are in common use in x-ray spectrochemical analysis: gas-filled proportional counters, scintillation counters, and solid-state semiconductor detectors. In each of these detectors, each absorbed x-ray photon is converted into a pulse of electric current having amplitude proportional to the photon energy. In these detectors, the x-rays to be detected are absorbed by, respectively, a gas atom, an iodine atom in a disk of thallium-activated sodium iodide—NaI(Tl), and a silicon atom in a layer of lithium-doped silicon—Si(Li).

At the photon energies involved in x-ray spectrochemical analysis ($\sim$1–50 keV), the detected x-rays usually interact with the detector by undergoing photoelectric absorption. Each absorbed x-ray photon expels a photoelectron from an atom in the active detector material. This photoelectron has kinetic energy (velocity) equal to the x-ray photon energy minus the binding energy of the electron in the atom. For our present purpose, we assume that this initial photoionization occurs in the outermost shell of the detector atom where the binding energy is very small, $\sim$30 eV at most, compared with the photon energies of x-ray spectral lines. Thus, for practical purposes, the photoelectron has substantially the same energy as the x-ray photon that expelled it—that is, in effect, the x-ray photon simply transfers its energy to the photoelectron. Inner-shell photoionization of detector atoms is considered below.

These photoelectrons may undergo any of several processes considered in detail in later sections. For the present, we consider only the simplest and probably the most common process in each type of detector: In the gas-filled detector, the photoelectrons expend their energy in ionizing detector gas atoms, thereby producing ion–electron pairs and losing ~30 eV for each pair. In the NaI(Tl) scintillation detector, the photoelectrons expend their energy in producing ~4100-Å (blue) light photons, losing ~3 eV for each. The photoelectrons excite valence-band electrons to higher levels; when these electrons return to their unexcited state, the energy is emitted as light photons. In Section 5.1.4, it is explained that most of these light photons enter a multiplier phototube where ~10% of them expel photo-electrons, most of which enter an electron multiplier. In the Si(Li) semi-conductor detector, the photoelectrons expend their energy in elevating silicon valence-band electrons to the conduction band, producing electron–hole pairs and losing 3.8 eV for each pair. Figure 5.1 and lines 1–4 of Table

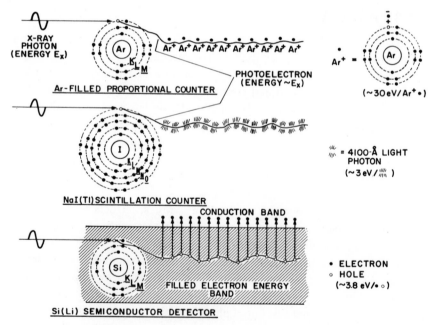

FIGURE 5.1. Simplified representation of the initial photoionization by the incident x-ray photon for Ar-filled proportional, NaI(Tl) scintillation, and Si(Li) semiconductor detectors. For outer-shell ionization, as shown, the photoelectron has substantially the same energy as the incident x-ray photon. The photoelectron expends its energy in producing Ar^+, e^- ion–electron pairs, 4100-Å light photons, and electron–hole pairs in the respective detectors.

TABLE 5.1. Proportionality in X-Ray Detectors

Line	Detector	Ar-filled proportional counter	NaI(Tl) scintillation counter	Si(Li) semiconductor detector
1	Initial photoelectric absorption[a,b]:	$Ar^0 \xrightarrow{E_x} Ar^+, e^-$	$I^0 \xrightarrow{E_x} I^+, e^-$	$Si^0 \xrightarrow{E_x} Si^+, e^-$
2	Energy of initial x-ray photoelectron:	$E_x - 30$ eV	$E_x - 30$ eV	$E_x - 8$ eV
3	Expenditure of energy of initial x-ray photoelectron:	Production of Ar^+, e^- ion–electron pairs	Production of 4100-Å light photons	Production of electron–hole pairs
4	Energy of process above (each event):	~30 eV	~3 eV[c]	3.8 eV
	Line λ (Å) E_x (eV)	Number of Ar^+, e^- ion-electron pairs	Number of 4100-Å light photons	Number of electron-hole pairs
5	V Kα 2.5 $\dfrac{12{,}396}{2.5} = 5000$	$\dfrac{5000}{30} = 167$	$\dfrac{5000}{3} = (1670)^d$ / 133	$\dfrac{5000}{3.8} = 1314$
6	Ge Kα 1.25 $\dfrac{12{,}396}{1.25} = 10{,}000$	$\dfrac{10{,}000}{30} = 333$	$\dfrac{10{,}000}{3} = (3330)^d$ / 266	$\dfrac{10{,}000}{3.8} = 2632$
7	Rh Kα 0.62 $\dfrac{12{,}396}{0.62} = 20{,}000$	$\dfrac{20{,}000}{30} = 667$	$\dfrac{20{,}000}{3} = (6670)^d$ / 532	$\dfrac{20{,}000}{3.8} = 5263$
8	Internal multiplication process:	Avalanching	Secondary-electron multiplication	None
9	Number of useful electrons collected[e]:	↑ ×10⁶	↑ ×10⁶	↑ ×1

[a] See Figure 5.1.
[b] K- and L-shell photoionization of argon, iodine, and silicon are disregarded here; see Section 5.1.3.3.
[c] See footnote d.
[d] Of these 1670, 3330, and 6670 light photons actually produced, ~0.9 reach the photocathode in the multiplier phototube of the scintillation counter (Figure 5.6); of these photoelectrons, ~0.1 expel photoelectrons; of these photoelectrons, ~0.9 reach the electron multiplier. Thus, $0.9 \times 0.1 \times 0.9 \approx 0.08$ (8 of each 100) of the 4100-Å photons are effective in producing a useful electron, and the effective light-photon energy (line 4) is $(100/8) \times 3$ eV = 40 eV. Then, for V Kα, Ge Kα, and Rh Kα, the numbers of effective 4100-Å photons are 133, 266, and 532, respectively.
[e] For each Ar^+, e^- pair, effective 4100-Å light photon, and electron–hole pair, respectively.

5.1 represent the initial photoionization and the process by which the photoelectron expends its energy for each of the three detectors.

In effect, in all three detectors, each detected x-ray photon expels and imparts its energy to a photoelectron, which, in turn, expends its energy in producing a certain number of *useful electrons*. In the gas-proportional, scintillation, and semiconductor detectors, these useful electrons are, respectively, those in the ion pairs, those entering the multiplier, and those in the electron–hole pairs.

Clearly, the higher the x-ray photon energy, the more energetic are the photoelectrons and the more numerous the useful secondary electrons. In fact, an important parameter of each detector is the *effective ionization potential*, or the energy required to produce each useful electron. Then the number n_0 of useful electrons is the x-ray photon energy E_x divided by the effective ionization potential $V_{i,\text{eff}}$,

$$n_0 = E_x/V_{i,\text{eff}} \tag{5.1}$$

For gas, scintillation, and semiconductor detectors, $V_{i,\text{eff}}$ is $\sim$20–30 eV (depending on the gas), $\sim$50 eV, and 3.8 eV, respectively.

Equation (5.1) is the basis for the proportionality of detector output pulse height and x-ray photon energy. Lines 5–7 in Table 5.1 show this relationship for x-ray lines of long (V $K\alpha$), intermediate (Ge $K\alpha$), and short (Rh $K\alpha$) wavelength. Note that these lines have wavelength ratio 4 : 2 : 1, photon energy ratio 1 : 2 : 4.

In the Si(Li) detector, the electrons produced by the initial photo-electron actually constitute the detector output pulse. However, in the gas-proportional and scintillation counters, these electrons are subjected to internal multiplication, so the output pulses contain more electrons than were produced initially. Lines 8 and 9 of Table 5.1 represent this feature.

In wavelength-dispersive spectrometry, detection occurs after dispersion of the secondary x-rays by the analyzer crystal. However, in energy-dispersive spectrometry, the undispersed x-rays emitted by the specimen are detected, and the detector output is subjected to electronic pulse-height dispersion; thus, the detector is itself an essential part of the dispersion process.

5.1.2. Pulse-Height Distributions

It has been explained that the output of an x-ray detector consists of a pulse of electric current for each absorbed x-ray photon and that the pulse height is proportional to the photon energy. The basis for this propor-

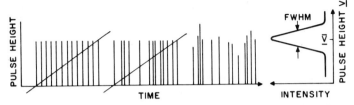

FIGURE 5.2. Amplitude and time distribution of detector output pulses and origin of the pulse-height distribution for monochromatic x-rays. The pulses have neither uniform time distribution nor exactly the same height. Rather, they have random time distribution and a Gaussian pulse-height distribution about a mean, as shown at the extreme right.

tionality is represented in Table 5.1, lines 5–7. In this section, we consider the detector output in more detail.

Figure 5.2 represents three concepts of the output of a detector receiving monochromatic x-rays. In the first, the output pulses have exactly the same height and uniform time distribution; in the second, they have the same height, but random time distribution. Neither of these concepts is correct. Rather, the pulses have both random time distribution and a statistical pulse-height distribution, as shown in the third pulse diagram.

The random time distribution arises from the random time distribution of x-ray photon emission. The pulse-height distribution arises from statistical variations in the processes from which the pulses arise. Even with strictly monochromatic x-rays incident upon a detector, the numbers of ion–electron pairs, light photons, or electron–hole pairs produced for each absorbed photon (Table 5.1, lines 5–7) are not exactly the same. Rather, they have a Gaussian distribution about a mean value $\bar{n}$ given by Equation (5.1) and a statistical deviation $\pm \bar{n}^{1/2}$. It is these mean numbers that are proportional to x-ray photon energy and provide the basis for proportionality of detector output pulse height and photon energy.

The spread in pulse height is represented by a plot of number of pulses of each height (intensity) *versus* pulse height, as shown at the right of the third pulse diagram in Figure 5.2. The plot has the form of a *peak* and is known as a *pulse-height distribution*. These curves are usually plotted with the intensity axis vertical; here, the unconventional orientation permits correlation with the pulse diagram.

The spread in the pulse-height distribution is represented by the *full width at half maximum* (FWHM) peak height or as the ratio of the FWHM to the mean pulse height, usually expressed in percent. High resolution of pulse-height distributions for different x-ray lines requires that the spread be as narrow as possible. The width of the pulse-height distribution for a detector is determined by the statistically limiting process, that is, that process in the production of each pulse that has the smallest population. The smaller the population, the greater will be the width (FWHM) of the pulse-height distribution because the statistical deviation in a mean population of $\bar{n}$ members is $\pm \bar{n}^{1/2}$.

In Table 5.1, lines 5–7, it is evident that the statistically limiting processes in the three types of detector are as follows. For the gas-proportional counter, it is the number of ion–electron pairs. For the scintillation counter, it is the number of photoelectrons that reach the electron multiplier (Table 5.1, footnote d). In the Si(Li) detector, it is the number of electron–hole pairs. Note that for a specified x-ray line, say, Ge $K\alpha$, the statistically limiting process is greatest for the Si(Li) detector (2632), smallest for the scintillation counter (266). Consequently, the width of the pulse-height distribution is narrowest for the Si(Li) detector, widest for the scintillation counter. The fact that this limiting population is multiplied in the scintillation counter, but not in the Si(Li) detector, does not change the width.

In a wavelength-dispersive spectrometer, the detector output pulses are amplified without change in relative pulse amplitude, then pass through a pulse-height selector to the counting components, the ratemeter and scaler, as shown in Figure 3.2. The pulse-height selector is described in Section 5.2.3; however, an elementary knowledge of its function is helpful in our consideration of detectors. Let it suffice here to state that the pulse-height selector passes all pulses having amplitude greater than a selected minimum, the *baseline*, and less than a selected maximum, the *window*. Average pulse height is proportional to detector potential and amplifier gain, as well as photon energy.

5.1.3. Gas-Filled Detectors

5.1.3.1. Structure

The basic structure of the gas-filled x-ray detector is shown in Figure 5.3. The active components are the cylindrical metal cathode, axial wire anode, and gas filling. An electrical insulator supports and permits external connection to the wire, and a window admits the x-rays to be detected. In operation, the cathode and anode are operated at negative and positive potential, respectively, by a dc power supply, as shown.

Gas-filled detectors are classified: (1) as ionization chambers, proportional counters, or Geiger or Geiger–Müller counters (Section 5.1.3.2);

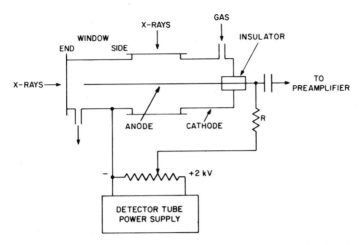

FIGURE 5.3. Structure of the gas-filled x-ray detector (proportional counter, Geiger counter, ionization chamber). The detector is shown to have both end and side windows for illustration only.

(2) as sealed or flow counters; (3) by window thickness and/or material; (4) as end- (axial-) or side- (radial-) window counters; and (5) by gas filling.

Gas-filled counters for use in the short-wavelength region have gastight windows of beryllium or mica, are permanently sealed after gas filling, and are known as *sealed counters.* Counters for use in the long-wavelength region must have "thin" plastic windows, such as 6- or 3-μm (0.00025- or 0.00012-inch) Mylar. These windows are not gastight, so when the detectors are in use, the gas must be passed through them continuously; inlet and outlet tubes are provided for this purpose, as shown in Figure 5.3. Such detectors are known as *flow counters.* These Mylar windows can withstand atmospheric pressure, so the gas is passed through the detectors at atmospheric pressure, and the detectors can be used in air, helium, or vacuum path. Counters for use in the ultralong-wavelength region must have "ultra-thin" windows, such as 1-μm polypropylene or 500–5000-Å Formvar. These are also flow counters. However, ultrathin windows cannot withstand atmospheric pressure, so these counters are used only in vacuum spectrometers with the detector gas at reduced pressure.

The detector may have a single end (axial) or side (radial) window or two diametrically opposite side windows, but never both end and side windows, as shown in Figure 5.3 for illustration. When pulse-height selection is to be used, the electric field in the detector must be as nearly uniform as possible: Side-window tubes are used to avoid the nonuniform field at the wire end, and the inside surfaces of mica, Mylar, and plastic windows are coated with a conductive layer of aluminum ~ 100 Å thick. The front and back side-window arrangement has two advantages. Unabsorbed x-rays that might excite photoelectron and secondary x-ray emission from the back cathode wall can emerge from the tube. Also, a thin-window counter for long-wavelength x-rays can be backed by a second detector for short-wavelength x-rays.

The gas filling is determined by the wavelength region. The shorter the wavelength, the higher must be the atomic number (or mean atomic number) of the gas if the x-rays are to be absorbed in the detector gas volume. For short wavelengths, xenon and krypton are used, for intermediate wavelengths, argon. For flow counters in the long-wavelength region, the most commonly used gas is "P10" (90 vol% argon, 10 vol% methane), but pure propane may be used for special applications. For the ultralong wavelengths, "P90" (90 vol% methane, 10 vol% argon), pure methane or ethane, and mixtures of neon or helium with methane or carbon dioxide are used.

5.1.3.2. Types of Gas-Filled Detectors; Gas Amplification

Consider an x-ray photon entering the detector shown in Figure 5.3 and passing through the active gas volume. Eventually, it undergoes *photoelectric absorption* by a gas atom and imparts its energy to a photoelectron, which produces ion–electron pairs along its path until its energy is spent (Figure 5.1, top). If no electric potential is applied, the ion pairs quickly *recombine*. However, if potential is applied, the negative electrons and positive ions *migrate* to the positive anode wire and negative cathode, respectively. If the potential is high enough, the electrons may *accelerate* to such velocity that they collide with gas atoms along their path, causing *ionization by collision* and producing more ion pairs. The additional electrons so formed may produce still more ion pairs, the electrons from which produce still more, and so on. Thus, a single primary electron may initiate a shower of hundreds to billions of electrons, a phenomenon known as *Townsend avalanching*. This avalanching is the basis of internal electron multiplication in gas-proportional counters.

The average number $\bar{n}_0$ of primary ion pairs produced by the photoelectron expelled by the detected x-ray photon is given by Equation (5.1). The *gas-amplification factor* or *gas gain* G is the number of electrons collected on the anode wire for each primary electron. If n is the total number of electrons arriving at the wire,

$$G = n/\bar{n}_0 = n(V_{i,\text{eff}}/E_x) \tag{5.2}$$

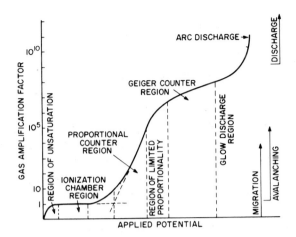

FIGURE 5.4. Gas-amplification factor as a function of applied potential for the gas-filled x-ray detector.

Figure 5.4 shows the gas-amplification factor as a function of detector potential. The curve is divided by its inflections into a number of regions. Let us consider the phenomena in the detector shown in Figure 5.3 as anode potential is increased from zero (potentiometer contact at extreme left) to progressively higher positive values.

At zero potential, the ion pairs quickly recombine, and no charge is collected on the anode. The gas amplification is zero. In the region of unsaturation, the applied potential is so low that some electrons still recombine with positive ions before they can reach the anode. The collected current is *unsaturated*; that is, fewer electrons are collected than were formed by the x-ray photon, and the gas amplification is <1. However, as the potential is increased, electron acceleration increases, and progressively more electrons escape recombination and reach the anode.

In the ionization-chamber region, the potential is just sufficient to overcome recombination completely, all electrons are collected (*saturation*), and the gas amplification is 1 and remains constant over a substantial interval of applied potential. However, no secondary ionization occurs in the gas or at the electrodes. Ionization-chamber detectors operate in this region; these detectors have little application in x-ray spectrochemical analysis.

The critical detector potential for onset of gas amplification is that at which electrons from the primary ion pairs are accelerated just to the ionization potential of the detector gas by the time they reach the anode wire. At any higher potential, the electrons are able to initiate some additional gas ionization along their path. In this intermediate region, detector potential is high enough to initiate mild avalanching, and the gas amplification becomes >1. This region is of no significance for x-ray detection.

In the proportional-counter region, the potential is high enough for marked avalanching to occur, and the gas amplification is 10^2–10^6. However, this avalanching is limited in extent. It is localized in a region within a few wire diameters of the anode, and, in general, each electron produced by the x-ray photon initiates only one avalanche, and the avalanches are largely free of any interaction. The number of avalanches is substantially the same as the number of primary ionizations, and, since all electrons are collected, the total collected charge is proportional to the x-ray photon energy.

As potential is increased still further, avalanching becomes more generalized. The number of avalanches per unit volume increases, and they begin to interact. The same electron initiates multiple avalanches. The region in which avalanching occurs spreads outward from the wire. Proportionality deteriorates as potential is increased. In this region of limited proportionality, gas amplification becomes a function of incident photon energy, and pulses of low initial amplitude are amplified more than those of high initial amplitude. The amplification is also strongly dependent on detector potential. This region, like the intermediate region, is also of little value for x-ray detection.

In the Geiger-counter region, the potential is so high that the avalanching becomes generalized. The gas ions and atoms are excited in the gas volume and at the cathode wall to emit ultraviolet radiation. Secondary electrons and low-energy x-rays are emitted when the electrons strike the anode, and electrons are expelled by ion bombardment of the cathode. These emitted electrons initiate new avalanches. The ultraviolet radiation permeates the entire gas volume, causing photoionization, the photoelectrons from which also initiate avalanches. Thus, in the Geiger region, the initial ionizing event merely triggers the general discharge of the entire tube. All vestiges of proportionality disappear, and the output pulses are of substantially the same high average amplitude—$\sim$1000 times the amplitude of pulses in the proportional region. The Geiger, or Geiger–Müller, counter operates in this region. These detectors have little application in x-ray spectrometry because of their lack of proportionality and their long dead time (Section 5.1.6.3). However, sometimes they are used on multichannel spectrometers, where they detect only a single wavelength.

In the glow-discharge region, the discharge is so severe that it does not quench after each incident photon. The photon triggers not one output pulse, but a continuous, sustained discharge. If the tube window is transparent, the gas volume can be seen to glow. Operation in this condition for even a short time is likely to ruin the tube. At still higher potentials, an arc discharge occurs between the cathode and anode, and the wire is likely to "burn out."

5.1.3.3. Phenomena in the Detector Gas Volume

It has been explained that: (1) the output of a gas detector consists of a pulse of electric current for each x-ray photon absorbed; (2) the numbers of primary ion pairs produced by the individual photons of a strictly monochromatic x-ray beam are not the same, but have a Gaussian distribution about a mean value; and (3) in proportional counters, these ion pairs initiate a complex gas-amplification process, resulting in an output pulse having 10^2–10^6 times as many electrons as were produced by the x-ray photon. Consequently, it is not surprising that the amplitudes of the detector output pulses also have a Gaussian distribution.

We are now in a position to consider the operation of the proportional counter in more detail. The diagrams in Figure 5.5 show five processes that an x-ray photon may initiate in the active volume of a proportional counter. Column 1 shows each process in a simplified schematic way. Column 2 shows the *individual* detector output pulse resulting from each process. Column 3 gives the output pulse height for each process in terms of incident x-ray photon energy E_x. Column 4 shows, for each process, the pulse-height distribution curve for a detector that is receiving a continuous monochromatic x-ray beam and in which only that process occurs. The "peaks" on these curves are *not* profiles of individual pulses; such profiles would be plotted on a grid of pulse height in volts *versus* time. Nor are they to be confused with the familiar 2θ scan of intensity *versus* 2θ angle. Rather, they show the numbers of pulses of each pulse height present in the detector output in unit time. It is also necessary to distinguish *pulse height*—the amplitude (in volts) of the individual pulse triggered by an individual x-ray photon—and *intensity*—the number of pulses per second.

Let us consider the five processes in detail. In Figure 5.5 and the following discussion, argon is assumed to be the detector gas, but the principles apply to any detector gas.

1. Transmission (Figure 5.5A). The x-ray photon may pass through the gas wholly unabsorbed, to be absorbed in the detector wall or to emerge through a rear window. Of course, no output pulse is produced. This process is more likely the shorter the x-ray wavelength and the lower the

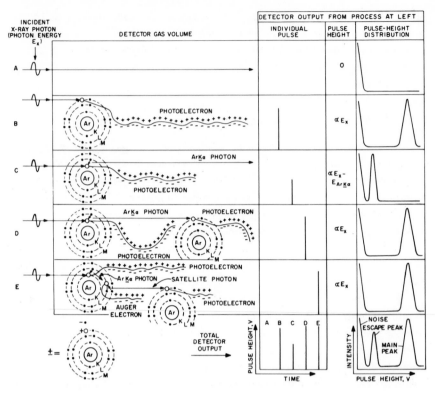

FIGURE 5.5. Phenomena in the active volume of an argon-filled proportional counter, showing, for each phenomenon, the photon–atom interaction, an individual pulse, the pulse height, and the pulse-height distribution; E_x is the energy of the incident x-ray photon.

A. The x-ray photon is not absorbed.

B. The x-ray photon undergoes photoelectric absorption in the outermost shell of a gas atom; the resulting photoelectron expends its energy producing Ar^+, e^- pairs.

C. The x-ray photon undergoes photoelectric absorption in the K shell of a gas atom; the resulting photoelectron expends its energy producing Ar^+, e^- pairs; the Ar $K\alpha$ photon "escapes" the gas volume unabsorbed.

D. The x-ray photon undergoes photoelectric absorption in the K shell; the resulting photoelectron behaves as in process C; the Ar $K\alpha$ photon behaves as the incident x-ray photon in process B.

E. The x-ray photon undergoes photoelectric absorption in the K shell; the resulting photoelectron behaves as in process C; the Ar $K\alpha$ photon undergoes internal absorption (Auger effect), producing an Auger electron and a satellite-line x-ray photon; the Auger electron expends its energy producing Ar^+, e^- pairs; the satellite photon behaves as the incident x-ray photon in process B.

absorption coefficient of the detector gas. In this way, short-wavelength x-rays pass through a two-window detector optimal for long-wavelength x-rays to a backing scintillation or sealed proportional counter.

2. Ion-Pair Production (Figure 5.5B). The x-ray photon may cause outer-shell photoionization of a detector gas atom, imparting its energy to the photoelectron, which in turn ionizes the detector gas atoms along its path to produce Ar^+, e^- pairs. Actually, the ion pairs are not uniformly distributed along the electron path, but become denser as the electron progressively loses energy. The output pulse height is proportional to the photon energy. If only this process occurred in a detector receiving a monochromatic beam, the output would consist of a single pulse-height distribution (peak) of average pulse height proportional to the photon energy.

It has been assumed so far that photoionization of the outer shells of the detector gas atoms is the only process by which x-ray photons are converted to electric charge. In fact, it would be preferable if this were true. However, if the x-ray photon has wavelength shorter than that of the argon *K*-absorption edge, it may undergo photoelectric absorption in the *K* shell of an argon atom (Figure 5.5C, D, E), expelling an argon *K* electron and expending the remainder of its energy in imparting kinetic energy (velocity) to the photoelectron. The *K*-excitation potential of argon (3.2 keV) is much higher than the ionization potential of the outermost electron in argon atoms (15.7 eV). Consequently, the incident x-ray photon loses much more energy in *K*-shell ionization than in outer-shell ionization. It follows that photoelectrons produced by *K*-shell ionization have much less energy ($E_x - 3.2$ keV) than those produced by the process in Figure 5.5B ($E_x - 15.7$ eV). The photoelectron expends its energy in producing Ar^+, e^- pairs as shown. However, the Ar $K\alpha$ (or $K\beta$) photon resulting from the filling of the *K*-shell vacancy can undergo one of three processes as follows.

3. Argon K Excitation Followed by Escape of the Ar Kα Photon (Figure 5.5C). The Ar $K\alpha$ (or $K\beta$) photon may escape the active volume of the detector. This is not an unlikely process because an element is relatively transparent to its own characteristic radiation. The output pulse height is proportional to the difference between the energies of the incident and Ar $K\alpha$ photons. If only this process occurred in a detector receiving a monochromatic beam, the output would consist of a single pulse-height distribution of correspondingly lower average pulse height than that in Figure 5.5B. This distribution is known as an *escape peak* because it is produced when some of the energy carried into the detector by the x-ray photon "escapes" the active volume.

4. Argon K Excitation Followed by Absorption of the Ar Kα Photon (Figure 5.5D). The Ar *Kα* photon may undergo photoelectric absorption in the outer shell of another argon atom, producing a photoelectron which expends its energy producing Ar^+, e^- pairs. To the extent that this occurs, the escape peak is avoided. The sum of the numbers of ion pairs formed by (1) the photoelectron produced by the incident x-ray photon and (2) the photoelectron produced by the Ar *Kα* photon equals the number formed in Figure 5.5B. The pulse height and pulse-height distribution are the same as in Figure 5.5B.

5. Argon K Excitation Followed by Auger Effect (Figure 5.5E). The Ar *Kα* photon may undergo absorption within the atom of its origin (Auger effect, Section 2.4) with resultant emission of an Auger electron and a satellite-line x-ray photon. The Auger electron expends its energy in producing Ar^+, e^- pairs, and the satellite photon undergoes photoelectric absorption, which, in turn, produces Ar^+, e^- pairs. The sum of the numbers of ion pairs formed by (1) the photoelectron produced by the incident x-ray photon, (2) the Auger electron, and (3) the photoelectron produced by the satellite photon equals the number formed in Figure 5.5B. The pulse height and pulse-height distribution are the same as in Figure 5.5B.

5.1.3.4. Detector Output; Escape Peaks

To summarize, the output of a proportional counter consists of a series of electric-current pulses of various heights randomly distributed in time, each pulse originating from absorption of a single incident x-ray photon. In addition, there is a series of very low amplitude detector noise pulses. For monochromatic incident x-rays having wavelength longer than that of the absorption edge of the detector gas, the output contains a single pulse-height distribution (peak) having mean pulse height proportional to the x-ray photon energy. For monochromatic incident x-rays having wavelength shorter than that of the absorption edge of the detector gas, the output may contain two pulse-height distributions (peaks). The *main* (*principal, natural,* or *photo*) peak has mean pulse height proportional to the x-ray photon energy. The *escape* peak has mean pulse height proportional to the difference between the photon energies of the incident x-rays and the spectral line of the detector gas. For incident x-rays containing more than one wavelength, the output contains a pulse-height distribution for each and an escape peak for each that can excite the detector gas. Since these peaks are relatively broad, they may overlap one another, and the main peak of the analyte line may be overlapped by a main or an escape peak of another element.

The intensity of the escape peak relative to the main peak is greater (1) the closer the wavelength of the x-rays lies to the short side of the absorption edge of the detector gas, and (2) the greater the fluorescent yield of the gas.

5.1.4. Scintillation Counters

The basic structure of the scintillation counter is shown in Figure 5.6. The counter consists of a scintillator affixed to the end of a multiplier phototube.

The scintillator consists of a disk of single-crystalline thallium-activated sodium iodide, NaI(Tl), $\sim$2.5 cm in diameter and 2–5 mm thick. The disk is aluminized around the edge and on the side that is to be toward the x-rays to reflect scintillation light radiating away from the detector and direct it toward the phototube. The aluminized disk is sandwiched between a beryllium window $\sim$0.2 mm thick on the x-ray side and a thin Lucite disk on the phototube side to act as a light coupler. The sandwich is then hermetically sealed and perhaps coated with opaque paint on the edge and x-ray side to exclude moisture and light, respectively. The packaged scintillator is affixed to the face of an end-window multiplier phototube just outside the photocathode. A layer of silicone is applied between the Lucite and phototube to improve optical coupling.

The multiplier phototube ("photomultiplier") consists of a photo-emissive cathode, a series of secondary-electron-emissive *dynodes*, and a

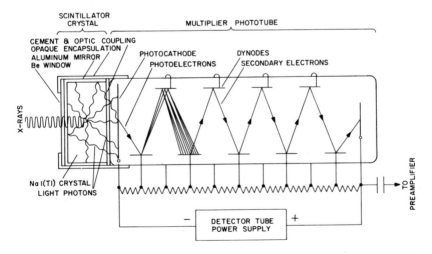

FIGURE 5.6. Structure of the scintillation counter.

collector sealed in an evacuated glass envelope and provided with a base having prongs to permit external connection of the electrodes. The tube is coated with optically opaque black paint to exclude external light and may be enclosed in a mu-metal shield to exclude stray magnetic fields. A regulated dc potential of 600–1000 V is applied to the tube with the photocathode negative. Each successive dynode is maintained at a higher positive potential by means of a potential divider usually consisting of a series of resistors mounted in the tube socket.

In operation, the x-rays are absorbed in the scintillator. Each x-ray photon transfers its energy to a photoelectron, which, in turn, expends its energy producing 4100-Å light photons (Figure 5.1). The more energetic the x-ray photon, the more light photons it can produce; this provides the basis for proportionality of detector output pulse height and x-ray photon energy in scintillation counters (Table 5.1, lines 5–7).

Of the 4100-Å light photons produced, $\sim$0.9 (90%) reach the photocathode of the multiplier phototube; this is the *scintillator efficiency* or *optical coupling efficiency*. Of these photons, $\sim$0.1 expel photoelectrons; this is the *photocathode efficiency*. Of these photoelectrons, $\sim$0.9 reach the first dynode; this is the *first dynode collection efficiency*. Thus, $0.9 \times 0.1 \times 0.9 = 0.08$, or 8 of each 100, of the 4100-Å photons are effective in initiating a secondary-electron cascade and contributing to the output pulse. The *effective light-photon energy* is then $(100/8) \times 3 \text{ eV} = 40 \text{ eV}$. The *statistically limiting process* in the scintillation counter, the process involving the least numerous population, is the collection of photoelectrons on the first dynode.

Each photoelectron arriving at the first dynode expels two to four secondary electrons. These secondary electrons are attracted to the second dynode, where more secondary electrons are emitted, etc. Thus, a progressively increasing shower of secondary electrons cascades through be multiplier. The collector receives the secondary-electron current, which constitutes the output pulse from the scintillation counter. This secondary-electron multiplication is the basis of internal electron multiplication in scintillation counters. The *gain G* of the electron multiplier is the number of electrons arriving at the collector for each photoelectron arriving at the first dynode:

$$G = k\sigma^n \qquad (5.3)$$

where k is the first dynode collection efficiency ($\sim$0.9), σ is the secondary-emission ratio (3–4), and n is the number of dynodes, usually 9–14. Thus, for a 10-stage multiplier having $\sigma \approx 4$ and $k \approx 0.9$, the gain is 0.9×4^{10}, or $\sim 10^6$.

The x-ray photon in the scintillation crystal may initiate one of five processes directly analogous to those that occur in the gas volume of a proportional counter (Section 5.1.3.3, Figure 5.5).

1. The x-ray photon may pass through the crystal unabsorbed, to be absorbed in the glass envelope or internal structure of the phototube. This process is unlikely except at very short wavelengths.

2. It may produce a burst of light photons (a scintillation) by the process described above.

If the x-ray photon has wavelength shorter than that of the iodine K-absorption edge, it may undergo photoelectric absorption in the K shell of an iodine atom, expelling an iodine K electron and expending the remainder of its energy in imparting kinetic energy to the photoelectron. The photoelectron expends its energy in producing light photons. However, the I $K\alpha$ (or $K\beta$) photon resulting from the filling of the K-shell vacancy can undergo one of the three processes listed below. Of course, analogous processes can occur when incident x-rays expel I L-shell electrons, but for simplicity the following discussion is in terms of I K-shell excitation.

3. The I $K\alpha$ (or $K\beta$) photon may escape the scintillation crystal. This is not an unlikely process, because an element is relatively transparent to its own characteristic radiation.

4. It may produce a burst of light photons by process 2 above.

5. It may undergo absorption within the atom of its origin (Auger effect) with resultant emission of an Auger electron and a satellite-line x-ray photon. The Auger electron expends its energy producing light photons. The satellite photon undergoes process 2 above.

The first of the foregoing processes produces no light photons. The second process produces a number of photons proportional to the x-ray photon energy. In the fourth process, the sum of the photons produced by the photoelectron and the I $K\alpha$ photon equals the number produced by the second process. In the fifth process, the sum of the photons produced by the photoelectron, the Auger electron, and the satellite-line photon equals the number produced by the second process. Thus, the second, fourth, and fifth processes produce a number of light photons proportional to the x-ray photon energy. However, in the third process, the number of light photons is proportional to the difference between the energies of the incident and I $K\alpha$ photons. This is, of course, the escape-peak phenomenon.

If there are n dynodes having a secondary-emission ratio of σ, the output pulse contains $0.9 \times 0.1 \times 0.9 \times \sigma^n$ electrons for each photon in the scintillation (see above). Thus, the output of the scintillation counter, like that of the proportional counter, contains a pulse-height distribution (peak) for each incident x-ray wavelength and may contain an escape peak for each incident wavelength shorter than the iodine K-absorption edge. The closer the incident wavelength lies to the K-absorption edge, the greater is the intensity of the escape peak relative to the main peak. Escape peaks are not nearly as troublesome with scintillation counters as with argon- and krypton-filled proportional counters. The iodine LIII absorption edge lies at 2.7 Å at the long-wavelength limit of the usual working range of the scintillation counter. The iodine K-absorption edge lies at 0.37 Å. Thus, troublesome escape peaks occur only for $K\alpha$ lines of elements having atomic number 57 (lanthanum) or higher, and these elements are usually determined by measurement of their L lines with gas-flow proportional counters. In general, all pulse-height distributions are much broader for a scintillation counter than for a proportional counter, and the detector noise pulses have much higher intensity.

5.1.5. Lithium-Drifted Silicon Detectors

The basic structure of the lithium-drifted silicon, Si(Li), detector is shown in Figure 5.7 and its associated liquid-nitrogen vacuum cryostat in Figure 3.3.

The detector consists of a single-crystalline disk of semiconductor-grade silicon having a compensated intrinsic (i-type) region sandwiched between positive and negative (p- and n-type) regions. It is thus a p–i–n-type diode. The compensated region is formed by diffusing ("drifting") lithium into p-type silicon to compensate the impurities and dopants already present. The thin p-type layer on the surface does not contribute to the detection process and is known as the *dead layer*. Gold-film electrodes

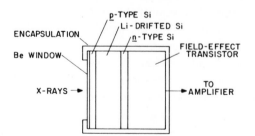

FIGURE 5.7. Structure of the lithium-drifted silicon, Si(Li), solid-state semiconductor detector with its integral field-effect transistor (FET) preamplifier.

~200 Å thick are evaporated on the faces of the Si(Li) wafer, which, together with its integral field-effect transistor (FET) preamplifier (see below), is then provided with a beryllium window ≤ 10 μm thick and encapsulated on the back side and rim. The window and encapsulation protect the detector from: (1) surface contamination, for which, being at liquid-nitrogen temperature (see below), it constitutes a likely condensation target; (2) extraneous light and light from optically fluorescent specimens; and (3) in electron-probe microanalyzers and scanning electron microscopes, scattered electrons. Alternatively, the window and encapsulation may be omitted from the detector itself and provided by the vacuum cryostat. For measurement of very long wavelengths, such as the K lines of carbon, nitrogen, and oxygen, provision may be made to tip open the window when the instrument is under vacuum and close it when air is admitted.

The two principal features of the Si(Li) detector are its area and thickness. Geometric efficiency—intercepted solid angle of x-rays—increases, but resolution decreases with increasing area; absorption efficiency increases with thickness. A typical size is 1.5 cm in diameter and 3 mm thick.

The detector is maintained at liquid-nitrogen temperature ($-196°C$, $77°K$) in its vacuum cryostat at all times, even when not in use. Refrigeration reduces noise and ensures optimal resolution when the detector is in use, and minimizes diffusion of the highly mobile lithium atoms at all times. A 20-liter dewar reservoir requires filling only once each week or two. The detectors can usually be "cycled"—that is, warmed to room temperature for a few days, then returned to liquid-nitrogen temperature—up to 10 times before performance deteriorates to an unacceptable extent. However, in no case may potential be applied to the detector when it is not refrigerated. Vacuum operation is required to prevent condensation of moisture on the detector and to permit operation at long wavelengths.

In operation, the x-rays are absorbed in the lithium-drifted layer. Each absorbed x-ray photon transfers its energy to a photoelectron, which, in turn, expends its energy producing electron–hole pairs (Figure 5.1). The more energetic the x-ray photon, the more electron–hole pairs it can produce; this provides the basis for proportionality of detector output pulse height and x-ray photon energy in Si(Li) detectors (Table 5.1, lines 5–7). The solid-state detector does not have internal electron multiplication analogous to gas amplification in gas counters and secondary-electron multiplication in scintillation counters.

The x-ray photon in the Si(Li) detector can initiate any of the five processes analogous to those described in Section 5.1.3.3 and Figure 5.5.

5.1.6. Detector Characteristics

The four most significant detector characteristics are quantum efficiency (which determines the useful wavelength region), plateau (which determines the maximum useful count rate), and resolution.

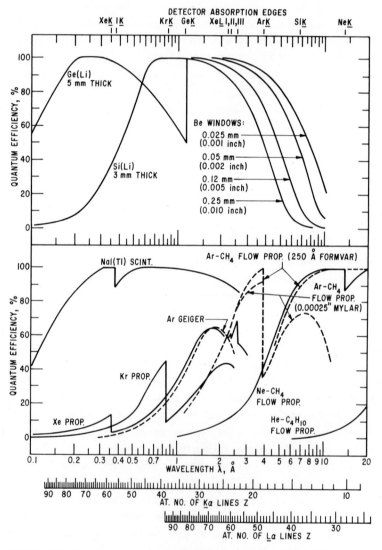

FIGURE 5.8. Typical quantum efficiencies of several x-ray detectors. The discontinuities correspond with absorption edges in the active detector volume (gas, scintillator, silicon).

5.1.6.1. Quantum Efficiency

Quantum efficiency is the fraction or percent of incident x-ray photons of a specified wavelength that produce pulses of countable amplitude. Figure 5.8 shows quantum efficiency curves for common detectors. These curves determine the useful spectral regions of the detectors.

5.1.6.2. Plateau

Suppose that monochromatic x-ray photons are arriving at the detector at some constant intensity, say 1000/s, and that amplifier gain and pulse-height analyzer baseline settings are constant. If, now, the detector tube potential is increased from zero, the observed detector output intensity follows the curve shown in Figure 5.9. The measured intensity, observed on the panel meter or recorder, just perceptibly rises above zero at the *starting potential*, becomes substantially constant at the *threshold potential*, remains substantially constant over a range termed the *plateau*, and finally rises again at the *discharge potential*. The *operating potential* is usually set near the center of the plateau. The interval between the threshold and operating potentials is termed the *overpotential*. The plateau curve is displaced to progressively lower potential the higher the amplifier gain, the lower the discriminator setting, and the shorter the wavelength (higher the photon energy) of the incident x-rays, as shown by the *dashed* lines in Figure 5.10.

The *slope* of the plateau curve is the inclination of the straight portion of the plateau curve expressed as the ratio of the change in measured intensity to the change in applied potential near the midpoint of the plateau. It may be expressed in counts per second per volt, or it may be expressed as percent change in measured intensity per volt. The slope increases and the

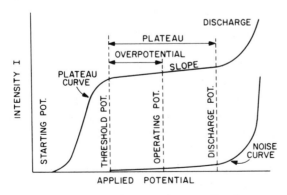

FIGURE 5.9. X-ray detector plateau and noise curves.

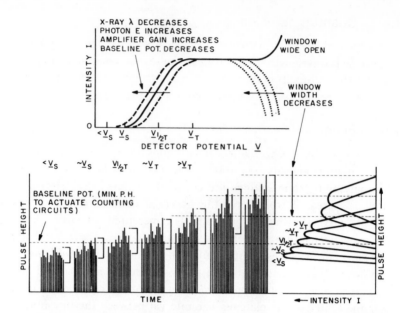

FIGURE 5.10. X-ray detector plateau phenomena. As detector potential increases, average pulse height increases and progressively more—and finally all—of the pulses exceed the minimum pulse height required to actuate the counting circuits, that is, the discriminator baseline. The plateau curve is shifted to lower potentials as x-ray wavelength decreases (photon energy increases), amplifier gain increases, and discriminator baseline potential decreases. If the pulse-height selector window is operative, increased detector potential eventually pushes pulses out the top of the window, and the observed intensity decreases; this occurs at progressively lower detector potential the narrower the window.

plateau shortens as the detector deteriorates with life, and these are good indicators of the condition of the detector.

The same procedure without incident x-rays gives the detector noise curve shown in Figure 5.9.

Figure 5.10 shows the origin of the plateau phenomenon in terms of detector output pulse height and pulse-height selector baseline—the minimum pulse height that is passed on to the counting circuits. At detector potentials below the starting potential V_S, none of the pulses in the distribution exceeds the baseline. As detector potential is increased, progressively more of the pulses exceed the baseline, until, at the threshold potential V_T, all pulses are counted and further increase in detector potential has no effect on the measured intensity. Figure 5.10 illustrates the plateau phenomenon in two ways, with plots of intensity *versus* pulse height and pulse height *versus* time. The former figure is shown in unconventional orientation to permit correlation with the latter.

It is also evident in Figure 5.10 why the plateau curve shifts toward lower potentials with increased amplifier gain, decreased baseline setting, and shorter wavelength. Increased amplifier gain increases *all* pulses proportionally so they exceed the minimum countable height at lower detector potential. Decreased baseline setting lowers this minimum passable pulse height. Shorter wavelength means higher photon energy and proportionally higher pulse amplitude.

The foregoing discussion applies to plateau curves made with the pulse-height analyzer in integral mode, that is, with only the baseline operative. If the analyzer is placed in differential mode with a window of finite width, increase in detector potential eventually results in increase of pulse height to be extent that pulses are "pushed" out the top of the window. In this case, the plateau curve falls off at higher detector potential. The narrower the window is, the lower is the potential at which this occurs and the shorter is the plateau itself, as shown by the *dotted* lines in Figure 5.10.

To summarize, a plateau curve is a plot of observed intensity *versus* detector potential made with monochromatic x-rays at constant incident intensity, amplifier gain, and discriminator setting—with or without the window. The plateau is the substantially straight, horizontal portion of the intensity *versus* potential characteristic of the detector tube, in which the counting rate is substantially independent of the applied potential.

5.1.6.3. Dead Time

There is a limit to the rate at which a detector can "count" incident x-ray photons. Figure 5.11 shows a detector output pulse (pulse 1) of normal amplitude and the following pulse (pulse 2) arising from a photon entering the detector at various intervals after initiation of pulse 1. The figure also shows the pulse-height selector baseline—the minimum countable pulse height (see the last paragraph of Section 5.1.2). The detector *dead time* is the time interval, after initiation of a pulse of normal amplitude, during which the detector is wholly insensitive and gives no response to a second incident x-ray photon. Just after the dead time, the detector responds to a photon, but the pulse is abnormally low. The longer after expiration the photon enters the detector, the more nearly normal is the pulse height. *Resolving time* and *recovery time* are the intervals, after initiation of a normal pulse, after which the detector produces pulses of countable and normal amplitude, respectively. To summarize, after the dead time, a

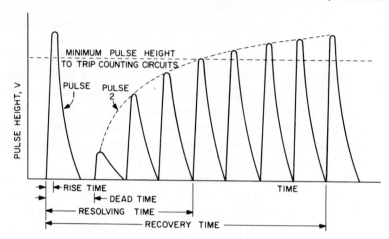

FIGURE 5.11. Amplitude of a detector output pulse as a function of the time interval after initiation of a preceding pulse of normal amplitude.

second pulse can occur; after the resolving time, a second pulse can be counted; after the recovery time, a second pulse has normal amplitude.

These phenomena are responsible for *coincidence loss*—x-ray photons that are not counted ("lost") because they enter the detector during the dead or resolving times of pulses from preceding photons. The nonlinear portions of the curves in Figure 5.12 result from coincidence loss.

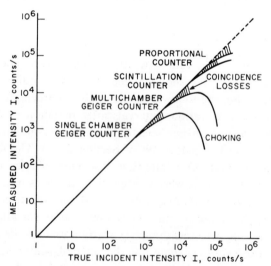

FIGURE 5.12. Linear counting range, coincidence loss, and choking for commonly used x-ray detectors.

Figure 5.13 shows one method of determining the true intensity from a detector subject to coincidence loss, the *multiple-foil method*. A stack of metal foils, usually aluminum or nickel, is placed in the x-ray path. The foils are removed one at a time, and x-ray intensity is scaled after each removal. In the figure, the plotted data shows noticeable departure from linearity at ~7000 counts/s and a true incident intensity of ~150,000 counts/s. The foils must be flat, free of pinholes, warp, and wrinkles, and of identical thickness. They must be placed precisely perpendicular to the x-ray beam, and secondary emission from the foils must be excluded from the detector.

The following equation may be used to correct for coincidence loss or, with the multiple-foil method, to calculate resolving time:

$$I_{\text{true}} = I_{\text{obs}}/(1 - I_{\text{obs}}T_R) \tag{5.4}$$

where I_{true} and I_{obs} are true and observed intensity (counts/s), respectively,

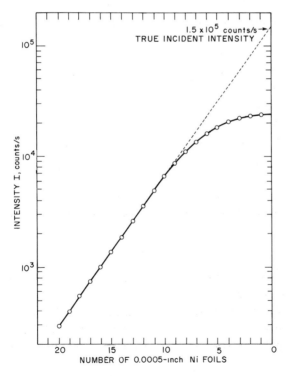

FIGURE 5.13. Multiple-foil method for evaluation of coincidence loss.

and T_R is resolving time (s). This equation is applicable to intensities measured at peaks, but not necessarily to integrated intensities, because the observed intensity, and therefore the coincidence loss, varies continuously during the integration. Also, the equation is valid only if the x-ray photons arriving at the detector have a wholly random time distribution. Consequently, the equation is much less satisfactory with full-wave excitation than with constant potential because with the former, x-rays are generated in the x-ray tube in bursts.

Coincidence losses in the preset-count method (Section 5.3.1) are calculated with the equation

$$I_{\text{true}} = N/(T - NT_R) \qquad (5.5)$$

where N is the accumulated count, T the counting time (s), and T_R the resolving time (s).

5.1.6.4. Resolution

The resolution of a detector is a measure of its ability to distinguish or recognize as separate two closely spaced pulse-height distributions and depends on their separation and breadth.

The theoretical resolution R of a detector can be expressed in terms of the *half-width* $W_{1/2}$ or $\Delta V_{1/2}$, or *full width at half maximum height* (FWHM), of the pulse-height distribution above background. These terms are synonymous and expressed in volts. The half-width of a Gaussian distribution is 2.35 σ intervals (Section 8.4.2, Figure 8.8), so that

$$R = \Delta V_{1/2} = W_{1/2} = \text{FWHM} = 2.35\sigma \qquad (5.6)$$

A more commonly used term is the *percent resolution*, or *relative resolution*, which is the half-width divided by the mean pulse height $\bar{V}$, expressed in percent, that is,

$$R = 100\Delta V_{1/2}/\bar{V} \, [=] \% \qquad (5.7)$$

This definition is shown in Figure 5.14. These equations can be put in terms of the x-ray photon energy E_x and the average number of electrons n in the electron population limiting the statistical precision of the detector. From Equation (8.22), $\sigma = n^{1/2}$, or, from Equation (8.23), $\%\sigma$ (or ε) = $100(n^{1/2}/n)$. Equations (5.6) and (5.7) then become, respectively,

$$R = 2.35n^{1/2} \qquad (5.8)$$

$$R = 235(n^{1/2}/n) = 235/n^{1/2} \, [=] \% \qquad (5.9)$$

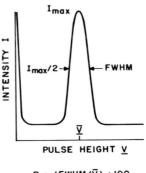

FIGURE 5.14. Detector resolution in terms of full width at half maximum height (FWHM) and mean pulse height $\bar{V}$.

For a gas proportional counter, n is the average number of primary ion–electron pairs generated by the photoelectron produced by the incident x-ray photon. For a semiconductor detector, n is the average number of electron–hole pairs generated by the photoelectron produced by the photon. For a scintillation counter, n is the number of photoelectrons collected by the first dynode of the multiplier phototube for each incident x-ray photon. If in all three cases one defines an *effective ionization potential* V_i as the average energy (eV) required to produce one electron in the statistically limiting process, then

$$n = E_x/V_i \tag{5.10}$$

where E_x is x-ray photon energy (eV). Equation (5.9) then becomes

$$R = \frac{235}{(E_x/V_i)^{1/2}} \: [=]\% \tag{5.11}$$

This equation is usually put in the form

$$R = \frac{235(E_x V_i)^{1/2}}{E_x} \: [=]\% \tag{5.12}$$

The resolutions of NaI(Tl) scintillation, Xe-filled proportional, and Si(Li) semiconductor detectors for Ag $K\alpha$ and $K\beta$ lines are compared in Figure 5.15. The resolutions of scintillation, proportional, and Si(Li) detectors are compared in Figure 5.16 with that of a wavelength-dispersive spectrometer having a LiF(200) crystal. It is evident that the crystal spectrometer is superior at photon energy $\lesssim 30$ keV, wavelength $\gtrsim 0.4$ Å (Cs $K\alpha$), substantially the entire useful spectral region of a 50-kV instrument.

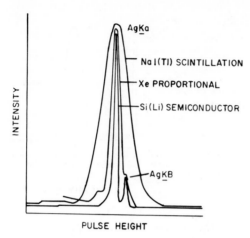

FIGURE 5.15. Comparison of the Ag $K\alpha,\beta$ pulse-height distributions for the NaI(Tl) scintillation counter, Xe proportional counter, and Si(Li) semiconductor detector. [P. G. Burkhalter and W. J. Campbell, *Oak Ridge National Laboratory Report* **ORNL-IIC-10**, vol. 1 (1967); courtesy of the authors.]

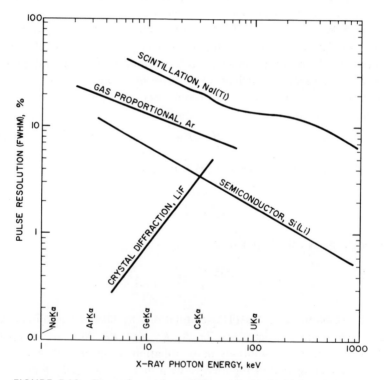

FIGURE 5.16. Comparison of resolutions of the NaI(Tl) scintillation counter, Ar proportional counter, and Si(Li) semiconductor detectors, and LiF analyzer crystal. [P. G. Burkhalter and W. J. Campbell, *Oak Ridge National Laboratory Report* **ORNL-IIC-10**, vol. 1 (1967); courtesy of the authors.]

Actually, the observed resolution of x-ray detectors based on ionization is substantially better than that indicated by statistics alone. Fano introduced the factor F (*Fano factor*) to correct the variance, that is, to bring theoretical statistical variance [Equation (8.13)] into agreement with observed variance, which becomes Fv or $F\sigma^2$. The Fano factor is an intrinsic material constant which reflects the ultimate detector resolution and correlates the observed resolution and average ionization energy. The observed resolution is then

$$R = \text{FWHM} = 2.35(FE_xV_i)^{1/2} \; [=] \; V \qquad (5.13)$$

where the notation is that of Equations (5.10) and (5.11). The term $(E_xV_i)^{1/2}$ is the half-width FWHM in electron volts to be expected for a Gaussian distribution of E_x/V_i electron–hole pairs. This value is reduced by the Fano factor F, which has a value <1. The Fano factor is evaluated by correlation of pulse distribution half-width and x-ray photon energy, and probably has a minimum value of ~ 0.125.

From Equation (5.11), it is evident that resolution is inversely related to x-ray photon energy,

$$R \propto 1/E_x^{1/2} \qquad (5.14)$$

For two x-ray photon energies E_1 and E_2 or wavelengths λ_1 and λ_2, detector resolutions are related by

$$R_1/R_2 = (E_2/E_1)^{1/2} = (\lambda_1/\lambda_2)^{1/2} \qquad (5.15)$$

Incidentally, the equations in this section are for resolution of the detector only. The resolution of the overall detector–readout system is given by introducing the noise contributions of the several components (detector, preamplifier, etc.) into Equation (8.17):

$$R = \{\textstyle\sum \sigma^2 + [2.35(FV_iE_x)^{1/2}]^2\}^{1/2} \qquad (5.16)$$

Mean pulse amplitude, half-width, and energy resolution are all related to photon energy E_x; they are proportional to E_x, $E_x^{1/2}$, and $E_x^{-1/2}$, respectively.

5.1.6.5. Comparison of Detectors

The five types of detector most commonly used in x-ray spectrometry are compared in Table 5.2.

TABLE 5.2. Characteristics of X-Ray Spectrometer Detectors

Property	Geiger	Sealed proportional	Flow proportional	NaI(Tl) scintillation	Si(Li) semiconductor
Window					
position	End	Side	Side	End	End
material	Mica	Mica[a]	Mylar[a]	Be	Be
thickness	10 μm	10 μm	6 μm	0.2 mm	
Gas	Ar–Br$_2$	Xe–CH$_4$	Ar–CH$_4$	—	—
Internal gain	10^9	10^6	10^6	10^6	0
Dead time (μs)	200	0.5	0.5	0.2	
Maximum useful count rate (s^{-1})	2×10^3	5×10^4	5×10^4	10^5	2×10^4
Background intensity (s^{-1})	2	0.5	0.2	10	
Resolution for Fe $K\alpha$ (%)	—	12	15	50	5
Useful wavelength region (Å)	0.5–4	0.5–4	0.7–10[b]	0.1–3	0.4–10[c]

[a] Aluminized.
[b] To ∼100 Å (Be $K\alpha$) with ultrathin window and special gas.
[c] To ∼50 Å (C $K\alpha$) with certain precautions.

In energy-dispersive spectrometry, the Si(Li) detector is used invariably because of its extremely high resolution, its high quantum efficiency over a wide spectral region, and its small size. The size permits its use in close proximity to the specimen, giving high geometric efficiency. The detector is limited to a maximum count rate of ∼20,000 counts/s. This limitation is particularly serious in energy dispersion, where the entire spectrum enters the detector at once—not just the analyte line.

In wavelength-dispersive spectrometry, the most commonly used detectors are the scintillation counter for the short-wavelength region and the argon–methane flow counter for the long-wavelength region. The two are often used in tandem—a two-window flow counter backed by the scintillation counter. A sealed xenon-filled proportional counter can be used in place of the scintillation counter. For the ultralong-wavelength region, flow counters having ultrathin windows and special gas fillings are used (Section 5.1.3.1). The scintillation counter has nearly 100% quantum efficiency over its entire wide useful spectral region. However, the resolution is the poorest by far of any x-ray detector and the detector noise the highest. The gas-filled detectors have better resolution than the scintillation counter and much lower noise. However, the quantum efficiency is lower and varies widely with wavelength.

The Geiger counter has little application in x-ray spectrometry because of its lack of proportionality and its long dead time. However, sometimes these detectors are used on the preset channels of multichannel spectrometers where they detect only a single wavelength.

5.2. READOUT COMPONENTS

5.2.1. Introduction

The fifth phase of an x-ray spectrometric analysis (Section 3.2) is readout, display, and measurement of the detector output data.

In Section 5.1, it is shown that the outputs of x-ray detectors of the types used on analytical x-ray spectrometers consist of series of pulses of electric current, one pulse for each photon detected. The pulses are randomly distributed in time. In proportional, scintillation, and Si(Li) detectors, the pulses have average amplitude proportional to the photon energy.

Even for strictly monochromatic x-rays, the pulses are not all of the same height, but have a Gaussian distribution about a *mean* height proportional to the photon energy. For incident x-ray beams containing more than one wavelength, that is, more than one spectral line, the detector output contains (1) a pulse-height distribution peak for each wavelength, and (2) another *escape* peak for each wavelength shorter than the absorption edge of the detector gas, the iodine in the NaI(Tl) scintillator, or the silicon in the Si(Li) detector. The electronic readout components count these pulses or measure their rate of production for analytical purposes. In addition to the pulses that arise from the discrete incident spectral wavelengths, the detector output also contains pulses arising from background and, especially in scintillation counters, pulses arising in the detector itself.

The detector and electronic readout components of a wavelength-dispersive x-ray spectrometer are shown in simplified block-diagram form in Figure 5.17 and schematically in Figure 5.18. The principal panel controls are shown along the bottom of Figure 5.17 and along the left edge of

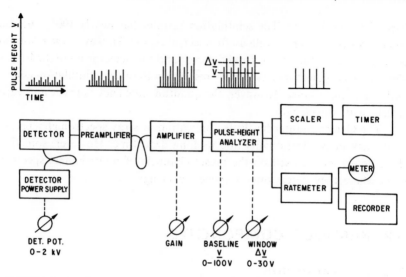

FIGURE 5.17. Electronic detection and readout components of an x-ray spectrometer—simplified block diagram. Simplified pulse diagrams are shown at the top on a grid of pulse height *versus* time. Panel controls involved in pulse-height selection are shown at the bottom.

Figure 5.18. Highly simplified diagrams of pulse height *versus* time are shown along the top of Figure 5.17 and at the upper right of Figure 5.18 for pulses of three amplitudes. The low pulses represent low-energy background and detector and amplifier noise. The medium and high pulses originate from two incident x-ray spectral lines having medium and high photon energies, respectively.

In Figure 5.18, the components are divided into three groups: the detector itself, the amplifier and pulse-height selection components, and the readout and display components. The detector and preamplifier are mounted on the spectrogoniometer. All the other components are mounted in some sort of relay rack or console, with the possible exception of the computer (if any), which may be separate, even in a location remote from the spectrometer. The detector power supply is included with the electronic components. It is a well-regulated ($\pm 0.05\%$ or so) 0–2-kV dc supply. Formerly, the readout components used electron tubes. However, most modern spectrometers have all-solid-state semiconductor readout components, mostly in the form of plug-in "NIM" (Nuclear Instruments Manufacturers) modular units having standardized dimensions, connections, potentials, etc. Conversion to solid-state modules has resulted in reduced size and power consumption and increased sophistication.

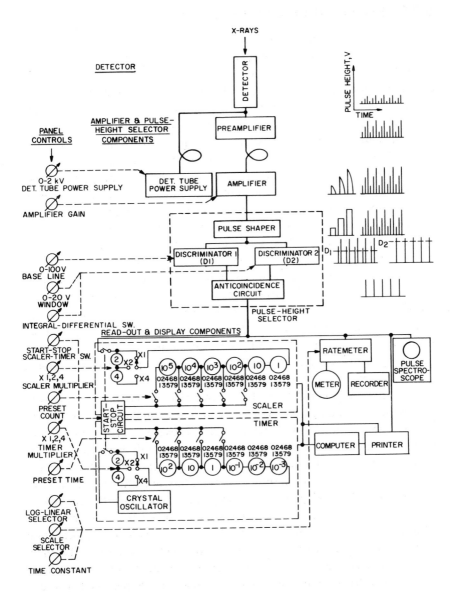

FIGURE 5.18. Electronic detection and readout components of an x-ray spectrometer. Simplified pulse diagrams are shown at the top right on a grid of pulse height *versus* time. Panel controls are shown along the left.

5.2.2. Preamplifier and Amplifier

The amplifiers amplify the detector output pulses to potentials high enough to pass the discriminator or pulse-height selector and actuate the ratemeter and scaler. In the older electron-tube units, the pulses are amplified to a maximum of 100 V, in NIM modules to 10 V. With the exception of sine-function amplifiers (see below and Section 5.2.3.6), the amplification is *linear*—that is, without change in relative pulse amplitudes.

Conventional preamplifiers are low-gain ($\sim 10 \times$) linear amplifiers mounted at the detector itself and connected by a relatively long cable to the main linear amplifier in the electronics console. The linear preamplifier provides optimal coupling to the cable and minimal attenuation along it. Sine-function preamplifiers amplify all detector output pulses, regardless of amplitude, to the same final amplitude to pass a fixed, preset pulse-height selector window. This is one way to effect automatic pulse-height selection (Section 5.2.3.6).

The main amplifier is a linear, low-noise, fast-response, video-type amplifier having gain that varies up to $\sim 10,000$. Alternatively, many amplifiers have a fixed gain of, say, 10,000, and provide for *attenuation* to lower gain. The amplifier is designed to amplify preamplifier output pulses to a maximum of 0.5–10 V in solid-state NIM modules, 5–100 V in electron-tube components. Geiger, proportional, and scintillation counters require low, intermediate, and high amplification, respectively.

The amplifier should have the following characteristics: (1) linearity of amplification, that is, equal gain for all pulse amplitudes; (2) stability of gain; (3) low noise; (4) rapid overload recovery, that is, rapid return to normal baseline after receiving pulses of abnormally high amplitude or frequency; (5) good thermal stability; (6) good baseline restoration (dc restoration), that is, rapid return to normal baseline after termination of a pulse; and (7) good pole–zero cancellation; sometimes the trailing or descending side of a pulse may undershoot the baseline, that is, momentarily fall to a potential less than that of the baseline; the pole–zero cancellation precludes this, or at least rapidly returns the undershoot to the baseline. Separate baseline restoration and pole–zero cancellation circuits or modules may be provided to effect these last two functions. Some amplifiers also provide the pulse-shaping function described in Section 5.2.3.1.

5.2.3. Pulse-Height Selector

Pulse-height *selection* (PHS) is the separation of one or a very few pulse-height distributions from that of the analyte line so that the analyte line can be measured without interference. This technique is used to supplement wavelength dispersion. Pulse-height *analysis* (PHA) is the separation of all the pulse-height distributions in the energy spectrum emitted and scattered by the specimen so that all its lines can be measured individually. This technique is used in—and, in fact, constitutes—energy-dispersive spectrometry.

Pulse-height selection is discussed in this section; pulse-height analysis and energy-dispersion in general are discussed in Chapter 6.

5.2.3.1. Instrument and Principle

A *discriminator* is an electronic circuit that passes all pulses having height greater than some preset minimum, and rejects all lower pulses. A *pulse-height selector*, or *pulse-height analyzer*, is an electronic circuit that passes on to the ratemeter and scaler all pulses having height greater than one preset level but lower than a second preset level, and rejects all pulses outside this *window* or *channel*. All x-ray spectrometers, have a discriminator to reject pulses having height, after amplification, of $\sim$0.5 V or less in solid-state NIM modules, $\sim$5 V or less in electron-tube components. This discriminator eliminates the detector and amplifier noise and some low-energy background. Although most modern x-ray spectrometers have pulse-height selectors, this component may be omitted, and, even if present, it may be bypassed.

The pulse-height selector is shown in Figure 5.18 enclosed in a dashline box. The particular type shown consists of four components, the functions of which are illustrated by pulse diagrams at the right of each component. It will be recalled that the low pulses represent detector and amplifier noise, and the medium and high pulses represent two x-ray spectral lines having medium and high photon energy, respectively. The four components are as follows.

The *pulse shaper* forms rectangular or square-wave pulses from the amplified pulse forms, as shown by the two groups of enlarged pulses, but does not alter their relative amplitudes. The shaped pulses are passed on to the inputs of both of the discriminators.

Discriminator 1 (the *lower-level discriminator*) passes all pulses having amplitude greater than level D_1. Thus, it passes the pulses of both spectral lines, but rejects the low-amplitude noise. A discriminator functioning in this way is always required, even when the pulse-height selector is absent or by-passed.

Discriminator 2 (the *upper-level discriminator*) passes all pulses having amplitude greater than level D_2. Thus, it passes only the pulses of the high-energy spectral line.

The *anticoincidence circuit* cancels all pulses that arrive simultaneously from both discriminators. Thus, it rejects the high-amplitude pulses, and passes on to the ratemeter and scaler only the pulses of intermediate amplitude.

To summarize, pulses of amplitude smaller than a selected minimum D_1 do not pass either discriminator. Pulses of amplitude greater than a selected maximum D_2 pass both discriminators and are eliminated by the anticoincidence circuit. Thus, only pulses having amplitude between the two levels pass on to be counted. The operation of the pulse-height selector is more simply illustrated by the simplified diagram in Figure 5.17.

The potential V corresponding to the lower discriminator level is usually referred to as the *baseline*, sometimes as the *window level* or *channel level*. It represents the minimum countable pulse height. The potential interval ΔV between the two levels is usually referred to as the *window* or *channel*, sometimes as the *window width* or *channel width*. The baseline may be varied from 0–10 V in solid-state NIM modules, from 0–100 V in electron-tube components. The window may be varied over the full baseline in NIM modules, from 0 to 20 or 30 V in electron-tube components. The lower level of the window always coincides with the baseline setting; that is, the window always "rests on" the baseline and moves with it.

The instrument just described is a *single-channel pulse-height selector*. It may be operated as a simple discriminator, passing all pulses having amplitude greater than a selected baseline V, or as a pulse-height selector, passing only pulses having amplitude greater than a selected baseline V, but less than the baseline plus window $V + \Delta V$.

5.2.3.2. Pulse-Height Distributions

An elementary discussion of the origin and nature of pulse-height distributions is given in Section 5.1.2. The remainder of Section 5.2.3 considers pulse distributions and their selection in detail.

If incident x-ray intensity is not high enough to cause abnormal pulse-amplitude effects (Figure 5.11), individual pulse height and mean height of a pulse distribution are determined by: (1) x-ray photon energy E_x (eV); (2) effective ionization potential $V_{i,\text{eff}}$ (eV) of the active detector medium (gas or scintillator), which determines the number of useful electrons produced by the initial ionization; (3) internal electron multiplication (gas or secondary) in the detector G_D; (4) detector potential; and (5) amplifier gain G_A. Mean pulse height $\bar{V}$ is given by

$$\bar{V} = (E_x/V_{i,\text{eff}})G_D G_A \qquad (5.17)$$

The gains are dimensionless, and G_D for the Si(Li) detector is 1. If the amplifier has fixed gain G_A reducible by attenuation L, (G_A/L) is used in the equation.

Detector potential does not appear in the equation explicitly, but largely determines G_D. It follows that for a specified detector, pulse height is determined by x-ray photon energy, detector potential, and amplifier gain, increasing for each of these parameters. These relationships are shown in Figure 5.19 and explained in its caption, which should be studied carefully. It is evident that: (1) as photon energy increases (wavelength decreases), mean pulse height increases and the pulse distribution falls at higher baseline potential, decreases in peak intensity, and widens; (2) at constant photon energy, as detector potential or amplifier gain increases, a pulse distribution moves to higher baseline potential, decreases in peak height, and widens. However, integrated intensity (area under the curve) remains constant. Decreasing photon energy, detector potential, and amplifier gain have the opposite effect.

Given mean pulse height $\bar{V}_1$ for x-rays of photon energy E_1, the mean pulse height $\bar{V}_2$ for x-rays of energy E_2 is given by

$$\bar{V}_1/\bar{V}_2 = E_1/E_2; \qquad \bar{V}_2 = \bar{V}_1(E_2/E_1) \qquad (5.18)$$

In general, if all other variables are constant, the pulse-height distribution has greater width (FWHM): (1) for scintillation counters than for proportional counters than for Si(Li) counters; (2) the higher the x-ray photon energy (shorter the wavelength); (3) the higher the detector potential; and (4) the higher the amplifier gain.

Figure 5.19 can also be used to illustrate the principle of pulse-height selection. Assume that the figure shows three pulse-height distributions of x-rays having low, medium, and high photon energy (long, medium, and short wavelength), respectively (case I in the caption). It is evident that with the window set as shown, the B peak is passed to the exclusion of A and C. If the baseline potential is varied, the pulse-height distributions remain fixed, but the window is moved and can be set to admit any one of them. If the detector potential and/or amplifier gain are varied, the window remains fixed, but the pulse-height distributions are moved, and any one can be moved into the window. Finally, the window width can be

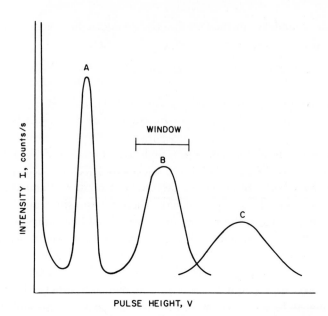

FIGURE 5.19. Effect of incident x-ray photon energy, detector potential, and amplifier gain on mean pulse height, width, and peak intensity of the pulse-height distribution. The integrated intensities (areas under the peaks) are equal. The figure represents any of three cases:

I. Detector potential and amplifier gain constant: peaks *A*, *B*, and *C* represent pulse-height distributions for x-ray lines of low, medium, and high photon energies (long, medium, and short wavelengths), respectively.

II. Photon energy and amplifier gain constant: peaks *A*, *B*, and *C* represent the same pulse distribution at successively higher detector potentials.

III. Photon energy and detector potential constant: peaks *A*, *B*, and *C* represent the same pulse distribution at successively higher amplifier gain.

varied to pass as many as possible of the pulses in the analyte-line distribution while excluding as many as possible of the pulses from neighboring peaks, which may, unlike the peaks shown here, overlap. For example, it is evident that the window width shown in Figure 5.19 just admits the entire *B* peak. If the window were moved to the *A* peak, it could be made narrower without loss of *A* pulses, but if it were moved to the *C* peak, many *C* pulses would be excluded unless the window were made wider. In practice, it is customary to move the pulse-height distribution to the window, rather than the window to the distribution. The window is usually set at a baseline of ~1–2 V in solid-state NIM modules, ~10–20 V in electron-tube components, where the pulse distribution is narrow, yet well resolved from the noise.

5.2.3.3. Pulse-Height Distribution Curves

Pulse-height distribution curves are best described in specific, rather than general, terms. A good illustrative example is the determination of ~1 wt% silicon in iron with an EDDT crystal. Fourth-order Fe $K\beta$ slightly overlaps Si $K\alpha$. Table 5.3 and Figure 5.20 show the origin of the problem, and Figures 5.21A and B show 2θ scans of the spectral region 103–113° 2θ occupied by the two lines, with and without pulse-height selection, respectively. Although the two lines are separated by 2° 2θ, the Fe $K\beta(4)$ line is so intense relative to Si $K\alpha$ that the tail of the iron line contributes intensity at the 2θ angle of the relatively weak silicon line. It must be emphasized that although the Fe $K\beta(4)$ intensity at the Fe $K\beta(4)$ peak (106° 2θ) is ~1600 counts/s, the Fe $K\beta(4)$ intensity *overlap* at the Si $K\alpha$ peak (108° 2θ) is only ~30 counts/s. Thus, if the spectrogoniometer is set to measure the Si $K\alpha$ line (108° 2θ), the detector or amplifier output contains five groups of pulses of various intensities and average amplitudes as follows: (1) detector-amplifier noise of very low amplitude and very high intensity; (2) continuous x-ray background having a wide range of amplitudes and an intensity, say, 160 counts/s; (3) Si $K\alpha$ pulses of low amplitude and

TABLE 5.3. Data for Si $K\alpha$ and Fourth-Order Fe $K\beta^a$

Line	λ (Å)	4λ (Å)	E_x (eV)	2θ, EDDT	I (counts/s) 106° Fe $K\beta(4)$	I (counts/s) 108° Si $K\alpha$
Si $K\alpha$	7.13	—	$\dfrac{12{,}396}{7.13} = 1739$	108°	0	200
Fe $K\beta(4)$	1.76	7.04	$\dfrac{12{,}396}{1.76} = 7045$	106°	1600	30
Background					160	160

^{a} See Figures 5.20 and 5.21A, B.

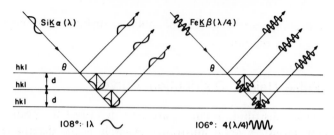

FIGURE 5.20. Partial overlap of fourth-order Fe $K\beta$ and first-order Si $K\alpha$ with an EDDT crystal (see Table 5.3). At nearly the same 2θ angle at which Si $K\alpha$ is diffracted in first order (one-wavelength path difference between successive planes), Fe $K\beta$ is diffracted in fourth order (four-wavelength path difference). If Fe $K\beta(4)$ is very intense relative to Si $K\alpha$, the tail of the Fe $K\beta(4)$ peak may contribute intensity at the Si $K\alpha$ peak.

intensity $\sim$200 counts/s; (4) Fe $K\beta$ pulses of relatively high amplitude and intensity $\sim$30 counts/s; and (5) Fe $K\beta$ escape-peak pulses (assuming an Ar-filled proportional counter); these are of no significance in the discussion to follow and are not considered further here.

There are several ways to display pulse-height distribution data, three of which are shown in Figure 5.22. In Figure 5.22B, *individual* pulses are displayed on a grid of pulse height *versus* time. Three of the groups of pulses listed above are distinguishable and are marked at the left of the drawing: detector–amplifier noise, Si $K\alpha$, and Fe $K\beta$. The background pulses are not

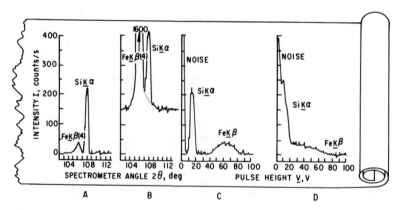

FIGURE 5.21. Ratemeter–recorder readout of intensity data. (A) 2θ scan (EDDT crystal) with pulse-height selection. (B) 2θ scan without pulse-height selection. (C) Differential pulse-height distribution curve. (D) Integral pulse-height distribution curve.

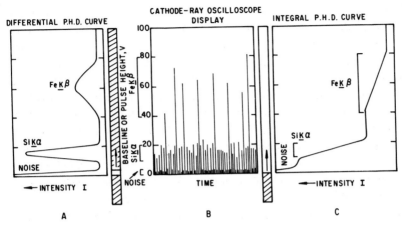

FIGURE 5.22. Function of the pulse-height selector. The center figure (B) represents the amplifier output on a grid of pulse height *versus* time and shows three groups of pulses: high-intensity, low-amplitude noise; intermediate-intensity, intermediate-amplitude Si $K\alpha$; and low-intensity, high-amplitude Fe $K\beta$. The left figure (A) shows the differential pulse-height distribution curve. Only pulses having amplitudes ("tops") lying in the open window in the shaded area pass to the ratemeter and scaler as the baseline, and with it the window, are moved up the baseline. The right figure (C) shows the integral pulse-height distribution curve. Only pulses having amplitudes above the shaded area pass to the ratemeter and scaler as the baseline is moved upward. Pulse-height distribution curves are shown conventionally with the intensity axis vertical; they are shown here in unconventional orientation to permit correlation with the center figure. For solid-state NIM modules, the baseline potential scale would be 0–10 V.

shown because they have a wide range of pulse heights and would obscure the other pulses. As expected, the high-energy Fe $K\beta$ distribution has higher average pulse height and is wider than the low-energy Si $K\alpha$ distribution. However, also as expected, at 108° 2θ, the Si $K\alpha$ pulses are more *numerous* ($\sim$200/s) than the Fe $K\beta$ pulses ($\sim$30/s). The x-ray photons incident on the detector, and therefore the output pulses, have a random distribution in time. Such a display is obtained by connecting the vertical input of a cathode-ray oscilloscope to the output of the linear amplifier. The displays at the output of the detector or preamplifier would differ only in linear amplification. Incidentally, if NIM modular units are used, the baseline pulse-height scale would be 0–10 V, not 0–100 V.

The single-channel pulse-height selector (PHS) or analyzer (SCA) and ratemeter–recorder may be used to display the pulse-height distribution in two ways.

A differential pulse-height distribution curve (Figure 5.22A) is recorded by: (1) setting the spectrogoniometer at 108° 2θ (Si $K\alpha$); (2) setting the window at an arbitrary small width ΔV, say 1 V; and (3) "scanning the baseline," that is, moving the baseline discriminator level V—and with it the window—from 0 to 100 V as intensity is recorded. At any instantaneous baseline setting, the window passes to the ratemeter only those pulses having height between V and $V + \Delta V$. The resulting curve of intensity *versus* pulse height, Figure 5.22A, shows two pulse-height distribution peaks and the detector-amplifier noise. The orientation is unconventional, to permit correlation with Figure 5.22B. At very low baseline potential, the high-intensity, low-energy noise passes the window, giving high response, which then falls to zero until the window passes through the Si $K\alpha$ distribution, giving a relatively high peak. The response then again falls to zero until the window passes the less intense Fe $K\beta$ distribution. The continuous x-ray background pulses are not shown in Figure 5.22B, and so background is also omitted in Figure 5.22A; it would appear simply as a finite—rather than zero—intensity between the noise, Si $K\alpha$, and Fe $K\beta$ peaks.

An *integral pulse-height distribution curve* (Figure 5.22C) is recorded by: (1) setting the spectrogoniometer at $108°\ 2\theta$ as before; (2) making the window inoperative or, in effect, "wide open"; and (3) scanning the baseline V from 0 to 100 V as intensity is recorded. At any instantaneous baseline setting, *all* pulses having height greater than V pass to the ratemeter. The resulting curve of intensity *versus* pulse height, Figure 5.22C, shows three *steps*, instead of the *peaks* of Figure 5.22A—the noise, Si $K\alpha$, and Fe $K\beta$. Again, the orientation is unconventional, to permit correlation. At very low baseline potential, *all* pulses—noise, Si $K\alpha$, Fe $K\beta$, and background— pass to the recorder, and the response is very high. At 3–5 V, the noise is discriminated out, and the intensity decreases. At a somewhat higher potential, the Si $K\alpha$ distribution is excluded, and at still higher potential, the Fe $K\beta$. Here again, if x-ray background were shown in the figure, it would appear simply as a finite—rather than zero—intensity after the Fe $K\beta$ step.

Although differential pulse-height distribution curves are more commonly used, both types of pulse-height distribution curve convey the same information, and the choice is largely a matter of personal preference. Such curves are made prior to undertaking analysis of a new sample type. They reveal escape peaks and overlapping of neighboring pulse-height distributions with that of the analyte, and permit the choice of baseline and window settings for the analysis.

Pulse-height distributions can be displayed in several ways. Individual pulses can be observed on a grid of pulse height *versus* time by connecting

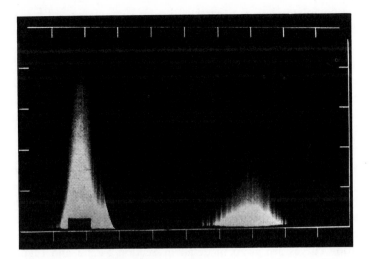

FIGURE 5.23. Pulse-spectroscope display of the Si $K\alpha$ (left) and Fe $K\beta$ (right) pulse-height distributions. The black bar on the Si $K\alpha$ peak is an electronic window marker.

the vertical deflection terminal of an oscilloscope to the output of the detector, preamplifier, or amplifier. Such displays resemble Figure 5.22B. If the oscilloscope time base is fast enough, the shapes of individual pulses can be observed, as shown in Figure 5.11 and at the top right in Figure 5.18. Differential and integral pulse-height distribution curves are observed on a grid of intensity *versus* pulse height on ratemeter–recorder charts, as shown in Figures 5.21C and D, respectively. Figure 5.23 shows a differential pulse-height distribution curve on a cathode-ray tube pulse spectroscope. The black bar on the Si $K\alpha$ peak is an electronic window marker, which moves along the baseline and varies in width as baseline and window controls are varied. The pulse distributions move along the baseline as detector potential and amplifier gain are varied. The great convenience of such instruments in setting the pulse-height selector is obvious.

5.2.3.4. Pulse-Height Selector Settings

Suppose that one or more analytes are to be determined in a specified sample type (matrix) for the first time. It is necessary to establish whether, with the goniometer set at 2θ for each analyte line λ_A, only pulse distributions (photo or escape) arising from that line are present in the detector output, aside from the noise. Extraneous pulse distributions arise principally from: (1) emitted matrix-element lines and scattered x-ray tube target lines

λ having complete ($\lambda = \lambda_A$), partial ($\lambda \approx \lambda_A$), or higher-order ($n\lambda \approxeq \lambda_A$) overlap with the analyte line (complete overlap cannot be dealt with by pulse-height selection); and (2) crystal emission lines arising from excitation of heavier constituent elements in the analyzer crystal by x-rays emitted or scattered by the specimen. The first group of lines is revealed by recording the wavelength spectrum (2θ scan) of the specimen, or at least of the spectral region of each analyte line. However, crystal emission occurs in all direction and enters the detector at all 2θ angles.

If it is possible that at the 2θ angle of the analyte line, the analyte pulse distribution may be partially overlapped by extraneous photo or escape distributions, the pulse-height selector settings (detector potential, amplifier gain, baseline, window) should be derived from full pulse-height distribution curves, as plotted on a recorder or observed on a pulse spectroscope. Techniques for recording integral and differential pulse-height distribution curves are given in detail in Section 5.2.3.3 and Figure 5.22, in simplified form in Figure 5.24A and its caption.

These curves are plotted with a specimen of the type to be analyzed, the instrument components [x-ray tube, crystal, collimator(s), detector] to be used in the analysis, and the goniometer set at 2θ for the analyte line. Detector potential is set at midplateau. With the pulse-height selector in integral mode and the discriminator (baseline) at $\sim$0.5 V for solid-state NIM modules, $\sim$5 V for electron-tube components, excitation (kV, mA) is set to give analyte-line intensity of up to 10,000 counts/s. Higher intensity is avoided to preclude pulse-height shift (Section 5.2.3.7). The curves are plotted on the ratemeter–recorder or perhaps photographed on a pulse spectroscope.

If it is decided to set the baseline and window on the analyte peak where it lies, the curves permit selection of these settings to admit as many pulses as possible from the analyte peak and as few as possible from other peaks and continuum. If it is decided to move the analyte peak to a narrow preset window at low mean pulse height, method B below may be used to establish the amplifier gain (or detector potential) to move the peak.

If it is certain that at the 2θ angle of the analyte line, the analyte pulse distribution is free from interference, pulse-height selector settings can be established conveniently and rapidly by one of the four simple methods shown in Figure 5.24 and summarized in its caption.

These methods are conducted at the same conditions as are used for recording pulse-height distribution curves (above). However, the specimen should be pure analyte, if practical. An oxide, other compound, or alloy may be used. However, no other element may be present that, at the 2θ

FIGURE 5.24. Methods for establishing pulse-height selector settings (detector potential, amplifier gain, baseline, window). The solid outlines represent pulse-height distributions where they fall along the baseline; the dashed outlines represent the distributions after moving with the amplifier gain (and/or detector potential) to a relatively narrow preset window in the low-pulse-height region. The reader is reminded that in integral mode, the pulse-height selector passes all pulses having amplitude greater than the baseline B, in differential mode, all pulses having amplitude within the window $B-W$.

Full pulse-height distribution curves may be plotted on a ratemeter–recorder chart (Section 5.2.3.3), as shown in detail in Figure 5.22, in simplified form in A above: The integral curve is plotted by moving the baseline B from 0 to maximum; the differential curve is plotted by moving the window $B-W$ from 0 to maximum baseline. Baseline and window settings are then derived from the plotted curves (Figure 5.22).

Four rapid simplified methods for establishing pulse-height selector settings are shown here:

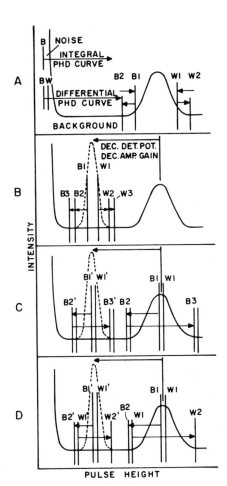

A. The baseline is moved to the low-energy side $B1$ of the pulse distribution where it lies, then the window is narrowed to the high-energy side $W1$.

B. The pulse distribution is moved with the amplifier gain (or detector potential) to a narrow, low-energy, preset window $B1-W1$, and the baseline and window adjusted.

C. A narrow window is moved over the pulse distribution to establish its lower and upper limits—either where it lies (solid outline, $B1-W1$) or after moving it to a preset window (dashed outline, $B1'-W1'$).

D. A narrow window is moved over the pulse distribution to establish its lower limit, then the window is widened to its upper limit—either where it lies (solid outline, $B2-W2$) or after moving it to a preset window (dashed outline, $B2'-W2'$).

angle for the analyte line, can produce an observable pulse distribution
—photo or escape— that may be confused with that of the analyte. The
action is best observed on a pulse spectroscope. However, the following
procedures provide for simply observing intensity on the panel ratemeter
or recorder (with the pen disengaged).

Before describing the four rapid methods, the reader is reminded that:
(1) in integral mode, the pulse-height selector passes all pulses having
amplitude greater than the baseline; (2) in differential mode, it passes all
pulses having amplitude within the window—that is, above the baseline,
but below the top of the window; (3) increased amplifier gain (and/or
detector potential) increases mean pulse amplitude, displacing the pulse
distribution upscale, widening it, and decreasing its peak intensity, but not
its integrated intensity; and (4) decreased gain (and/or detector potential)
decreases mean amplitude, displacing the distribution downscale, narrowing
it, and increasing its peak intensity, but not its integrated intensity.

Method A (Figure 5.24A). The baseline and window may be set to
bracket the pulse distribution where it lies as follows. With the pulse-height
selector in integral mode with a low baseline, it is likely that the entire
analyte pulse distribution is passed, and its intensity is indicated on the
meter and recorder. While intensity is observed on the meter or recorder,
the baseline is moved to progressively higher potential until intensity
decreases sharply, indicating that the baseline has begun to discriminate
the low-amplitude side of the pulse distribution (*B*1). The baseline is then
backed off (*B*2) to allow some latitude for shift and distortion. The selector
is now switched to differential mode with the window at maximum width.
The width is decreased progressively until intensity decreases sharply,
indicating that the window has begun to exclude the high-amplitude side
of the distribution (*W*1), then increased for latitude (*W*2).

If, when the baseline is first increased, the observed intensity sharply
decreases immediately, the pulse distribution lies partly below the original
baseline. In this case, amplifier gain (or detector potential) must be increased to move the pulse distribution to higher average amplitude. Similarly, if, when the window width is first decreased, the observed intensity
sharply decreases immediately, the pulse distribution lies partially above the
original (maximum) window. In this case, amplifier gain (or detector potential) must be decreased to move the pulse distribution to lower average
amplitude.

Method B (Figure 5.24B). Alternatively, the pulse distribution may be
moved to a narrow preset window and the baseline and window adjusted

as follows. The selector is set in differential mode with the baseline at some arbitrary low potential ($B1$) and a narrow window ($W1$): say, 1–2-V baseline, 0.5–1-V window for solid-state NIM modules, 10–20-V baseline, 5–10-V window for electron-tube components. At these conditions, the pulse distribution is likely to fall wholly or partly outside the window (solid-line peak), and the panel ratemeter and recorder indicate low intensity accordingly. While intensity is observed on the meter or recorder, the amplifier gain (or detector potential) is varied until the observed intensity is maximum, indicating that the peak of the pulse distribution has moved into the window (dash-line peak), as indicated by the long arrow in the figure. The baseline is now reduced decrementally while the upper level of the window is held fixed at its original absolute potential. This means that for each decrement in baseline potential, the window width is *increased* an equal increment in potential. This is repeated until no further increase in intensity is observed, indicating that the low side of the pulse distribution now lies entirely in the window ($B2$); the baseline is then decreased a little farther ($B3$). The window is now widened until no further increase in intensity is observed, indicating that the high side of the distribution now lies entirely in the window ($W2$), and is then widened a little more ($W3$).

Incidentally, in the foregoing discussion, it must be borne in mind that a decrease in baseline potential and/or an increase in window width usually results in an increase in intensity; this is true even in a region where there is no pulse-height distribution, and is due to admission of increased numbers of background pulses. Similarly, an increase in baseline and/or decrease in window width usually results in a decrease in intensity due to exclusion of background pulses. However, these intensity variations are usually very small and gradual, and the operator soon learns to distinguish them from the large, sharp intensity changes when a pulse distribution moves in or out of the window, or vice versa.

Method C (Figure 5.24C). A narrow window may be moved over the pulse distribution to establish its lower and upper limits, as follows. The selector is set in differential mode with the baseline at some arbitrary low potential ($B2'$) and a narrow window: say, 0.5–1-V baseline, 0.1–0.2-V window for solid-state NIM modules, 5–10-V baseline, 1–2-V window for electron-tube components. At these conditions, the pulse distribution is likely to fall wholly or partly outside the window, and the panel ratemeter and recorder indicate low intensity accordingly. Let the solid-line peak represent this distribution. (Assume that the dash-line peak is absent; this peak is considered below.) While intensity is observed on the meter or recorder, the baseline, and with it the window, is moved to progressively higher potential until intensity increases sharply, indicating that the window has begun to admit pulses at the low end of the peak ($B2$). The baseline is increased further until intensity is maximum, indicating that the window ($B1$–$W1$) now lies at the peak, then still further until intensity no longer decreases, indicating that the window has passed the high end of the peak ($B3$). The baseline setting is then $B2$, and the window width is set to correspond to $B3$. Somewhat wider settings are used for latitude.

Method D (Figure 5.24D). A narrow window may be moved over the pulse distribution to establish its lower limit, and the window opened to its higher limit, as follows. The procedure is identical with that for method C above, except that when baseline setting $B2$–$W1$ is established, the *window* is widened until intensity no longer increases, indicating that the entire peak has been bracketed ($W2$).

Alternatively, and perhaps preferably, methods C and D may be modified to move the pulse distribution to a narrow preset window ($B1'$–$W1'$): say, ~2-V baseline, 0.1–0.2-V window for solid-state NIM modules, ~20-V baseline, 1–2-V window for electron-tube components. While intensity is observed on the meter or recorder, amplifier gain (or detector potential) is adjusted to give maximum intensity, indicating that the peak of the pulse distribution has moved to the selected baseline potential and is more or less centered in the window. In Figures 5.24C and D, this is depicted by the movement of the solid-line peaks downscale to become the dashed-line peaks, as indicated by the arrow at the top of each figure. Methods C and D are then conducted in exactly the same way as before, but on the new peaks.

In methods C and D, especially their modified versions, if on decreasing the baseline from the peak position ($B1$–$W1$ or $B1'$–$W1'$), intensity continues to decrease right down to the point where it increases again due to detector amplifier noise, it means that the low end of the pulse distribution overlaps the noise. In this case, amplifier gain (or detector potential) must be increased to move the distribution upscale.

Regardless of which of the four foregoing procedures is used, insofar as possible, the baseline and window should be set to pass the entire analyte pulse-height distribution and its escape peak, if any, with some latitude.

All settings—kV, mA, detector potential, amplifier gain, baseline, and window—are recorded and always used thereafter to determine the specified analyte in the specified type of sample. The foregoing procedure must be followed for each other analyte to be determined with the aid of the pulse-height selector. If the analyte having spectral line of longest wavelength is evaluated first, it may be that the other analyte lines simply require progressively lower amplifier gains to move their pulse distributions successively to the window. This preliminary work is done only once for a given sample type. However, it is prudent to check conditions occasionally, especially if the detector plateau curve shows a change. The great convenience of the pulse spectroscope (Figure 5.23) for this preliminary work is obvious.

If the pulse-height selector is used only to reduce background rather than to exclude another pulse distribution, the operator may want to determine whether the selector is really beneficial. A simple test follows. After the baseline and window are set by one of the methods described above, the goniometer is moved just off the analyte-line peak and the adjacent continuous background intensity is observed on the panel ratemeter or recorder, or scaled. Then the pulse-height selector is set at integral mode and the baseline discriminator at 0.5–2 V for solid-state NIM modules, 5–20 V for electron-tube components, and the background intensity is measured again. If this intensity is substantially the same as before, the pulse-height selector is not doing any good and may just as well be operated as a simple discriminator.

5.2.3.5. Applications and Limitations

In many analyses, the pulse-height selector is not required, perhaps not even beneficial. When sharp, intense, spectral lines occur in a sparse spectrum having low continuous background, there is little, if any, advantage in use of the selector. However, for the following applications, it may be extremely beneficial, even indispensable. (1) The pulse-height selector discriminates the analyte line λ and *harmonic overlap*—higher orders of other lines that diffract at the same crystal 2θ setting as λ, $n(\lambda/n)$, that is, second-order $\lambda/2$, third-order $\lambda/3$, etc.; some examples are given in Table 5.4. However, the selector cannot discriminate pulse distributions from similar wavelengths, such as As $K\alpha$ and Pb $L\alpha_1$ (both 1.18 Å). (2) It reduces continuous background in the longer-wavelength region, where the continuum is mostly higher orders of the shorter-wavelength "hump" of the primary continuum scattered by the specimen. (3) It decreases (improves) the detection limit by reducing background. (4) It permits higher sensitivity in the long-wavelength region, where it discriminates higher-order interfering lines so that coarse collimation can be used. (5) It removes certain types of radioactive background from radioactive specimens. (6)

TABLE 5.4. Discrimination of Harmonic Overlap by the Pulse-Height Selector

Line	λ (Å)	$n\lambda$	E_x (keV), $12.4/\lambda$
V $K\alpha^a$	2.50	—	4.96
Ge $K\alpha$	1.25	$2\lambda = 2.50$	9.92
Hf $L\alpha_1{}^a$	1.57	—	7.90
Zr $K\alpha$	0.79	$2\lambda = 1.58$	15.70
K $K\alpha^a$	3.74	—	3.32
Ge $K\alpha$	1.25	$3\lambda = 3.75$	9.92
Si $K\alpha^a$	7.13	—	1.74
Fe $K\beta$	1.76	$4\lambda = 7.04$	7.04

^{a} Analyte line.

Finally, it permits energy-dispersive analysis for specimens that produce only one or a few widely spaced pulse distributions.

The resolution of a pulse-height selector used with a proportional counter is far superior to that with a scintillation counter—but is still none too good. Pulse distributions for the same lines of elements differing in atomic number by ± 1 are not resolved; those differing by ± 2 or 3 are resolved only by sacrifice of a substantial fraction of the total intensity, with consequent loss in counting efficiency; those differing by ± 4 are resolved with little loss of counting efficiency. The ability of a counter (proportional or scintillation) to resolve pulse distributions of neighboring elements is substantially constant throughout the periodic table. Only lithium-drifted silicon and germanium detectors can resolve lines of adjacent elements.

The application of the pulse-height selector in reduction of background is discussed in the last two paragraphs of Section 5.3.2.3. The selector removes only background having wavelength substantially different from that of the analyte line.

The effectiveness of the pulse-height selector is limited by: (1) the difference in mean pulse height of the pulse-height distributions to be resolved; (2) their half-widths; (3) their profiles, that is, whether Gaussian or distorted; (4) their relative intensities; (5) the presence or absence of escape peaks and their character with respect to the first four properties; and (6) the continuous background pulses. The pulse-height selector is more effective: (1) the more widely separated, narrower, and more nearly Gaussian the distributions; (2) the more intense the analyte distribution and the less intense the interfering one; (3) the fewer, weaker, narrower, and more separated the escape peaks of diverse elements; and (4) the lower the continuous background.

5.2.3.6. Automatic Pulse-Height Selection

Because mean pulse height is proportional to x-ray photon energy, for each wavelength to be measured, different settings are required for the baseline and window, or for the amplifier gain and/or detector potential. Such resetting is very inconvenient when several analyte lines are to be measured on a number of specimens, and is intolerable in automatic instruments and in recording of 2θ scans. Accordingly, methods have been devised to couple the pulse-selection function to the θ–2θ drive of the spectrogoniometer so as to vary the pulse-selection conditions automatically and synchronously with 2θ, that is, with wavelength. With such automatic pulse-height selectors, a 2θ scan gives a pure first-order spectrum.

The methods for effecting this function are of two basic types—those that vary the pulse height and those that vary the baseline, or, in effect, those that move the pulses to the window and those that move the window to the pulses. The two approaches are best discussed in terms of Equation (5.17), $\bar{V} = (E_x/V_{i,\text{eff}})G_D G_A$, where $\bar{V}$ is mean pulse height; E_x is x-ray photon energy; $V_{i,\text{eff}}$ is energy to produce one useful electron in the detector; G_D is internal electron multiplication in the detector; and G_A is external amplifier gain.

In the first method, the baseline and window are constant, and, as 2θ varies, the amplifier gain G_A is varied to amplify the pulses to different degrees and move them into the preset window. The same effect would result from variation of detector potential, and, thereby, of G_D. However, G_D is highly dependent on the specific detector, and this method is avoided. For a specified detector, $V_{i,\text{eff}}$ is constant if the detector potential and therefore G_D are constant. Then, from Equations (5.17) and (2.21), the Bragg law,

$$V \propto E_x G_A, \qquad E_x \propto 1/\lambda, \qquad \lambda \propto (2d \sin \theta)/n \qquad (5.19)$$

Then,

$$G_A \propto V(2d/n) \sin \theta \qquad (5.20)$$

For a specified analyzer crystal and order n of λ, $2d/n$ is constant. Then,

$$G_A \propto V \sin \theta \qquad (5.21)$$

Consequently, if G_A is varied sinusoidally with θ, all pulse-height distributions occur at V. Amplifiers designed to effect this function are known as *sine-function amplifiers*.

However, if the crystal ($2d$) or spectral-line order n is changed, all pulse-height distributions are displaced to some new constant value V'. If the original pulse-height selector settings are to be retained, the average pulse height must be moved from V' to V by changing the amplifier gain as a function of $2d/n$. Consequently, in semiautomatic and automatic spectrometers, coupling of the amplifier gain and 2θ necessitates provision for automatic change of the amplifier gain also with selection of the analyzer crystal and order. In the following examples, Ni $K\alpha$ [1.659 Å, 7.47 keV, 48.67° 2θ (LiF)] is used for illustration, and it is assumed that the amplifier gain–2θ coupling is calibrated for LiF; that is, when the goniometer is at 48.67° 2θ, the amplifier gain is set automatically to amplify pulses from 1.659-Å (7.47-keV) photons to the correct amplitude to pass the preset pulse-height selector.

Suppose now that second-order Ni $K\alpha$ [2λ 3.318 Å, 2θ (LiF) 111.00°] is to be measured. But when the goniometer is set at 111.00° 2θ, the amplifier gain is set automatically to amplify pulses originating from 3.318-Å (3.47-keV) photons, not Ni $K\alpha$ photons. Since these photons are only half as energetic as Ni $K\alpha$ photons, the amplifier will assume automatically twice the gain it would have as 48.67° 2θ (Ni $K\alpha$). Thus, provision is required to halve the amplifier gain automatically when the order selector is set at $n = 2$, or, in general, to change the gain by $1/n$ when the order selector is set at n.

Similarly, if Ni $K\alpha$ is to be measured with, say, LiF(420), PET, or gypsum crystals, the peak is diffracted at a different 2θ angle for each, as shown in line 3 of Table 5.5. Incidentally, in order to show the relationships more simply, these particular crystals were chosen because they have $2d$ spacings respectively $\sim 1/2$, ~ 2, and ~ 4 times that of LiF(200). However, at each of these 2θ angles, the amplifier gain will be set automatically to amplify pulses originating from photons of the wavelength that would be diffracted at that angle by LiF(200) (line 4 of the table). The photon energies for these wavelengths are given in line 5 and are different from that for Ni $K\alpha$. Thus, the gain would be different from that required for Ni $K\alpha$. The photon energies are in the same ratio as the $2d$ values (line 6), and the required amplifier gain is proportional to $1/2d$ (line 7). Thus, provision

TABLE 5.5. Relationship of Required Amplifier Gain and Crystal 2*d* Spacing for Automatic Pulse-Height Selection

Crystal	LiF(420)	LiF(200)	PET	Gypsum
2*d* (Å)	1.80	4.028	8.742	15.185
2θ for Ni $K\alpha$ (deg)	134.37	48.67	21.88	12.54
λ, wavelength diffracted at above 2θ by LiF(200) (Å)	3.712	1.659	0.764	0.440
E_x, photon energy at above wavelength (keV)	3.34	7.47	16.22	28.17
R, ratio of 2*d* (or E_x) to 2*d* (or E_x) for LiF(200)	0.447	1.00	2.17	3.77
Required relative amplifier gain for λ, $1/R$	2.24	1.00	0.461	0.265

must be made to vary the amplifier gain automatically as a function of 2*d* when the crystal changer is actuated to select a different crystal.

A disadvantage of the sine amplifier arises when it is applied over very wide x-ray wavelength regions. The sine amplifier maintains constant pulse height at the pulse selector input, regardless of 2θ, by varying the amplifier gain while leaving detector potential constant. Now for measurement of F $K\alpha$ to Cl $K\alpha$, relatively high detector potential is preferable to take advantage of the relatively noise-free internal amplification of the detector. Conversely, at shorter wavelengths, relatively lower potential is preferable to avoid running the detector into the nonproportional region. Consequently, at least one manufacturer supplies a spectrometer in which detector potential, rather than amplifier gain, is linked to the 2θ drive, so that potential varies inversely as $\sin\theta$.

In the other method of automatic pulse-height selection, as 2θ varies, the pulse-height distributions of the spectral lines are allowed to undergo their natural variation in mean height and in width. The baseline, and perhaps also the window width, are varied to receive them. Again, for a specified detector and potential, $V_{i,\text{eff}}$ and G_D are constant. Then, if G_A is also constant,

$$V \propto E_x \propto 1/\lambda \propto 1/\sin\theta \qquad (5.22)$$

Consequently, if, as θ varies, the baseline is varied in proportion to $1/\sin\theta$, the baseline follows the pulse-height distributions.

As x-ray photon energy increases, the window width required to pass the pulse distribution increases (Figure 5.19), and if the width remains constant, some analyte-line pulses may be excluded. Similarly, as photon energy decreases, the required window width decreases, and if the width remains constant, some unnecessary background pulses or pulses from an adjacent distribution may be admitted. Thus, an added refinement to any of the types of automatic pulse selector described above would be an arrangement to vary the window width to correspond with the variation in the width of the pulse distribution with 2θ. For instruments in which amplifier gain and detector potential are linked to the 2θ drive, the baseline and window width should be made to vary as $1/\sin\theta$ and $1/(\sin\theta)^{1/2}$, respectively. One commercial instrument provides two window widths—one narrow, one wide—which are automatically selected on selection of the flow-proportional and scintillation counter, respectively, and which can be selected manually for low and high photon energies, respectively.

5.2.3.7. Pulse-Height Selection Problems

If the detector is receiving a monochromatic x-ray beam and the pulse-height selector is not used, each pulse resulting from an incident x-ray photon is counted, provided only that it is higher than the discriminator baseline. It does not matter whether the pulse is a photo or escape pulse, or whether the pulse-height distribution has shifted or become distorted. However, when the pulse-height selector is used, many conditions become troublesome that would otherwise be insignificant. These pulse-height selector problems may be classified in four categories: (1) pulse-height distribution shift; (2) pulse-height distribution distortion; (3) additional pulse-height distributions arising from the measured x-ray line; and (4) additional pulse-height distributions arising from other x-ray lines.

1. Pulse-Distribution Shift. This effect is shown in Figure 5.25A. Suppose that a pulse-height selector window is set to pass the distribution represented by the solid outline. If the distribution shifts, as shown by the dashed outline, those pulses in the shaded area now lie outside the window and are not counted, so the measured intensity decreases proportionally. Such shifts are always caused by changes in amplification, either internal (detector) or external (amplifier). Proportional counters are more prone to shift than scintillation counters. In proportional counters, the effect is greater for flow counters (compared with sealed counters), high detector potential, high-atomic-number detector gas, small and/or nonuniform

FIGURE 5.25. Effect of shift and distortion of the pulse-height distribution. (A) Shift. (B) Distortion. (C) Intensity effect—shift caused by dead-time effect. The detector is assumed to be free from dead-time losses up to 10,000 counts/s. At lower intensities, the detector output intensity decreases proportionally and without change in pulse height. At higher incident intensities, the detector output intensity increases at a progressively slower rate than the incident intensity, and the entire pulse-height distribution shifts to smaller average pulse height.

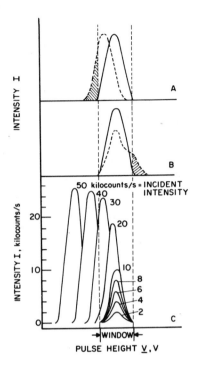

anode wire diameter, and stainless steel (compared with tungsten) anode wire. Modern solid-state amplifiers usually have very stable gain, so shift usually arises from change in internal detector amplification. The most common causes are the following.

A *detector-potential drift* of only 1% may cause a change of $\lesssim 10\%$ in mean pulse height. The obvious remedy is high regulation ($\pm 0.05\%$ or less) of the detector power supply.

The *detector gas-density effect* arises from variations in ambient (room) temperature and/or pressure. Typically, a change of 1% in either may cause a change of $\lesssim 5\%$ in mean pulse height, and typical laboratory variations of 3% in (absolute) temperature and 7% in pressure may cause shifts of $\lesssim 20\%$ and $\lesssim 40\%$, respectively. Fortunately, temperature and pressure changes tend to compensate one another. The best remedies are temperature-controlled ($\pm 0.5°C$) spectrometer chambers and discharging flow counters into gas-density compensating manostats rather than directly into the atmosphere.

The *detector-gas composition effect* arises from slight changes in the, say, argon/methane concentration ratio as the gas cylinder empties. The methane concentration in P10 gas (90 vol% argon, 10 vol% methane) may decrease by 0.5 vol% between full and near-empty cylinders, and this may cause a change of 10–20% in mean pulse height.

The *intensity effect* arises when the incident intensity is so high that photons enter the detector between the resolving and recovery times (Figure 5.11), so the pulses are above the baseline and counted, but not of normal height. The effect is illustrated in Figure 5.25C. The outline marked 10 represents the pulse-height distribution for the maximum intensity at which no significant decrease in mean pulse height occurs, say, 10,000 counts/s. If now the incident intensity is increased incrementally to 20, 30, 40, and 50 kilocounts/s, the pulse distribution undergoes two changes: (1) The intensity of the entire distribution increases, but each 10-kilocount increase gives a progressively smaller increase in integrated detector output intensity (area under the curve); and (2) the entire distribution shifts to progressively lower mean pulse height, so that an increasing portion of the pulse distribution moves out of the window. Conversely, if the incident intensity is decreased decrementally to 8, 6, 4, and 2 kilocounts/s, the distribution decreases in integrated intensity in proportion to the incident intensity and remains constant in mean pulse height. The effect may be serious in flow counters at intensities >20,000 counts/s, in sealed proportional counters >50,000 counts/s. However, scintillation counters may remain substantially free from the effect up to 10^5 counts/s. The most obvious remedy is to reduce incident analyte-line intensity by reducing excitation conditions (kV, mA), using a less intense line ($K\beta$ instead of $K\alpha$), or using a higher-order line. If a tandem detector is used, the signal may be taken from the *less* favorable one for the analyte line.

   2. Pulse-Distribution Distortion. This effect is shown in Figure 5.25B. Again, suppose that a pulse-height selector window is set to pass the distribution represented by the solid outline. If the distribution becomes distorted, that is, departs from its Gaussian profile as shown by the dashed outline, those pulses in the shaded area now lie outside the window and are not counted. Distortion is caused by heterogeneity of the electric field in the detector arising usually from window effects, anode-wire end effects, and dirt, corrosion, and erosion along the wire. The effect is most common in flow counters, and end-window counters are more prone to it than side-window counters. The effect is certain to occur if the inside surfaces of Mylar and other plastic windows are not aluminized or carbonized. The anode-wire damage is caused by dirt and traces of moisture and/or oxygen in the gas, and by reduction or polymerization of the hydrocarbon (methane, etc.) component of the gas on the anode wire. Remedies include specification of high-purity detector-grade gas, use of filters and/or chemical absorbers for oxygen and moisture in the gas line, and substitution of carbon dioxide for methane in the gas mixture.

Because of the shift and distortion effects, it is wise to set the pulse-height selector window somewhat wider than the minimum width required to pass the distribution.

3. Additional Pulse Distributions Arising from the Measured Line. In addition to its own photo pulse-height distribution peak, the analyte line may initiate *escape peaks*, and *double* and *sum peaks*.

The origin of *escape peaks* (and of the term) is explained in Sections 5.1.3.3 and 5.1.4, and Figure 5.5. The intensity of the escape peak relative to the photo peak increases with: (1) proximity of the incident wavelength to the short-wavelength side of the absorption edge of the detector medium (gas, iodine, silicon); (2) fluorescent yield of the detector medium; (3) opacity (high absorption coefficient) of the detector to the analyte line; (4) transparency (low absorption coefficient) of the detector to its own line; and (5) decreasing detector thickness. As analyte-line wavelength decreases (photon energy increases), the separation of photo and escape peaks decreases, making it easier to include them both in the window. As wavelength increases (photon energy decreases) and approaches the long-wavelength limit of the detector's useful region, the escape peak is likely to lie outside this limit, where, although not counted, at least it causes no interference.

In general, escape peaks are not a serious problem in analytical x-ray spectrometry. In argon proportional counters, elements of atomic number 18 (argon) and below do not excite the argon K-absorption edge, and for elements of atomic number 28 (nickel) and above, the escape and photo peaks are not resolved. Thus, escape peaks may be troublesome only for elements 19–27 (potassium through cobalt). Even for these, the low argon K fluorescent yield (~ 0.1) results in relatively weak escape peaks, and the potassium and calcium escape peaks may well be lost in the detector noise. However, if these escape peaks do interfere with an analyte photo peak, use of propane detector gas may provide the remedy. For example, when an argon detector is used for determination of fluorine in the presence of calcium, the Ca K escape peak interferes with the F $K\alpha$ photo peak. The interference does not occur in propane detectors.

In xenon proportional counters and NaI(Tl) scintillation counters, the LIII absorption edges (both ~ 2.6 Å) lie at or beyond the usual spectral regions of the detectors. Thus, x-ray lines most efficient in exciting these edges give very low-energy escape peaks that are lost in the detector noise.

Although such escape peaks are lost for counting purposes, at least they cannot interfere with other, useful, pulse distributions. In both detectors, the K edges (both $\sim$0.36 Å) are excited only by the K lines of elements of atomic number $\sim$57 (lanthanum) or above. The K lines of these elements are excited relatively inefficiently by a 50-kV generator, and the elements are usually determined by use of their L lines. However, the K fluorescent yields of xenon and iodine are high ($\sim$0.9), and, when escape peaks do occur, they have two to three times the intensity of the photo peak.

In the krypton proportional counter, the K edge (0.86 Å) is excited by elements of atomic number $\sim$38 (strontium) and above, and the K fluorescent yield is relatively high ($\sim$0.6). Thus, the krypton detector is the most prone to escape peaks of all the common detectors, but even it has a wide useful range.

Escape peaks, like photo peaks, arise from the x-ray spectral line being measured, so, in general, the pulse-height selector window should be set to admit both. However, this is not feasible if: (1) the escape peak has such low pulse height that it is lost in the detector noise; (2) the two peaks are too widely separated; (3) too many background pulses are admitted; or (4) a photo or escape peak of another element is wholly or partially admitted. Sometimes, even the last three of these difficulties can be surmounted if two pulse-height selectors are available and set for the photo and escape peaks, respectively. In this way, background and interfering peaks between the two peaks of interest are rejected.

Escape peaks may be troublesome in two ways. If, for any of the reasons given above, it is not feasible to include both the photo and escape peaks in the window of a single-channel analyzer, only one can be measured. Then the measured intensity is reduced by the intensity represented by the rejected peak. Also, escape peaks complicate the energy spectrum and increase the probability of overlapping.

Double or *sum peaks* arise when n identical x-ray photons enter the detector simultaneously, or very nearly so, giving a pulse of n times normal amplitude. If this happens often enough, a second pulse distribution appears in the detector output at mean amplitude n times that of the normal peak, known as a *double* ($n = 2$) or *sum peak*. If the selector is set to admit the normal peak, the number of photons detected but not counted is twice the number of double pulses, thrice the number of triple pulses, etc., that is, n times the number of n-sum pulses. Even without the selector, there is a loss equal to the number of double pulses, twice the number of triple pulses, etc., that is, $n - 1$ times the number of n-sum pulses. Moreover, two or more simultaneous pulses of amplitude too low to be counted—that is, below the baseline—may sum to give a countable pulse and an erroneous net gain of one pulse. Sum pulses may be difficult to understand in view of the dead-time phenomenon. However, the problem is resolved if one considers that, in effect, two concurrent photons are equivalent to a single photon of twice the energy. A photon entering the detector after the quench process has begun is, of course, subjected to the dead-time effect.

4. Additional Pulse Distributions Arising from Extraneous Lines. There are several sources of such distributions, but the only one that causes problems specifically with pulse-height selection is *crystal emission.* Scattered primary and emitted secondary x-rays incident upon the analyzer crystal excite the elements in the crystal to emit their characteristic spectra. If the crystal is purely organic (EDDT, PET, etc.), these spectra are of little significance in conventional x-ray spectrometry. If the crystal contains a heavier element, such as aluminum (topaz), silicon (quartz, silicon), phosphorus (ADP), sulfur (gypsum), chlorine (NaCl), or potassium (KHP), the spectra still do not interfere if high-Z elements are being determined in air path, or if low-Z elements are being determined in helium or vacuum *without* pulse-height selection. In the former case, the air absorbs the crystal emission before it reaches the detector. In the latter case, since the crystal emission radiates in all directions, it merely contributes to the continuous background. However, if the pulse-height selector is used, that portion of the crystal emission that does enter the detector gives rise to a pulse-height distribution that may interfere with the measured distribution.

5.2.4. Ratemeter–Recorder

The ratemeter provides readout and display of x-ray intensity data in instantaneous analog mode—like an automobile speedometer. It indicates the mean value of the current consisting of the individual pulses from the discriminator or pulse-height selector accumulated over a certain selected short time interval determined by the time constant. This mean current is proportional to the instantaneous x-ray intensity incident on the detector and is displayed as intensity or count rate on a panel meter or strip-chart recorder. The ratemeter may have linear or logarithmic response. Linear response reduces tailing of the peaks and is preferable for estimation of intensity. Logarithmic response may be advantageous for recording spectra having both intense and weak peaks. Wavelength-dispersed spectra recorded in linear and logarithmic mode are shown in Figure 7.2. The ratemeter–recorder combination is used principally to record the x-ray spectrum for qualitative or semiquantitative analysis, or preparatory to quantitative analysis. Its use for these purposes is described in Section 7.1.1.

Other applications of the ratemeter–recorder panel meter system in x-ray spectrometry include the following: (1) quantitative analysis in certain applications where high accuracy is not required; (2) rapid sorting of specimens on the basis of presence or absence of a key element, or high or low concentration of such an element; (3) mapping of analyte-line intensity-distribution profiles in a line across the specimen to evaluate concentration distribution or thickness distribution of thin films; (4) line and raster scans in electron-probe microanalysis (Figure 11.6); (5) in selected-area analysis and electron-probe microanalysis, for searching over a specimen for regions containing a specified analyte; the spectrometer is set at the analyte line, and the specimen is moved as intensity is observed on the panel ratemeter or recorder; (6) alignment and "peaking" of the spectrogoniometer (Section 4.2.4); (7) slow scanning of peaks to reveal their profiles and possible spectral overlap; (8) recording of a peak while a continuous or stepwise integrated count is made (Section 5.3.1); (9) evaluation of short- and long-term instrument stability by recording spectral-line intensity as a function of time; (10) recording pulse-height distribution curves (Sections 5.2.3.3, 5.2.3.4); and (11) use as a sensitive, fast-response intensity meter to be observed as various instrument conditions are varied.

5.2.5. Scaler–Timer

The *scaler* or *count register* provides readout and display of x-ray intensity data in integrated or accumulative digital mode—like an automobile odometer.

The *timer* or *timer register* is a clock, but it keeps time by counting the positive pulses from the sine-wave output of a high-frequency crystal oscillator. Their rate of arrival is determined by the frequency of the oscillator and is, say, 1000 Hz, that is, 1000 positive pulses per second. Their time distribution is perfectly uniform, and, at 1000 Hz, they are spaced at 1-ms intervals.

Both the scaler and timer, then, are electronic binary or decimal counters, and if there is any essential difference between them, it is that the scaler must be capable of counting at extremely high rates, whereas the timer need not count faster than $\sim$1000/s. Thus, the scaler and timer are "fast" and "slow" counters, respectively.

Figure 5.18 shows, very schematically, a scaler and timer, each having a six-decade decimal counter and provision for twofold or fourfold binary multiplication.

A *binary circuit* may be regarded as one that passes every second pulse it receives; that is, it blocks pulses 1, 3, 5, ... and passes pulses 2, 4, 6, ... Usually, provision is made to indicate which condition the circuit is in. When it has received and blocked a pulse, a small neon lamp glows until the next pulse is received and passed.

A *decimal circuit* may be regarded as one that passes every 10th pulse it receives, blocking—or rather, storing—pulses 1–9. Provision is made to indicate how many pulses are accumulated in the circuit. In the EIT decimal indicator tube, a fluorescent spot advances sequentially from 0 to 9 on a numerical scale. Modern instruments display the count and time numerically using neon Nixie or solid-state light-emitting diode (LED) indicators.

In Figure 5.18, suppose that both the scaler and timer multiplier switches are at $\times 1$, that all of the preset-count and preset-time switches are open, and that both start–stop switches are simultaneously closed (start). The analyte-line pulses from the pulse-height selector pass directly to the first scaler decade, and the pulses from the crystal oscillator pass directly to the first timer decade.

The first nine analyte-line pulses are accumulated in and indicated by the first scaler decade. Pulse 10 passes to the second decade, causing it to indicate 1 and returning the first decade to zero. Pulses 11–19 accumulate in the first decade, and pulse 20 passes to the second. This sequence continues until both decades are filled by pulse 99. Pulse 100 then passes to the third decade, returning the first two to zero. Thus, the individual decades indicate, respectively, individual pulses, 10's, 100's, 1000's, etc., as shown.

Meanwhile, the timer is counting positive pulses from the crystal oscillator in the same way, but since these pulses arrive at exact 1-ms intervals, the timer is, in effect, counting milliseconds of elapsed time. However, the decades are labeled 0.001, 0.01, 0.1, 1, ... *seconds*, rather than 1, 10, 100, 1000, ... *milliseconds*.

Since the scaler and timer were started together, if they are now stopped together at any arbitrary time, the accumulated count and counting time may be read on the scaler and timer, respectively. The two instruments are stopped by opening the start–stop switches.

If it is intended to make a preset-count measurement of, say, 10^4 counts, the switch following the 10^3 scaler decade, that is, between the 10^3 and 10^4 decades, is closed. The scaler and timer are started simultaneously. The action is the same as before up to the 9999th pulse. However, the 10,000th pulse not only passes to the 10^4 decade, but also passes through the switch and actuates the stop circuit, opening both start–stop switches and stopping the scaler and timer. The counting time is then read on the timer.

Similarly, if it is intended to make a preset-time measurement of, say, 100 s, the switch following the 10 timer decade is closed. The first 99,999 millisecond pulses fill the first five timer decades. The 100,000th pulse (100.000 s) stops the scaler and timer. The accumulated count is then read on the scaler.

Now, suppose that the scaler multiplier switch is set at $\times 2$. The analyte pulses no longer pass directly to the decimal counter, but first pass through a binary stage (marked "2" in Figure 5.18), which passes only every second pulse. The action is the same as before, but the accumulated count in the scaler must be multiplied by 2. Similarly, if the multiplier switch is set at $\times 4$, the incoming pulses pass through two binary stages in series, so that only every fourth pulse reaches the counter, and the accumulated count must be multiplied by 4. Binary multipliers are also shown on the timer. Thus, the scaler and timer in Figure 5.18 have capacities of 1, 2, and 4×10^6 counts and 1, 2, and 4×10^3 s, respectively.

Provision is also made to clear the scaler and timer, that is, reduce all decades to zero, prior to starting a new measurement.

Another way to measure intensity is to allow the pulses to accumulate in a capacitor for a certain interval, then measure the accumulated charge. This method is inexpensive but has the disadvantage that the total accumulated count is not known, so that calculation of counting error is less convenient than for a scaler. This method is used in Applied Research Laboratories automatic x-ray spectrometers.

A printer may be provided and programmed to print out the analytical data—sample number; symbol of analyte; and line, background, and net intensities—on a paper strip.

5.2.6. Computer

Computers are widely used in modern wavelength-dispersive x-ray spectrochemical analysis in three ways: (1) X-ray intensity data can be copied or printed out, then input to a computer for calculation of analytical concentration. (2) The intensity data can be input to the computer directly; analytical concentrations are then computed from the stored data. (3) The computer can be programmed to control automated spectrometers, including indexing specimens into measuring position, and for each specimen, selection of 2θ angles for a number of analyte lines and backgrounds, selection of excitation conditions (kV, mA), crystal, collimator(s), detector, pulse-height selector settings, etc. The use of computers in energy-dispersive spectrometers is discussed in Section 6.1.3. The computer may be "dedicated" (an integral part of the spectrometer), a separate laboratory computer, or a large, central, time-shared computer. The more sophisticated the functions and the larger the volume of data the computer must deal with, the larger its capacity must be.

5.3. X-RAY INTENSITY MEASUREMENT

5.3.1. Measurement Techniques

In principle, the *integrated intensity* under the analyte-line peak should be measured. In energy dispersion, where all x-ray photons are admitted to the detector simultaneously, integrated intensity is measured readily. But in wavelength dispersion, where the crystal diffracts only a discrete wavelength to the detector, integrated intensity measurement is inconvenient and time-consuming. However, the *peak intensity*, at the 2θ angle of maximum intensity, is representative of the integrated intensity and is measured conveniently. So in wavelength dispersion, it is almost universal practice to measure peak intensity, except for certain special cases. Another advantage of measurement at the peak is the relative insensitivity of the measured intensity to slight error in 2θ setting.

Before any x-ray intensity measurements are undertaken, all excitation, detection, and readout components should be allowed to warm up for 30–60 min. The x-ray tube should be operated at or near the excitation conditions to be used. The operating potential should be applied to the detector, and for flow counters, the gas should be flowing at its proper rate.

Intensity can be measured with the ratemeter panel meter when only estimates are required—for example, in distinguishing 35–65 and 50–50 (wt%) nickel–iron alloys.

Intensity is frequently measured with the ratemeter–recorder for semiquantitative analysis and for quantitative analysis when high accuracy is not required. Three ratemeter–recorder techniques are commonly used for x-ray spectrochemical intensity measurements; two of these require recording of the entire analyte peak for each sample and standard. The least accurate technique is measurement of peak height above background. A more accurate technique is measurement of the area under the peak, but above the background, with a planimeter or by counting squares on the recorder chart paper. In the most accurate technique, the goniometer is set at the 2θ angle for the analyte line and left undisturbed; the recorder traces analyte peak intensity for several centimeters from each sample and standard. Background is recorded the same way at a 2θ angle adjacent to the peak. A straight line is drawn through the mean of the recorder "noise." In all three methods, measurements are made above background, and calibration curves or mathematical relationships are established from the standard data.

For all three recorder methods, the ratemeter should be in linear, rather than logarithmic, mode and should be accurately calibrated. Also, insofar as practical, all samples and standards should be recorded at the same full-scale intensity setting and other instrument settings, and the analyte peaks from all samples and standards should have as nearly as possible the same intensity. In the two methods for which peaks are recorded, the 2θ scanning rate should be slow and must be the same for all specimens. The time constant should be fast enough to indicate true peak intensity, but slow enough to dampen excessive recorder fluctuations.

However, intensity is usually—and most accurately—measured digitally with the scaler–timer components in one of five ways: the preset-time, preset-count, integrated-count, ratio, and monitor methods; the first two methods are the most commonly used.

In the *preset-time method*, the timer is set for a specified time T long enough to permit accumulation of enough counts for the required statistical precision. The timer and scaler are started simultaneously, operate until the preset time has elapsed, then stop simultaneously. The number of counts accumulated N is read on the scaler. Intensity $I = N/T$. If the same preset time is used for all samples and standards, the accumulated counts N may be used in calculations and calibrations instead of intensities. If this is done, and background corrections are made, the same preset time must be used for them also.

In the *preset-count method*, the scaler is set for a specified number of counts N large enough to give the required statistical precision. The timer and scaler are started simultaneously, operate until the preset count has accumulated, then stop simultaneously. The elapsed time T is read on the timer. As before, intensity $I = N/T$. If the same preset count is used for all samples and standards, the elapsed times may be used in calculations and calibrations instead of intensities. If this is done, and background corrections are made, the same preset count must be used for them also.

The *integrated-count method* may be conducted in either continuous or stepwise mode. In the continuous mode, the spectrometer is set at a 2θ angle on one side of the peak to be measured (*a* in Figure 9.2). The 2θ drive, timer, and scaler are started simultaneously, operate until the entire peak has been scanned (*a—b* in Figure 9.2), then are stopped simultaneously. The accumulated count and elapsed time are read. The integrated count or intensity may be used. In the stepwise mode, the spectrometer is moved manually or with a step scanner in small, equal-2θ increments, and a preset-time or preset-count measurement is made at each position. The integrated-count method is used when an accurate peak profile is required—to deal with spectral overlap, for example.

It is evident that the preset-time, preset-count, and integrated-count methods are simply ways of defining the duration of the counting time. In the three techniques, the count is continued for the duration of, respectively, a preset time, the time to accumulate a preset count, and the time to scan through the analyte peak. Two other ways to define the counting time are the ratio and monitor methods.

In the *ratio method*, the time to accumulate a specified count from the standard defines the time during which the analyte-line count is accumulated from the sample. In effect, a preset count N_S is made from the standard, during which an "up-and-down" timer "winds up" from time 0 to T_{N_S}, then "runs down" as counts accumulate from the sample. The standard count may be: (1) the analyte line measured from a calibration standard or from pure analyte; (2) an internal standard line measured from the sample itself; (3) a scattered target line or primary-continuum wavelength measured from the sample; or (4) a scattered target line or continuum wavelength measured from a separate scatter specimen. These methods are widely used on automatic spectrometers.

When a line having very low intensity is to be measured, a very long counting time is required to accumulate enough counts to give reasonable statistical precision. During such long times, the count is also affected by fluctuations in excitation and other instrument conditions. Thus, correlation of the measurements from a series of specimens may be meaningless. The *monitor method* provides one way to deal with this condition. A small portion of the primary x-ray beam is diverted, usually by a small scatter target, directly into a second detector–scaler system. The spectrometer and monitor counts are started simultaneously and terminated when the monitor has accumulated a preset count. All samples and standards are counted for the time for the monitor to accumulate this same preset count.

5.3.2. Background

5.3.2.1. Nature

Background is the intensity that would be measured at the 2θ angle of the analyte line (wavelength dispersion) or in the analyte-line window (energy dispersion) from a specimen identical with the analytical specimen in all respects except that the analyte is absent. Background consists of continuum and possibly wholly or partly overlapping spectral lines. It is significant in three ways: (1) Usually it must be subtracted from the analyte and internal-standard peak intensities. (2) It limits the minimum detectable amount of analyte, which may be defined as that amount that gives a net analyte-line intensity of three times the square root of the background intensity. (3) Finally, it may be used to compensate certain types of error. A practical valid method of background measurement is one of the two most frequent obstacles to a quantitative x-ray spectrochemical analysis. (The other obstacle is availability of suitable standards.)

The principal sources of background intensity I_B are: (1) primary continuous and spectral-line x-rays from the source (x-ray tube or radio-isotope) scattered by the specimen $I_{\text{pri,sc}}$; (2) specimen spectral emission scattered by the crystal $I_{C,\text{sc}}$; and (3) crystal emission $I_{C,\text{em}}$ excited by the scattered primary and emitted secondary x-rays incident on the crystal. Then,

$$I_B = I_{\text{pri,sc}} + I_{C,\text{sc}} + I_{C,\text{em}} \qquad (5.23)$$

5.3.2.2. Measurement

If a calibration curve of analyte-line intensity *versus* concentration is plotted, it may well be unnecessary to measure and correct for background. Peak intensities from the samples are applied to a curve of peak intensities from the standards. If such a curve is linear, its intercept on the intensity axis is the background intensity (see Figure 9.1).

Figure 9.2 shows a composite typical x-ray wavelength spectrum (2θ scan) showing emitted and scattered x-rays that may be used for or interfere with an analysis. The significance of each feature is given in the caption. The analytical aspects of the figure are discussed in Section 9.2.1. Here, the figure serves to illustrate several ways to measure background. In the following discussion, the notation is that of Figure 9.2.

If the analyte line A is superimposed on uniform continuous background, background is measured adjacent to the peak, B_{adj}.

If the analyte line A' is superimposed on nonuniform continuous background, background is measured at equal-2θ intervals on the low-2θ B_L' and high-2θ B_H' sides of the peak. Then,

$$B_P' = (B_L' + B_H')/2 \qquad (5.24)$$

If the background profile is strongly nonuniform in the region of the analyte peak, it may be necessary to measure background at two or more angles on each side of the peak, plot the profile, and read the background at the peak from the curve.

If the analyte line A'' is superimposed partly or wholly on a scattered x-ray tube target line T, background is measured in one of three ways: (1) B_P'' may be measured from a blank at the 2θ angle for the analyte line. (2) From measurements on blanks, the intensity ratio can be established between the interfering line T and another scattered target line that can be measured without interference, say T_R; then measurement of T_R on each sample and standard permits calculation of T. (3) A substantially constant relationship usually exists between adjacent background B_{adj}'' and the background at the peak B_P'', which, in this case, includes T. Then B_P'' may be calculated from measured B_{adj}'':

$$B_P'' = kB_{adj}'' \tag{5.25}$$

The constant k is evaluated from measurements at the 2θ angles of B_P'' and B_{adj}'' on suitable blanks containing no analyte.

If the analyte line A''' is superimposed partly or wholly on a line of a specimen matrix element M, background, including M, may be measured from a blank. Alternatively, the intensity ratio can be established between the interfering line M and another line of the same matrix element that can be measured without interference M'. Then measurement of M' on each sample and standard permits calculation of M.

If the peak intensity is measured by a continuous or stepwise integrated count $(a-b)$, background may be measured by an integrated count on an equal-2θ interval of adjacent uniform background $(a'-b')$, or by a count at B_{adj} for a time interval equal to that required for the integrated peak count. If the background is nonuniform, this equal-time interval may be divided between equal-2θ intervals on the high- and low-2θ sides of the peak.

Reference is made above to the use of blanks. A blank is usually defined as a specimen identical with those to be measured in composition and physical form, except that the analyte is absent. In practice, if the analyte is a trace constituent, or if its atomic number is near the average atomic number of the remaining matrix, this definition may apply rigorously. However, if the analyte is a minor or major constituent having atomic number significantly different from that of the matrix, its omission may result in a change in scattering power of the remaining matrix. The change may be to higher or lower scattered intensity, depending on the relative values of the atomic number of the analyte and effective atomic number of the matrix. In such cases, a correction factor may be applied, or the analyte may be replaced in the blank with an equal concentration of an element having about the same scattering power, but no interfering spectral lines. An element having atomic number one or two higher or lower than the analyte may be suitable. Sometimes one can use for a blank a pure chemical element having atomic number about equal to the effective atomic number of the specimens—but having no spectral lines at the analyte-line wavelength.

For specimens consisting of microgram or milligram amounts of analytical material supported on thin Mylar film, Millipore or Nuclepore filters, filter paper, ion-exchange membrane, etc., almost invariably background can be measured from clean substrate at the 2θ angle of the analyte line.

In the measurement of the low intensities from trace elements, long counting times may be required with consequent loss of precision, which is due mostly to fluctuations in the x-ray generator. In such cases, the error is decreased substantially by simultaneous measurement of line and background intensities. This is no problem with multichannel spectrometers, but is impossible with unmodified single-channel instruments. However, the counting time can be alternated between peak and background.

The foregoing discussion is in terms of wavelength dispersion, but analogous techniques are applicable to energy dispersion. However, several other sophisticated electronic methods of background measurement are applicable to energy dispersion that have no analog in wavelength dispersion (Section 6.1.3.3, Figures 6.2F, 6.4, 6.6B).

5.3.2.3. Reduction

In general, background intensity is reduced by the following conditions:

1. X-ray tube target atomic number is as low as possible.
2. Excitation conditions (kV, mA) are as low as possible.
3. Filters are used to remove target lines or to reduce continuum intensity in the region of a specific analyte line.
4. The specimen compartment is evacuated or flushed with helium to eliminate air scatter.
5. The specimen compartment is fitted with baffles, primary and/or secondary apertures, or a collimated specimen mask to minimize background from air in the specimen compartment, the surface outside the specimen area of interest, or other structures in the specimen compartment.
6. Matrixes of low atomic number are avoided, such as solutions, fusion products, and glass disks.

7. Small specimens and small selected areas on large specimens are measured in a collimated specimen holder or by selected-area techniques.

8. Very small specimens are supported at the ends of fine plastic or quartz fibers in free space.

9. Milligram to submicrogram quantities of analytical material are supported on thin substrates; background, and therefore detection limit, decreases successively for filter paper, Millipore, 6- and 3-μm (0.00025- and 0.00012-inch) Mylar, and ultrathin Formvar or collodion substrates.

10. Such thin-film substrates are supported on open frames or stretched over the tops of empty liquid cells or other cylindrical supports; primary-beam apertures or special support-cell geometries are used to prevent primary-beam irradiation of the inside cell walls to avoid scatter from the walls; and primary x-ray beam traps are used behind the specimen (opposite the x-ray tube).

11. Low-Z binders or diluents and plastic cell windows are avoided for powder specimens.

12. Liquid specimens are frozen or held in open-top cells, again to avoid cell-window covering.

13. It is possible for primary continuum x-rays to scatter from depths greater than that from which analyte-line x-rays can emerge from the specimen—that is, the effective layer thickness for scattered background can exceed that for the emitted analyte line. In such cases, peak-to-background ratios can be increased by reduction of specimen thickness to about the effective layer thickness for the analyte line. Further reduction in thickness results in decreased analyte-line intensity.

14. High-order lines are used; these lines lie at larger 2θ angles where background is lower.

15. The crystal contains no element(s) of relatively high atomic number to produce secondary emission.

16. Collimator foils are treated to minimize reflection and scatter.

17. Fine collimation is used.

18. A slit is placed at the detector window (in curved-crystal spectrometers).

19. The crystal axis is not parallel to the specimen plane; this discriminates the polarized component of the scattered background.

20. A gas-filled detector is used, rather than a scintillation counter, to avoid multiplier phototube noise.

21. If a scintillation counter is used, the multiplier phototube is selected for low noise.

22. The detector is operated at relatively low potential.

23. The amplifier is operated at relatively low gain.

24. Pulse-height selection is used.

Probably the most intelligent procedure is to determine whether reduction of background is beneficial for the particular analysis at hand. If so, then the various ways to reduce it are tried in order of decreasing effectiveness, using peak-to-background ratio as a criterion of effectiveness.

It is evident that most of the techniques that reduce background also reduce analyte-line intensity, and therefore may give little or no increase in peak-to-background ratio, and may even reduce it. For a specified specimen, peak-to-background ratio is mostly a matter of excitation conditions.

The pulse-height selector discriminates against only that portion of the background that has wavelength substantially different from that of the analyte line. At a specified spectral line (2θ), that component of the background originating from primary radiation scattered by the specimen (or by the mask or other structure in the specimen compartment) *and* diffracted by the crystal has the same *product* of order and wavelength $n\lambda$ as the analyte line. However, the pulse-height selector is still applicable if the *wavelengths* are not the same.

The effectiveness of the pulse-height selector in reducing background varies with the spectral region. In the low-2θ (short-wavelength) region, the background has high intensity and is principally first-order scattered continuum, which is why it has the characteristic "hump." It is also essentially "monochromatic" in the sense that there is only one wavelength at each increment of 2θ. In the high-2θ (long-wavelength) region, the background has low intensity and is principally higher-order scattered continuum. Thus, the effect of the pulse-height selector in rejecting background is least at low 2θ, greatest at high 2θ.

SUGGESTED READING

ADLER, I., *X-Ray Emission Spectrography in Geology*, Elsevier, Amsterdam (1966); Chap. 4, pp. 43–60.

BERTIN, E. P., *Principles and Practice of X-Ray Spectrometric Analysis*, 2nd ed., Plenum Press, New York (1975); Chaps. 6 and 7, pp. 219–314.

BIRKS, L. S., *X-Ray Spectrochemical Analysis*, 2nd ed., Interscience, New York (1969); Chap. 5, pp. 43–55.

BOWMAN, H. R., E. K. HYDE, S. G. THOMPSON, and R. C. JARED, "Application of High-Resolution Semiconductor Detectors in X-Ray Emission Spectrography," *Science*, **151**, 562–568 (1966).

DOWLING, P. H., C. F. HENDEE, T. R. KOHLER, and W. PARRISH, "Counters for X-Ray Analysis," *Philips Tech. Rev.* **18**, 262–275 (1956–57); *Norelco Rep.* **4**, 23–33 (1957).

GEDKE, D. A., "The Si(Li) X-Ray Energy Analysis System: Operating Principles and Performance," *X-Ray Spectrom.* **1**, 129–141 (1972).

HEATH, R. L., "Application of High-Resolution Solid-State Detectors for X-Ray Spectrometry—a Review," *Advan. X-Ray Anal.* **15**, 1–35 (1972).

JENKINS, R., *Introduction to X-Ray Spectrometry*, Heyden, London (1974); Chap. 6, pp. 60–75.

JENKINS, R., and J. L. DE VRIES, *Practical X-Ray Spectrometry*, 2nd ed., Springer-Verlag New York, New York (1969); Chaps. 3 and 4, pp. 47–89.

KILEY, W. R., and J. A. DUNNE, "X-Ray Detectors," *Amer. Soc. Test. Mater. Spec. Tech. Publ.* **STP-349**, 24–40 (1964).

LIEBHAFSKY, H. A., H. G. PFEIFFER, E. H. WINSLOW, and P. D. ZEMANY, *X-Rays, Electrons, and Analytical Chemistry*, Wiley–Interscience, New York (1972); Chap. 2, pp. 58–126.

PARRISH, W., "X-Ray Intensity Measurement with Counter Tubes," *Philips Tech. Rev.* **17**, 206–221 (1956).

PARRISH, W., and T. R. KOHLER, "Use of Counter Tubes in X-Ray Analysis," *Rev. Sci. Instrum.* **27**, 795–808 (1956).

Chapter 6

Energy-Dispersive
X-Ray Spectrometry

6.1. ENERGY-DISPERSIVE X-RAY SPECTROMETRY

6.1.1. Introduction

The distinction drawn in this book between energy-dispersive and nondispersive methods is given in the last paragraph of Section 3.2. Excitation for energy dispersion by high- and low-power x-ray tubes, radioisotopes, and ion beams is discussed in Sections 4.1.4, 4.1.3, and 4.1.2, respectively. The advantages and limitations of energy and wavelength dispersion are compared in Section 3.5.4. This section (6.1) is concerned with only the analytical techniques of energy dispersion.

Energy-dispersive x-ray spectrochemical analysis can be applied on standard commercial x-ray fluorescence spectrometers and diffractometers using proportional and scintillation counters. However, because of the poor resolution of these detectors, such application is feasible only for specimens having only one heavier element in a light matrix—or at most three or four such elements having widely separated pulse-height distributions.

For fluorescence spectrometers, the goniometer is fixed at $0°\,2\theta$ and the crystal is removed; coarse collimators may be used in the source and detector positions. The specimen is placed in the same position as for wavelength dispersion. In earlier instruments having tabletop goniometers, like that shown in Figure 3.2, the distance from specimen to detector may be reduced by mounting the detector after the source collimator. For diffractometers, the goniometer is fixed at $90°\,2\theta$, and coarse collimators or wide slits may be used. The specimen is placed in the same position as for diffractometry so that lines joining the specimen with the x-ray tube and detector windows make 45° angles with the specimen plane and are mutually perpendicular.

However, most energy-dispersive analysis by far is done on modern systems of the type shown in Figure 3.3 and consisting of the following components: (1) excitation source—high- or low-power x-ray tube, secondary target, electron beam, proton or other ion beam, or radioisotope; (2) specimen presentation system; (3) Si(Li) detector in a liquid-nitrogen vacuum cryostat; (4) multichannel analyzer; (5) computer and memory unit; and (6) cathode-ray tube, X–Y recorder, digital, and/or printout display units. The computer-memory unit may be "dedicated," that is, part of the spectrometer itself, a separate laboratory unit, or a central time-shared computer. If only simple qualitative analysis is required, this component may be omitted.

Commercial Si(Li), multichannel-analyzer, energy-dispersive spectrometers are applicable usually to all elements down to atomic number 11 (sodium), although with special provision, they may be used down to atomic number 6 (carbon). Sensitivity *in favorable cases* may be <1 ppm. Counting times may vary from <1 min to 15 min or more, depending on the mode of excitation and whether major, minor, or trace constituents are present. Specimens may have any form, but supported specimens are most commonly used. These spectrometers are extremely convenient, flexible, and versatile, and usually also rapid. This chapter is intended to acquaint the reader with the variety of their functions.

Incidentally, whereas x-ray spectrochemists using wavelength-dispersive spectrometers become accustomed to thinking of x-ray spectral lines and absorption edges in terms of their *wavelengths* in angstroms, those using energy-dispersive instruments think in terms of their *energies* in kiloelectron volts. For example, in wavelength terms, one would say that if Mo $K\alpha$ (λ 0.71 Å) is to be excited efficiently, the primary beam must be rich in wavelengths shorter than that of the Mo K absorption edge ($\lambda_{\mathrm{Mo}K_{\mathrm{ab}}}$ 0.62 Å). In energy terms, one would say that if Mo $K\alpha$ (E_x 17.4 keV) is to be excited, the excitation radiation must be substantially more energetic than the Mo K critical absorption energy or excitation potential ($V_{\mathrm{Mo}K}$ 20.0 keV).

6.1.2. Multichannel Pulse-Height Analyzer

In the *multichannel pulse-height analyzer* (MCA), the baseline in Figure 5.22 is, in effect, divided into 200, 400, 800, 1200, or even more individual equal, narrow, fixed windows or channels, each connected to its own accumulator or memory. The analyzer receives the amplified detector output and sorts the incoming pulses by height, directing each to its proper "box" along the baseline. The process continues long enough to permit accumulation of a suitably large count—say, 10^5—in the channel corresponding to the highest peak. The number of counts in each channel can be displayed during accumulation on a cathode-ray oscilloscope, or read out at the end of the counting time on an X–Y recorder or printer. A 400-channel analyzer, in effect, step-scans the baseline in 400 incremental steps and scales the intensity at each step—but it does it "all at once" rather than sequentially.

    Figure 6.1 shows the function of a multichannel analyzer in a highly simplified schematic way, using only 30 channels for ease of illustration. Figure 5.22B, showing noise, Si $K\alpha$, and Fe $K\beta$ pulse distributions, is reproduced at the left of Figure 6.1. However, instead of "scanning the baseline" sequentially in one of the two ways described in Section 5.2.3.3, the baseline is in effect divided into 30 equal pulse-height intervals or windows (*channels*), each having its own accumulator. The instrument accumulates pulses in all channels simultaneously until the channel corresponding to the pulse height of the highest intensity (channel 4, average pulse height of the Si $K\alpha$ pulse distribution) has accumulated some arbitrarily large preset count. Then the number of counts accumulated in each channel is rapidly read out electronically—not mechanically, as in Figure 6.1—and displayed on an X–Y recorder or cathode-ray tube, as shown, or printed out digitally. If an instrument having 200 or more chan-

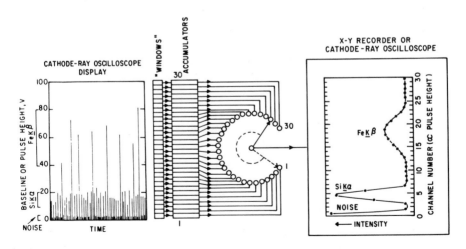

FIGURE 6.1. Schematic representation of the multichannel analyzer (see Figure 5.22).

nels were used, the display would appear substantially as in Figure 5.22A. Alternatively, the accumulation may proceed for a preset time, or until a *group* of channels ("window") accumulates a preset count. The window may consist of a group of channels bracketing a specified peak, for example, channels 2–7 (Si $K\alpha$) in Figure 6.1.

It must be emphasized that the energy spectrum shown in Figure 6.1 results from multichannel analysis of the x-rays diffracted at 108° 2θ (Si $K\alpha$) by an EDDT crystal. A more likely application of the multichannel analyzer would be to eliminate the crystal and disperse directly the x-rays emitted by the specimen. This energy spectrum would show very intense Fe K and very weak Si K pulse-height distributions. Moreover, the broad pulse distributions shown in Figure 6.1 result from conventional scintillation or proportional counters. If a lithium-drifted silicon [Si(Li)] detector is used, very narrow pulse distributions are obtained with consequent resolution of pulse distributions for photon energies as close as 200 eV or less.

The multichannel analyzer is calibrated to correlate channel number and photon energy with known spectral lines, often from radioisotopes, [55]Fe, for example.

A multichannel analyzer to be used with a Si(Li) detector in an energy-dispersive x-ray spectrometer system should have the following characteristics: (1) 400 channels or more; a 400-channel analyzer used to cover the photon-energy range 0–20 keV (Mo $K\alpha$, any $L\alpha_1$) would have 50 eV/channel; (2) count capacity per channel 10^6 or more; (3) good integral linearity—that is, proportionality of channel number and energy; (4) good differential linearity—that is, uniformity of channel width; and (5) low dead time.

The principal advantages of the multichannel analyzer all result from the fact that the entire energy spectrum is accumulated simultaneously: It provides the most rapid way to display an energy spectrum and therefore to make a qualitative x-ray spectrometric analysis. Analyses can be terminated as soon as the spectrum is seen to be adequate for the purpose at hand. Also, instrumental instability has minimal effect on the simultaneously accumulated spectral-line counts.

The applications of the multichannel analyzer in x-ray spectrometry include the following: (1) rapid qualitative and quantitative energy-dispersive analysis; (2) preliminary "setup" work for crystal spectrometers; (3) improvement of spectral resolution by expansion of regions of particular interest; (4) rapid comparison analyses; (5) unfolding of overlapped pulse-height distributions; (6) determination of optimal excitation conditions quickly and without the long times required for a 2θ scan or for a baseline scan on a single-channel pulse-height analyzer; and (7) energy-dispersive x-ray diffractometry–spectrometry (Section 6.3).

Alternatively, the instrument can be used as a *multichannel scaler* to accumulate in successive channels pulses having the *same* average pulse height from a series of related individual measurements. The series may be simple replicates, an evaluation of short- or long-term instrument stability, or a point-by-point line or raster scan across the specimen by an x-ray "milliprobe" or electron-probe microanalyzer. A striking example of this last application is shown in Figure 11.9.

6.1.3. Instrument Functions

6.1.3.1. General

The x-ray energy spectrum is accumulated in the multichannel analyzer, as already explained. Usually a cathode-ray tube display is provided, in which case the display is continuous and one can watch the spectrum accumulate—that is, watch the channels filling. The accumulation continues until: (1) a preset time has elapsed; (2) a preset count (100–10^6) has accumulated in a selected channel or window (a group of consecutive channels spanning a selected individual pulse-height distribution peak); or (3) the accumulating spectrum observed on the cathode-ray tube display is satisfactory to the operator, who then terminates the counting. At the end of the counting period, the spectrum can be recorded in any of three ways, immediately or after processing (see below): (1) the display on the cathode-ray tube can be photographed with a Polaroid camera; (2) the accumulated spectrum can be plotted on an X–Y recorder; or (3) the number of counts in each channel can be printed out on a paper tape. The accumulated spectrum can also be input to the computer memory and processed electronically in a wide variety of ways with great speed and convenience.

The remarkable functioning of the Si(Li) detector/multichannel analyzer/memory/computer/cathode-ray tube display system may be considered in five categories: (1) spectrum display and labeling; (2) spectrum processing; (3) qualitative and comparison analysis; (4) quantitative analysis; and (5) automation. The remainder of this section (6.1) is devoted to a brief discussion of these functions. Qualitative and comparison analyses are discussed in Sections 7.1.2.2 and 7.2.2, respectively, and are not considered further here. Although all the features described in Sections 7.1.2.2 and 7.2.2 and below may not be available on any one commercial instrument, most are available on all instruments and all are available on one or

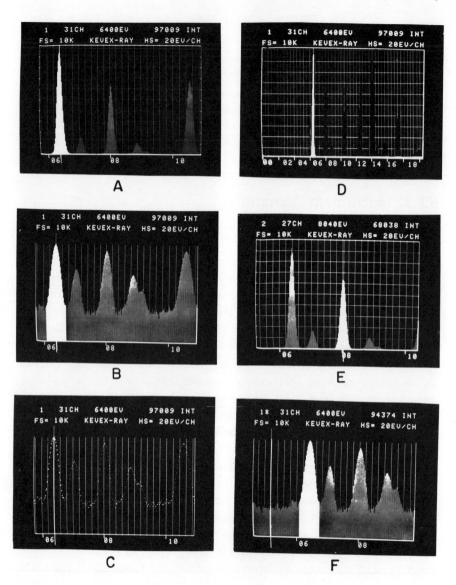

FIGURE 6.2. X-ray energy spectra of a test specimen containing Fe, Cu, Sr, Mo, and Pb. [Courtesy of Kevex Corp.] The alphanumeric data has the following significance. *Upper top* (from left): number of brightened "window"; width of selected window, number of channels (CH); energy (EV) of peak centroid in selected window; integrated (INT) total accumulated count in the selected window. (This position can also display net count in the selected window or total accumulated count or count rate for the entire

another. This discussion is illustrated by Figures 6.2–6.6, 7.1, and 7.10, which have detailed captions and should be studied carefully, together, in sequence, more or less independently of the text. However, specific figures are cited at appropriate places in the following text.

6.1.3.2. Spectrum Display and Labeling

The incoming count data is displayed on the cathode-ray tube on a grid of accumulated count *versus* channel number as a bar (histogram) or dot graph, and on a linear or logarithmic count scale. The vertical length of each bar or position of each dot represents the instantaneous accumulated count in one of the 200, 400, or more channels. Channel number is proportional to pulse height and x-ray photon energy. The sophisticated electronic circuitry provides many aids to facilitate visualization and interpretation of the displayed spectrum.

Markings are generated on the display in four forms: alphanumeric characters, bright-line cursors, an energy scale, and a rectilinear bright-line grid over the entire display (Figures 6.2 and 6.3). The horizontal (channel,

---

spectrum). *Lower top* (from left): number of counts for full vertical scale (FS); horizontal scale (HS) in eV/channel (EV/CH). *Bottom*: energy scale in keV.

A. X-ray energy spectrum in the region 5–11 keV, showing peaks at energies (keV) as follows (left to right): Fe $K\alpha$ (6.40), Fe $K\beta$ (7.06), Cu $K\alpha$ (8.04), Cu $K\beta$ (8.90), Pb Ll (9.18), and Pb $L\alpha_1$ (10.55). In this spectrum and those in B and C below, a window (No. 1) 31 channels wide is set on the Fe $K\alpha$ peak (6400 eV) and brightened, and a bright-line peak-centroid marker is visible above and below the peak.

B. Spectrum A displayed on a three-decade logarithmic vertical scale.

C. Spectrum B with dot instead of bar display.

D. X-ray energy spectrum in the region 0–20 keV, showing the peaks in spectrum A plus the following (left to right): Pb $L\beta_1$ (12.61), Sr $K\alpha$ (14.14), Pb $L\gamma_1$ (14.76), Pb γ_3 (15.21), Sr $K\beta$ (15.83), Mo $K\alpha$ (17.44), and Mo $K\beta$ (19.60).

E. Spectrum A slightly expanded horizontally and with a bright-line centroid marker and brightened window (No. 2) 27 channels wide set on the Cu $K\alpha$ peak (8040 eV); integrated count (68038 INT) in the window is displayed at the top right.

F. Spectrum B slightly expanded horizontally with Fe $K\alpha$ background subtracted numerically. The asterisk (*) after the window number (1) indicates that the top right number (94374) is a net count rather than a total count as in spectrum B (97009). The background is obtained by multiplying the number of counts in the channel marked by the bright line by 31, the number of channels in the Fe $K\alpha$ window, and subtracting the product from the total accumulated count in the window (97009).

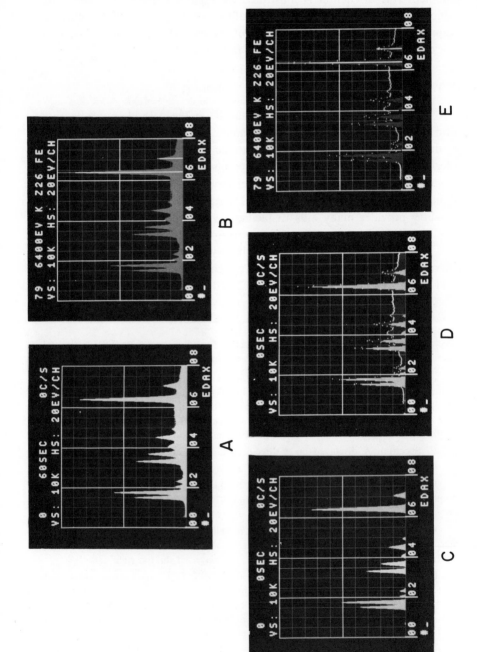

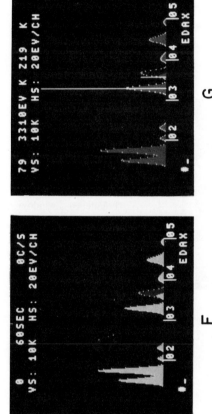

FIGURE 6.3. X-ray energy spectra of a test specimen containing Al, Si, P, S, K, Ca, Ti, and Fe. The alphanumeric data has the same significance as in Figure 6.2 with the following exceptions and additions: VS (vertical scale) is used instead of FS (full scale); in A and F, 60 SEC indicates the counting time; in B and E, K Z 26 FE indicates that the bright-line markers label the K lines of Fe (Z 26); in G, K Z 19 K indicates that the bright-line markers label the K lines of K (Z 19). [Courtesy of EDAX International, Inc.]

A. Entire x-ray energy spectrum in the region 0–8 keV, showing peaks at energies (keV) as follows (left to right): Al $K\alpha$ (1.49), Si $K\alpha$ (1.74), P $K\alpha$ (2.01), S $K\alpha$ (2.31), K $K\alpha$ (3.31), Ca $K\alpha$ (3.69), Ca $K\beta$ (4.01), Ti $K\beta$ (4.51), Fe $K\alpha$ (4.93), Ti $K\alpha$ (6.40), and Fe $K\beta$ (7.06).

B. Spectrum A with bright-line markers on Fe $K\alpha$ and Fe $K\beta$.

C. Spectrum A with background subtracted.

D. Spectrum A (dots) superimposed on spectrum C (bars).

E. Similar to spectrum D with bright-line markers on Fe $K\alpha$ and Fe $K\beta$.

F. The 1–5-keV region of spectrum A expanded with the Ca $K\alpha$ peak (dots) stripped to uncover the underlying K $K\beta$ peak (bars).

G. Spectrum F with bright-line markers on K $K\alpha$ and K $K\beta$.

energy) scale is the series of bright-line ticks, each marked numerically in photon energy (keV) along the bottom of each display. The horizontal scale can be varied from 0–1 to 0–20 keV. The horizontal scale (HS) factor in electron volts per channel is displayed alphanumerically at the top of each display (HS: 20 EV/CH in Figures 6.2 and 6.3). The vertical (count) scale can be varied from 100 to 10^6 counts full scale, linearly or logarithmically (Figures 6.2A and B). The *vertical*-scale (VS) or *full*-scale (FS) count is displayed alphanumerically at the top of each display (VS: 10K, that is, 10,000 counts, in Figure 6.3, FS = 10K in Figure 6.2, FS = 500 in Figure 6.4).

A single vertical bright-line cursor can be moved over the display. Whenever it is stopped—at a peak centroid, for example—the photon energy

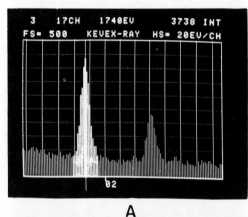

A

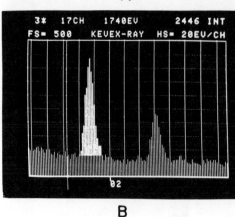

B

FIGURE 6.4. Background subtraction. X-ray energy spectrum showing Si $K\alpha$ and Cl $K\alpha$ peaks at 1.74 and 2.62 keV, respectively. The alphanumeric data has the same significance as in Figure 6.2. [Courtesy of Kevex Corp.]

A. A brightened window (No. 3) 17 channels wide and a bright-line peak centroid marker are set on the Si $K\alpha$ peak (1740 eV); integrated count (3738 INT) in the window is displayed at the top right.

B. Spectrum A with Si $K\alpha$ background subtracted. The asterisk (*) after the window number (3) indicates that the top right number (2446) is a net count rather than a total count as in spectrum A (3738). The background is obtained by multiplying the number of counts in the channel marked by the bright line by 17, the number of channels in the Si $K\alpha$ window, and subtracting the product from the total accumulated count in the window (3738). Note that only the net count in each window channel is brightened. Compare with Figure 6.2F.

(keV) corresponding to that channel is displayed alphanumerically (Figure 7.10A). The positions of the $K\alpha$, $K\beta$, $L\alpha_1$, $L\beta_1$, $L\beta_2$, $L\gamma_1$, $M\alpha$, and $M\beta$ lines of a specified element can be marked with bright-line ticks and the atomic number and/or symbol of the marked element displayed alphanumerically. All these lines that lie in the displayed spectral region are so marked by entering the atomic number of the element. The α, β, and γ ticks are long, medium, and short, respectively, suggestive of their relative intensities (Figures 6.3B and G, and Figure 7.10B).

Any portion of the displayed spectrum can be expanded up to the full width of the display for closer examination and better separation of partially overlapped peaks. The energy scale expands and contracts with the spectrum (compare Figures 6.2D and A, and Figures 6.5D and A, B, C).

The displayed spectrum, say 0–20 keV, can occupy the full width of the display (say, 400 channels), or the left or right half (channels 1–200 or 201–400), or the left, center, or right third (channels 1–133, 134–266, or 267–400). Comparison spectra may be displayed in the unused half (compare Figures 7.1A and C) or thirds. Alternatively, if the analytical spectrum occupies all 400 channels, the comparison spectrum can be superimposed on the analytical spectrum, one as bars, the other as dots (compare Figures 7.1A and B, and Figures 6.3D and A, C). Either spectrum can be displaced vertically or horizontally to facilitate comparison.

One can define the group of consecutive channels comprising the pulse-height distribution peak arising from the measured line of each analyte. Such groups of channels are known as *windows* and their central channels as *peak centroids*. The centroid represents the photon energy and mean pulse height of its peak and has the highest accumulated count of any channel in its window. The windows of analytes widen with increasing photon energy, that is, contain increasing numbers of channels. The windows for all analytes of interest are assigned serial numbers and defined in terms of the channel numbers of their high and low edges or of their centroids. All the data is input to the computer. Thereafter, if a specified analyte is to be measured, its window number is entered. Thereupon, the instrument automatically accents (brightens) all bars or dots in that window, marks its centroid with a bright-line cursor, and displays alphanumerically the window number, atomic number and/or symbol of the analyte, number of channels in the window, photon energy at the centroid, and integrated count in the window (Figures 6.2E and 6.4B).

The display data can be color coded and displayed on a color television monitor to facilitate visual conception and interpretation (Figure 6.6).

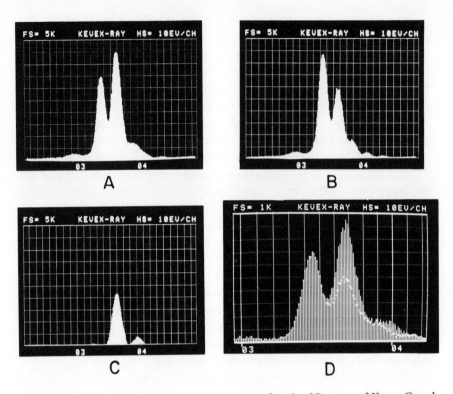

FIGURE 6.5. Stripping of overlapped energy-spectral peaks. [Courtesy of Kevex Corp.]

A. X-ray energy spectrum of a specimen containing Ca and Sn, showing overlapped Ca K and Sn L peaks at energies as follows (left to right): Sn $L\alpha_1$ (3.44); Sn $L\beta_1$ (3.66) and Ca $K\alpha$ (3.69); Sn $L\beta_2$ (3.90) and Ca $K\beta$ (4.01); and Sn $L\gamma_1$ (4.13).

B. Sn L spectrum from pure tin.

C. Spectrum B, adjusted to appropriate amplitude, subtracted from spectrum A, leaving Ca $K\alpha$ and Ca $K\beta$ peaks.

D. Spectra A (as bars) and B (as dots) expanded horizontally, superimposed, and displayed on a logarithmic vertical scale.

6.1.3.3. Spectrum Processing

The analytical spectrum can be processed in several ways as the progress of the processing is observed on the display. The system can be programmed (see below) to perform a specified series of spectrum processings automatically. The accumulation of the spectrum from the next sample can proceed

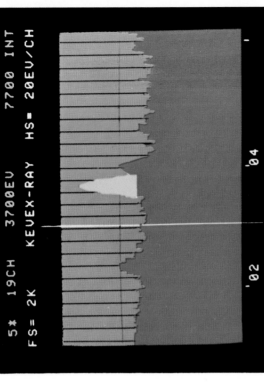

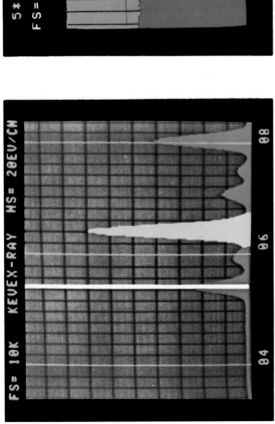

FIGURE 6.6. Color-coded x-ray energy spectra. [Courtesy of Kevex Corp.] (A) Color-coded x-ray energy spectrum in the region 3.5–8.5 keV, showing peaks at energies (keV) as follows (left to right): Cr $K\alpha$ (5.41), Cr $K\beta$ (5.95), Fe $K\alpha$ (6.40), Fe $K\beta$ (7.06), Ni $K\alpha$ (7.47), and Cu $K\alpha$ (8.04). A bright-line marker is set on the Cr $K\alpha$ peak and a window on the Fe $K\alpha$ peak. The color code is red for the general spectrum, yellow for the window, white for the marker. (B) Color-coded logarithmic x-ray energy spectrum, showing a Ca $K\alpha$ peak at 3.69 keV. The alphanumeric data have the same significance as in Figure 6.2. A window (No. 5) 19 channels wide is set on the Ca $K\alpha$ peak (3700 eV). The asterisk (*) after the window number (5) indicates that background has been subtracted from the window and the top right number (7700) is integrated net counts in the window. The background was obtained from the number of counts in the channel marked by the bright line in the same way as in Figures 6.2F and 6.4B. The color code is red for the general spectrum, yellow for the window and alphanumeric data pertaining to it, and white for markers and other alphanumeric data. Note that only the net count in each window channel is coded yellow. Compare with Figures 6.2F and 6.4B. The color code is red for the general spectrum, yellow for the window and alphanumeric data pertaining to it, and white for markers and other alphanumeric data.

while the previous spectrum is being processed in the computer. The analytical spectrum can be corrected for coincidence loss and pulse pileup, smoothed, stripped, and/or reconstructed.

Coincidence-loss correction is facilitated by the simultaneous accumulation of the spectrum, which permits application of novel and ingenious ways to compensate these losses. The *Harms dead-time correction* circuit continuously compares the number of pulses actually arriving from the detector—regardless of amplitude—with the number of pulses actually stored in the multichannel analyzer after baseline restoration and pulse pileup rejection. The circuit stores a number of pulses equal to the difference. This number corresponds to the pulses lost due to coincidence effects. At the end of the counting time, the circuit apportions these counts among the channels in the multichannel analyzer, incrementing each in proportion to its share of the total accumulated count. The *Lowes live-time correction*, in effect, estimates the dead time for each pulse from its half-width, keeps account of these dead times, then extends the counting time to allow accumulation of enough additional pulses to compensate for those lost by coincidence effects.

Pulse pileup is the superimposition of pulses on the trailing edges of immediately preceding pulses so that apparent pulse amplitude is the true amplitude plus the contribution of the tail of the preceding pulse. Pileup occurs at high intensity, principally in Si(Li) detectors because their long pulse decay times may exceed their dead and resolving times (Section 5.1.6.3, Figure 5.11). The pulse-height analyzer treats these pulses according to their apparent—abnormally high—amplitudes. Pileup appears on the energy spectrum as abnormally high and prolonged trailing on the low-energy sides of intense peaks. The condition is dealt with by pulse-pileup correction or rejection circuits.

Smoothing reduces the channel-to-channel fluctuations due to counting statistics and is especially useful when very weak peaks are present that might otherwise be "lost in the noise."

Stripping consists of subtraction of certain data from the analytical spectrum after the spectrum has been accumulated, or even while it is actually being accumulated. The data may actually be subtracted (Figures 6.3A, C, and D), leaving only the net spectrum in the display, or the net data—in each channel, that portion of the bar histogram that presents net counts—can be brightened Figures 6.4B and 6.6B). The subtracted data may be: (1) the spectrum contributed by the specimen support or matrix; (2) background; and/or (3) overlapped interfering peaks. The subtracted spectral data may be accumulated in unused portions of the multichannel analyzer from analyte-free specimen support or matrix, blanks, pure interfering elements, etc. Alternatively, the subtracted data may be entered from punched or magnetic tapes or stored in the computer memory, if it has sufficient capacity; interfering peaks to be subtracted can actually be generated electronically.

Background for specimens supported on Mylar film, Millipore filters, filter paper, ion-exchange membrane or paper, etc., may usually be corrected by subtraction of spectral data from clean support. Background for trace and, usually, minor constituents may usually be corrected by subtraction of spectral data from analyte-free blanks (Section 5.3.2.2). Background for minor and major constituents in more massive specimens is usually more difficult to deal with. One instrument provides two windows in the adjacent background region bracketing the selected analyte peak; the accumulated count from these windows is summed, halved, then subtracted from the peak count.

Alternatively, the integrated count can be measured in one or more channels adjacent to the peak window on each side; background is then stripped from each channel in the window assuming linearity of background profile between the measured adjacent backgrounds. In still another instrument, one of the windows is set to just bracket the peak of interest; let us assume that the required window width is, say, 40 channels. A single channel adjudged by the operator to be representative of the background under the peak is selected in the spectrum adjacent to this window and indicated on the display by a bright-line background marker. The instrument multiplies the number of counts in this background channel by the number of channels in the analyte peak window (40) and subtracts the product from the integrated analyte-peak count in the window. The net count is displayed numerically, and the net peak can be brightened on the display (Figures 6.2F, 6.4, and 6.6B). A different representative background can be selected for each peak.

In subtracting interfering peaks (Figures 6.3F and 6.5), if the subtracted data is accumulated directly from a blank, it may usually be subtracted directly. Otherwise, a scale factor is applied to adjust the subtracted peak heights. Alternatively, the progress of the subtraction can be observed on the cathode-ray tube as the amplitude of the subtracted data is varied, and, sometimes, the correct amplitude can be established visually. When only a few overlapping peaks are involved, the more accurate method of ratio subtraction can be used. The method requires that for each interfering (overlapping) peak, there be another peak of the interfering element in the analytical spectrum free of interference. The ratio of the overlapping and free peaks is established from measurements on pure element. Thereafter, the height of the peak to be subtracted is established from the height of the free peak measured on the sample, and the ratio.

Reconstruction is the opposite of stripping. The analytical spectrum is displayed in, say, channels 1–200. A series of peaks is generated electronically of elements known or likely to be present in the specimen, and is displayed in channels 201–400. The generated peak heights and widths are then adjusted to obtain a close visual match to the analytical spectrum.

6.1.3.4. Quantitative Analysis

The analytical spectrum can be processed for quantitative analysis in many ways, the following being typical: (1) Each peak is corrected for dead time. (2) Each peak is corrected for pulse pileup and background. (3) Each peak is corrected for, or stripped of, spectral-line interference as described above. (4) The channels to be integrated for each peak to be measured and its adjacent background are selected; the selection is facilitated by brightening the dots or bars in the selected channels as mentioned above. (5) The corrected integrated intensities are then printed out. (6) Alternatively, the intensities can be converted to analytical concentrations by the computer programmed to follow the method selected by the operator. This may be a simple intensity *versus* concentration function, in which case intensity data from standards must be accumulated, entered from tape, or stored. Or the method may make Lucas–Tooth and Pyne, Lachance–Traill, or other absorption–enhancement corrections, in which case constants for the elements present are entered, or, if the computer has sufficient capacity, constants for all elements are permanently stored. No knowledge of the mathematics of the method or of the electronics of the instrument is required, and the operation is largely automatic.

It is explained above that a quantitative analysis involves selection of the peaks to be actually measured, then for each, correction for dead time, pulse pileup, background, and spectral-line interference, and selection of the channels to be integrated for peak and background. The operator establishes these conditions more or less by trial and error while observing the progress on the display screen. Once the sequence of operations is established, the operator runs through the sequence once with the instrument in the "learn" mode. Thereafter, the instrument repeats the sequence on all subsequent samples of the series. Up to 50 steps can be learned and programmed in this way. If a multiple specimen holder is provided, the system can be programmed to index each specimen into position in turn, accumulate its spectrum, process the data, and print out the results. Any accumulated spectrum or analytical procedure program can be recorded on magnetic or punched tape or in the computer memory, if its capacity is adequate.

6.2. NONDISPERSIVE X-RAY SPECTROMETRY

It is unfortunate that the term *nondispersive* analysis is commonly—although inappropriately—used as a synonym for *energy-dispersive* analysis because in the latter, there is no actual spacial dispersion of the spectrum. I prefer to reserve the term "nondispersive" for those analytical methods in which the spectral lines are not really separated on the basis of either wavelength or photon energy. Such methods include selective excitation, selective filtration, and selective detection.

In *selective excitation*, the analyte line is excited selectively, that is, to the exclusion, or substantially so, of the lines of other elements in the specimen, usually by potential discrimination or monochromatic excitation. In *potential discrimination*, the x-ray tube is operated at a potential below the excitation potentials of the interfering lines. The method is feasible if the excitation potentials of the interfering lines are much higher than that of the analyte line *and* high enough so that an x-ray tube operating below them still excites the analyte efficiently. For example, the method is applicable to determination of tin-plating thickness on steel (Sn and Fe K-excitation potentials are 28 and 7 kV, respectively), but not nickel-plating thickness (Ni K-excitation potential is 8.3 kV).

In *monochromatic* or *tertiary excitation*, the primary beam may be used to irradiate a secondary target (radiator or "fluorescer") having strong spectral lines on the short-wavelength side of the analyte absorption edge. The resulting spectral emission excites the analyte line. The method may be useful when intensity can be sacrificed for a "clean" spectrum, that is, one free of continuum; the only continuum background arises from double scatter of the primary continuum from both the secondary target and the specimen. The analyte-line intensity is very low, but its measurement by the nondispersive method is feasible. Monochromatic x-radiation can also be obtained by filtering the continuum from the primary beam, leaving substantially pure target-line radiation, and from radioisotope sources. The filtered primary radiation is more intense than either secondary-target or radioisotope sources. However, secondary targets provide substantially any required wavelength, whereas the other two sources are much more limited.

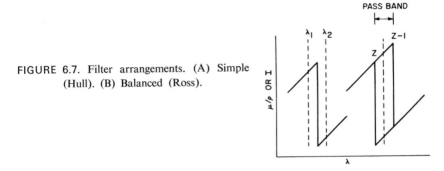

FIGURE 6.7. Filter arrangements. (A) Simple (Hull). (B) Balanced (Ross).

One of the ways to isolate a specified spectral line from a beam containing many such lines is by use of filters (Figure 6.7). The choice of filter material and thickness depends on three criteria: (1) degree of attenuation of the interfering x-rays; (2) remaining intensity of the wanted x-rays after adequate attenuation of the interfering x-rays; and (3) possible increase in background intensity by secondary excitation of the filter element(s).

The principle of the *simple (Hull) filter* is shown in Figure 6.7A. The filter is a thin foil or other layer of an element having a K or LIII absorption edge at a wavelength just shorter than that of the analyte line λ_2. Thus, the filter has low absorption for the analyte line. All lines of shorter wavelength, such as λ_1, are strongly absorbed. The method is very simple, but gives only rough discrimination, and none for interfering lines on the long-wavelength side of the filter edge. The intensity of the passed wavelength is lower the more nearly complete the discrimination—that is, the thicker the filter. The method is most useful for removing the shorter of two interfering lines bracketing the filter absorption edge. Two examples of simple filters follow. In the analysis of brass (copper–zinc alloy), a filter of nickel ($\lambda_{K_{ab}}$ 1.49 Å) 24 μm thick almost completely suppresses the Zn $K\alpha$ line (λ 1.44 Å) while reducing the Cu $K\alpha$ (λ 1.54 Å) intensity only to ~0.35 of its original value. In the analysis of tantalum–tungsten alloys, a filter of nickel 20 μm thick reduces the W $L\alpha_1$ line (λ 1.48 Å) to ~0.02 of its original intensity while reducing the Ta $L\alpha_1$ line (λ 1.52 Å) only to ~0.44 of its original intensity.

The principle of the *balanced (Ross) filter* is shown in Figure 6.7B. A balanced filter consists of two separately mounted thin foils or other layers of two elements of adjacent atomic number selected so that the wavelength to be measured lies in the *pass band* between their absorption edges. For example, nickel ($\lambda_{K_{ab}}$ 1.49 Å) and cobalt ($\lambda_{K_{ab}}$ 1.61 Å) filters can be used to measure Cu $K\alpha$ (λ 1.54 Å). The thicknesses of the filters are adjusted so that they transmit equal intensities at all wavelengths except those between their absorption edges. This is done by actual adjustment of their thickness or area density (mg/cm²), or by stretching or tilting the filters in their frames. Another way is to add to a filter having slightly low absorption a thin filter of a third element having very low absorption, no absorption edge, and no spectral line in the wavelength region of interest—aluminum, for example.

Intensity is measured first with one filter in the x-ray beam, then the other. The difference between the two intensities is due entirely to x-rays having wavelengths between the two absorption edges, principally the analyte line. Alternatively, the x-rays emitted and scattered by the specimen can be received by two identical detectors simultaneously, each covered by one of the balanced filters. Again, the difference in the two detector outputs is due to the wavelengths in the filter pass band.

The balanced-filter method can be made substantially specific for any wavelength, but the filters are difficult to balance, and a different pair of filters is required for each analyte line. The resolution of a balanced filter varies from $\sim$300 eV at 1.5 keV (Al $K\alpha$) to $\sim$700 eV at 10 keV (Ge $K\alpha$) to $\sim$2 keV at 70 keV (Hg $K\alpha$). Scintillation counters, despite their poor resolution, can be used with balanced filters because resolution is determined by the pass band of the filter rather than by the detector.

Selective detection, or "gas discrimination," takes advantage of the dependence of quantum efficiency of the detector (Figure 5.8) on wavelength to realize high detection efficiency for the analyte line—or low efficiency for interfering lines. The method is somewhat inconvenient to apply to specific spectral lines, but may be used in special cases. Gas discrimination is most effective in the long-wavelength region. Helium, neon, methane, and propane detector gases are relatively efficient at these wavelengths, but have very low absorption for higher-order, shorter-wavelength lines of elements of higher atomic number.

6.3. ENERGY-DISPERSIVE X-RAY DIFFRACTOMETRY–SPECTROMETRY

The condition for diffraction of x-rays by a crystal lattice can be expressed in the form of the Bragg law (Section 2.3.1),

$$n\lambda = 2d_{(hkl)} \sin \theta \qquad (6.1)$$

where n is the diffraction order; λ is the wavelength of the diffracted x-rays; $d_{(hkl)}$ is the interplanar spacing of the diffracting planes (hkl); and θ is the angle between the incident x-rays and the diffracting planes. Since in x-ray diffraction it is always a measurement of d that is required, the many diffraction methods fall into two fundamental groups on the basis of the Bragg law—those that vary λ and those that vary θ. However, more practically, x-ray diffraction methods are classified in the four groups given in Table 6.1—Laue, moving-crystal, Debye–Scherrer, and divergent-beam.

In the *Laue methods*, a collimated polychromatic x-ray beam irradiates a fixed single crystal. Each set of planes (hkl) in the crystal presents a certain fixed angle $\theta_{(hkl)}$ to the incident beam and "selects" from the beam and diffracts the particular wavelength that satisfies the Bragg law for its d and θ.

In the *moving-crystal methods*, a collimated, monochromatic x-ray beam irradiates a single crystal. If the crystal were fixed, as in the Laue case, only by chance would any of the planes happen to present the required θ to the incident beam to satisfy the Bragg law. However, if the crystal is rotated or made to undergo some other appropriate motion, various sets of planes are brought in turn to the required θ's to satisfy the Bragg law for their respective values of $d_{(hkl)}$ and the incident wavelength.

In the *Debye–Scherrer* (or *powder*) *methods*, a collimated, monochromatic x-ray beam irradiates a polycrystalline specimen. Any specified set of planes (hkl) can diffract the incident wavelength only from those crystallites so oriented as to present that plane to the incident beam at the required θ. However, if the number of crystallites in the beam is large and they are randomly oriented, each plane is likely to have some crystallites properly oriented to permit it to diffract.

In the *divergent-beam methods*, a spherically divergent point source of monochromatic x-rays is generated at or just above the surface of a fixed, thin, radiolucent single crystal. A solid cone of x-rays is transmitted through the crystal. If there were no diffraction, a photographic plate a short distance below the crystal would be *uniformly* exposed by the transmitted x-rays. Actually, certain rays of the diffracted cone strike each fixed crystallographic plane at the correct angle for diffraction. The intensities of the rays in the divergent cone are increased in the directions of the diffracted rays, and decreased in the directions of the corresponding incident rays prior to diffraction. In this way, a pattern of dark and light circles, parabolas, and hyperbolas (*Kossel lines*) is formed on the photographic plate.

The method of *energy-dispersive x-ray diffractometry–spectrometry* combines features of the Laue and Debye–Scherrer x-ray diffraction methods with energy-dispersive x-ray spectrometry. In fact, the method may be regarded as an energy-dispersive Laue method (because polychromatic x-rays are used) or as an energy-dispersive Debye–Scherrer method (because polycrystalline specimens are used). The originators of the method propose the term *energy-dispersive spectrometric powder diffractometry* (SPD). The technique is also known as the "Giessen–Gordon method." The instrument arrangement is shown in Figure 6.8 and is essentially that of a conventional diffractometer. An x-ray tube, Si(Li) detector, and a plane polycrystalline specimen are arranged with the incident and diffracted beams making equal angles θ_1 with the specimen surface. However, the arrangement differs from that in diffractometry in that the geometry is fixed and the x-ray tube irradiates the specimen with polychromatic, rather than monochromatic (filtered target-line) x-rays.

Diffracted x-rays from a specified set of planes (*hkl*) can be detected only from those crystallites that happen to be oriented so that those planes lie parallel to the specimen plane—that is, present angle θ_1 to the incident and diffracted rays. Those crystallites for which this is the case "select" from the incident beam and diffract the wavelength that satisfies the Bragg law for their $d_{(hkl)}$ and fixed angle θ_1. Other crystallites permit other planes to diffract other wavelengths in the same way.

Actually, the specimen may also be cylindrical with its axis at the point of intersection of incident and diffracted rays and perpendicular to their plane. In this case, diffracted rays from a specified set of planes are detected only from those crystallites that happen to be oriented so that their plane normals are coplanar with and bisect the angle $(180° - 2\theta_1)$ between the incident and diffracted rays.

The x-rays entering the detector consist of: (1) diffracted rays of various wavelengths, each diffracted by a different set of planes (*hkl*); (2) emitted

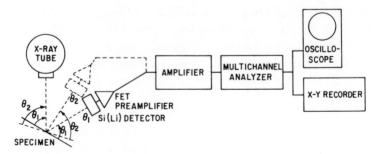

FIGURE 6.8. Instrument arrangement for energy-dispersive x-ray diffractometry–spectrometry.

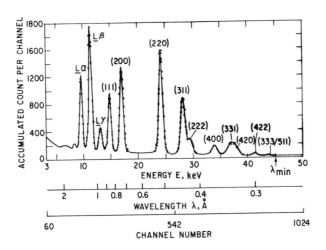

FIGURE 6.9. Energy-dispersive x-ray diffractometry-spectrometry spectrum of polycrystalline platinum. [B. G. Giessen and G. E. Gordon, *Science* **159**, 973 (1968); courtesy of the authors and the American Association for the Advancement of Science.]

x-rays of various wavelengths arising from secondary excitation of the elements in the specimen; and (3) scattered primary x-rays—continuum and target lines. The detector output is amplified and fed to a multichannel analyzer, the accumulated output of which is displayed on a cathode-ray tube or X–Y recorder. Figure 6.9 shows a diffraction-spectrometry energy spectrum of polycrystalline platinum foil. Each peak is a narrow pulse-height distribution, and the distributions of the Pt L lines and of the several wavelengths diffracted by the crystal planes are shown.

In Figure 6.9, the Pt L peaks constitute a Pt L energy spectrum identical with that obtainable from an energy-dispersive x-ray spectrometer. However, the diffraction peaks differ from those obtainable on a conventional diffractometer: In Figure 6.9, each peak represents a *different* wavelength diffracted from a different Pt (*hkl*) plane at the *same* angle θ_1. In a diffractometer record, each peak represents the *same* wavelength (the filtered target line) diffracted from a different plane at a *different* angle $\theta_{(hkl)}$.

The *emission* peaks can be identified by simple inspection if: (1) only one or a very few elements in the specimen emit lines giving pulse-height distributions in the energy region displayed; (2) these elements are known; and (3) there is no overlap of emission and diffraction pulse distributions. These conditions prevail in Figure 6.9. If there is difficulty in identifying the emission lines, a second energy spectrum is recorded at a different angle θ_2 (Figure 6.8) 5–15° away from θ_1. The pulse-height distributions of the *emitted* and *scattered* x-rays are the same in both spectra, but the distributions of the *diffracted* x-rays are different at the two angles. The emitted and scattered peaks may be eliminated by visual comparison of the two spectra, or the two multichannel analyzer outputs may be fed directly to a computer and the peaks common to both—that is, the emission and scatter peaks—electronically subtracted.

The interplanar spacing $d_{(hkl)}$ of the crystal plane giving rise to each diffraction peak is readily calculated from the Bragg law and Equation (1.17):

$$\lambda = 12.396/E_x \tag{6.2}$$

where E_x is photon energy (keV). Substitution in the Bragg equation gives

$$d_{(hkl)} = \frac{12.396/E_x}{2 \sin \theta} = \frac{6.198}{\sin \theta} \frac{1}{E_x} \tag{6.3}$$

For any specified angle θ, $\sin \theta$ may be included in the constant, giving simply $d = k/E_x$. The E_x is obtained from the calibrated scale of the multichannel analyzer.

If accurate relative intensities of the diffraction peaks are required, the detector response at each wavelength in the primary beam must be established and used to correct the observed relative intensities.

Applications and advantages of the method arise from the following three features: (1) the extreme rapidity with which the pattern is recorded, as quickly as 1 s; (2) simultaneous analysis for both elements and chemical species; and (3) elimination of the need for scanning over a wide, continuous angular range. The method is applied to studies of continuous dynamic phenomena, short-lived phenomena, very large objects, and specimens in restricted enclosures, such as high-pressure and high-temperature vessels.

The relationships, in terms of the Bragg law, among the various methods considered in this section are summarized in Table 6.1. It is evident that in the Laue, Debye–Scherrer, and divergent-beam methods, all planes diffract simultaneously, whereas in the moving-crystal method, in general, only one plane diffracts at a time. In all these methods, the diffracted rays from different planes travel in different directions. In energy-dispersive diffractometry–spectrometry, all planes diffract simultaneously and in the same direction.

TABLE 6.1. Comparison of X-Ray Diffraction, Spectrometry, and Energy-Dispersive
Diffractometry–Spectrometry

	$n\lambda$	$=$	$2d$	$\times$	$\sin\theta$
Diffraction, Laue	Variable		Calculated		Fixed
Diffraction, moving-crystal	Fixed		Calculated		Variable
Diffraction, Debye–Scherrer	Fixed		Calculated		Variable
Diffraction, divergent-beam	Fixed		Calculated		Variable
Spectrometry	Calculated		Fixed		Variable
Diffractometry–spectrometry	Variable		Calculated		Fixed

SUGGESTED READING

BERTIN, E. P., *Principles and Practice of X-Ray Spectrometric Analysis*, 2nd ed., Plenum Press, New York (1975); Chap. 8, pp. 315–403.

BIRKS, L. S., *X-Ray Spectrochemical Analysis*, 2nd ed., Interscience, New York (1969); Chap. 6, pp. 57–69.

GIESSEN, B. C., and G. E. GORDON, "X-Ray Diffraction—New Technique Based on X-Ray Spectrography," *Science* **159**, 973–975 (1968).

KIRKPATRICK, P., "Theory and Use of Ross Filters," *Rev. Sci. Instrum.* **10**, 186–191 (1939); **15**, 223–229 (1944).

ROSS, P. A., "New Method of Spectroscopy for Faint X-Radiations," *J. Opt. Soc. Amer.* **16**, 433–437 (1928).

RUSS, J. C., *Elemental X-Ray Analysis of Materials—EXAM Methods*, EDAX International, Inc., Prairie View, Ill., 89 pp. (1972).

RUSS, J. C., ed., "Energy-Dispersion X-Ray Analysis: X-Ray and Electron-Probe Analysis," *Special Technical Publication* **STP-485**, American Society for Testing and Materials, Philadelphia, 285 pp. (1971).

SALMON, M. L., "Practical Applications of Filters in X-Ray Spectrography," *Advan. X-Ray Anal.* **6**, 301–312 (1963).

WOLDSETH, R., *X-Ray Energy Spectrometry*, Kevex Corporation, Burlingame, Calif., 151 pp. (1973).

Chapter 7

Qualitative
and Semiquantitative Analysis

Qualitative analysis is the *detection* or *identification* of the constituent elements in the sample, *semiquantitative* analysis is the *estimation* of their approximate concentrations, and *quantitative* analysis is the accurate *determination* of their concentrations. (The common expression *quantitative determination* is redundant because a determination is necessarily quantitative.) In any of these three types of analysis, the interest may be in all constituent elements or only one or more specified elements.

X-ray emission spectrometry is extremely well suited to qualitative and semiquantitative analysis. It is convenient, rapid, and nondestructive, and applicable to major, minor, and trace constituents in any sample that can be accommodated by the instrument. Milligram samples are sufficient— even microgram samples in favorable cases. Identification of constituent elements is facilitated by the simplicity of x-ray spectra. The scope of the method with respect to atomic number, sensitivity, and specimen form, and the advantages and limitations of the method in general are given in Chapter 3.

This chapter is devoted mostly to qualitative and semiquantitative x-ray spectrometric analysis as applied on standard commercial x-ray wavelength-dispersive fluorescence spectrometers and energy-dispersive spectrometers.

7.1. QUALITATIVE ANALYSIS

A qualitative x-ray spectrometric analysis consists in excitation and recording of the x-ray spectrum of the sample, identification of the peaks, and, perhaps, classification of each element as a major, minor, or trace constituent on the basis of the intensities of its peaks.

7.1.1. Recording the Spectrum

In energy-dispersive x-ray spectrometry, the x-ray spectrum emitted and scattered by the specimen is directed on an x-ray detector, usually of the lithium-drifted silicon type, the output of which is dispersed by electronic pulse-height analysis. The entire dispersed spectrum is accumulated in a multichannel analyzer simultaneously in a few minutes. The accumulated spectrum is read out on an $X-Y$ recorder or on a cathode-ray tube display, which may be photographed. Typical energy-dispersive x-ray spectra photographed from cathode-ray tube displays are shown in Figures 7.1 and 7.10. The photographs constitute plots of intensity *versus* photon energy and are therefore x-ray energy spectra.

In wavelength-dispersive x-ray spectrometry, the x-ray spectrum emitted and scattered by the specimen is directed on an analyzer crystal, where it is dispersed by means of a "2θ scan": The crystal and detector arm are rotated about a common axis at relative speed ratio 1 : 2. Thus, the secondary beam and normal to the detector window both always present angle θ to the crystal surface, and the detector is always in position to receive any x-rays diffracted by the crystal. As the crystal rotates, it presents a continuous succession of θ angles to the secondary beam and, one by one, in accordance with the Bragg law, diffracts each wavelength in the secondary beam to the detector. The detector output is amplified and applied to the ratemeter, the output of which displaces the recorder pen. The scanning–recording process usually requires 15 min to 1 h, depending on scanning speed and the scope of the scanned region. The recorder chart drive is synchronizad with the goniometer 2θ drive so that chart divisions are directly convertible to 2θ angles. The result is a chart having the form of those shown in Figure 7.2. The chart is a plot of intensity *versus* 2θ and thereby intensity *versus* wavelength, and is therefore an x-ray wavelength spectrum.

In wavelength-dispersive flat-crystal x-ray fluorescence spectrometry, the instrument components and conditions to be considered in recording the spectrum are as follows. *Excitation conditions*: x-ray tube target, potential (kV), current (mA), and primary-beam filter; *spectrogoniometer conditions*: collimators, crystal, radiation-path medium, detector, and goniometer scanning speed; *readout conditions*: pulse-height selection, ratemeter sensitivity (intensity for full-scale recorder deflection), ratemeter time constant (response time), ratemeter mode (linear or logarithmic), recorder response time (time to reach full scale), and recorder chart speed. If there is interest in only one or a few elements, the instrument may be arranged for optimal recording of the spectra of these elements. If little or nothing is known of

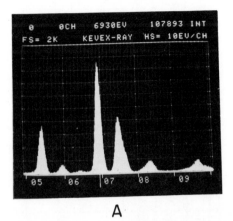

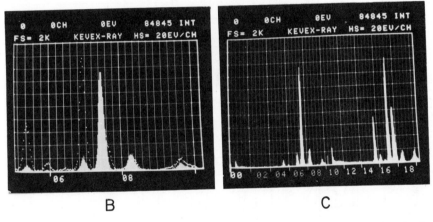

FIGURE 7.1. Energy-dispersed x-ray spectra (photographs of cathode-ray tube displays). These three photographs also illustrate comparison analysis (Section 7.2.2).

A. X-ray energy spectrum of a sample in the region 4–10 keV, showing peaks at energies (keV) as follows (left to right): Cr $K\alpha$ (5.41), Cr $K\beta$ (5.95), Fe $K\alpha$ (6.40), Co $K\alpha$ (6.93), Ni $K\alpha$ (7.47), W $L\alpha_1$ (8.40), W $L\beta_1$ (9.67).

B. Spectrum A (dots) compared with a superimposed reference spectrum (bars).

C. Spectrum A compressed into the right half of the display compared with the reference spectrum in B compressed into the left half of the display. The energy scale is correct for the left spectrum, 10 keV high for the right spectrum.

It is evident that the comparison spectrum in B and C is not the same as the sample spectrum; that is, the reference material is not the same as the analytical sample material.

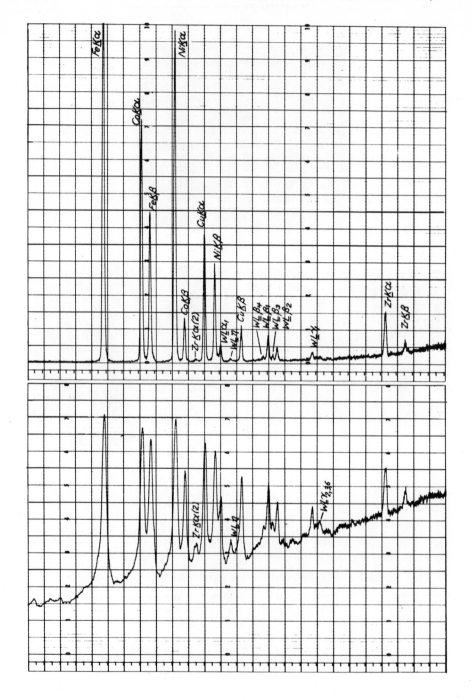

the sample and maximum information is required, the following general considerations may guide selection of the components and conditions.

Tungsten is the most commonly used target for all elements, but chromium is preferable for K lines of elements having atomic number $\lesssim 22$ (titanium). A dual tungsten–chromium target permits convenient change-over. If the sample has low mean atomic number, it will strongly scatter the primary x-rays, and a primary-beam filter may be used to exclude x-ray tube target lines. Nickel–brass and titanium filters are optimal for tungsten L and chromium K lines, respectively.

A 50-kV unit should be operated at maximum potential to ensure efficient excitation of elements having high excitation potentials. The current setting must be a compromise between adequate intensity from trace elements and relatively low scattered continuous background.

The collimators are selected to give a compromise between high intensity and resolution. No crystal gives as favorable a combination of high intensity, high resolution, and wide useful region (all elements down to potassium) as LiF(200). LiF(220) may be used to better disperse the K lines of elements 22–26 (titanium to iron). PET and KHP (or RHP) may be used down to aluminum and oxygen, respectively. A crystal changer is very convenient.

Air path can be used for lines having wavelength <4 Å ($_{19}$K $K\alpha$, $_{48}$Cd $L\alpha$), but elements having lines >2.5 Å ($_{22}$Ti $K\alpha$, $_{58}$Ce $L\alpha$) must be present in relatively high concentration. Helium or vacuum path is beneficial at wavelength >1 Å ($_{35}$Br $K\alpha$, $_{88}$Ra $L\alpha$) and should always be used >2.5 Å. Vacuum is preferable to helium >6 Å ($_{15}$P $K\alpha$, $_{40}$Zr $L\alpha$).

A combination detector consisting of a scintillation counter and flow-proportional counter having a 3-μm (0.00012-inch) Mylar window and using P10 gas (90 vol% argon, 10 vol% methane) gives satisfactory response for all spectral lines down to that of fluorine.

The goniometer 2θ scanning speed should be slow enough to permit response to weak peaks, yet fast enough to permit the spectrum to be recorded in reasonable time. Commonly used speeds are 0.5–2° 2θ/min. The weaker the peaks of interest—that is, the lower the concentration of the elements of interest—the slower must be the scanning speed.

FIGURE 7.2. Wavelength-dispersed x-ray spectra (ratemeter–recorder charts of 2θ scans) of an Alnico alloy [53Fe–17Ni–13Co–9.5Al–6Cu–0.5Ti–0.5Zr (wt%)]. The upper chart was recorded with linear ratemeter mode, the lower chart with logarithmic mode. Compare these spectra with the optical-emission spectrum of Alnico in Figure 3.4.

Automatic pulse-height selection gives a pure first-order spectrum, thereby eliminating high-order spectral-line interference and reducing high-2θ continuous background. However, for highly crowded spectra, it is preferable to record the better resolved higher-order spectra.

If there are several major constituents, the full-scale deflection is usually set so that the strongest peak is just on-scale. If there is only one major constituent, or even two or three, it may be preferable to set the scale so that the strongest minor-constituent peak is just on-scale and let the major peak(s) run off-scale. In either case, the peak to be used to set the scale may be known or found by making a rapid manual 2θ scan. It must be remembered that the response is smaller when the goniometer scans through a peak than when it is stationary at the peak, and response decreases the faster the scanning speed, the longer the ratemeter time constant, and the slower the recorder response time. If the scale is made too small, background "noise" fluctuations become severe, and weak peaks may be obscured.

The ratemeter time constant should be short enough to permit response to weak peaks, yet long enough to damp background fluctuations. Typical values are 1–4 s, but, in general, the time constant is selected for compatibility with the 2θ scanning speed and full-scale deflection. If minor and trace constituents as well as major constituents are of interest, that is, if very weak as well as very strong peaks are to be recorded, operation of the ratemeter in logarithmic, rather than linear, mode may be preferable. Recorder chart speed is selected to give a chart of convenient length, but there is no point in using a relatively slow scanning speed and short time constant in order to show weak peaks, then running over them with a fast chart speed. In general, a chart speed that gives $1° 2\theta$/cm is usually satisfactory. The charts in Figure 7.2 were recorded at $2° 2\theta$/min and 1 cm/min or $2° 2\theta$/cm, giving a $52° 2\theta$ spectrum on a 26-cm chart. It is evident in the figure that in the logarithmic spectrum, the peaks of lower intensity are accentuated, but so are the background and the "tails" ("skirts," "wings") of all peaks.

Inspection of the recorded spectrum may reveal questionable peaks that are due to poor resolution from an adjacent peak or from background noise. These spectral regions may be rerun at conditions to accentuate the questionable peak.

Energy-dispersive x-ray spectrometers have no spectrogoniometer, so the process of obtaining energy spectra is greatly simplified. However, many modes of excitation are applicable to energy-dispersed spectra. These are discussed in Chapter 4.

7.1.2. Identification of Peaks

Figure 7.2 shows the x-ray wavelength spectrum of an Alnico alloy [53Fe–17Ni–13Co–9.5Al–6Cu–0.5Ti–0.5Zr (wt%)] recorded with linear and logarithmic ratemeter modes. Certainly the linear spectrum is preferable for general purposes, and is always used when intensities are to be estimated from the chart. However, the very weak Zr $K\alpha(2)$, W $L\eta$, and W $L\gamma_1$ peaks are accentuated on the logarithmic spectrum.

The peaks on ratemeter recorder charts are identified by means of scales or, usually, tables. The ruler-like or transparent-overlay scales are marked with the positions of the strongest lines of all elements and, when laid on the charts, permit direct identification of the peaks. A different scale is required for each combination of analyzer crystal, 2θ scanning speed, and chart speed. Such scales are easily prepared by the x-ray spectrochemist or by the laboratory drafting department. Tables require reading of the 2θ angles of each peak on the chart. The 2θ angle is derived from the chart from the known 2θ-scanning and chart speeds, from a scale previously prepared from these parameters, or from the degree marker. The degree marker is a small pen actuated by a microswitch in the goniometer and makes ticks along the edge of the chart at $1°$, $0.5°$, or $0.25°$ intervals. Degree marker traces are shown along the bottom edges of the chart in Figure 7.2. The peaks are then identified by applying the 2θ values to tables.

7.1.2.1. X-Ray Spectrometer Tables

Tables for wavelength-dispersive x-ray spectrometry are of two forms: The *line-to-2θ* table enables one to find at what 2θ angle a specified order of a specified line is diffracted by a specified crystal; the *2θ-to-line* table enables one to identify a peak from the 2θ angle at which it is diffracted by a specified crystal. The 2θ-to-line table is also useful for predicting possible spectral-line interference. The analyte line is located in the table, and a search is made for $1°$ or $2°$ 2θ on either side of the line for lines of other elements known to be present in the specimen and for x-ray tube target lines.

The most comprehensive x-ray spectrometer table now available is that published by the American Society for Testing and Materials (ASTM). The table lists substantially all K, L, M, and N lines (including satellite lines) and absorption edges having wavelength up to ~ 160 Å for elements 4–98 (beryllium to californium)—some 3400 lines and absorption edges in all.

El	Line	C	W	I	Z	R	KeV	Lambda 10⁻	P	A	Lsteer (Film)	Llaur (Film)	Chlor (001)	KAP (100)	RbAP (100)	Mica (002)	Gyps (020)	ADP (101)	EDDT (020)	PET (002)	Graph (002)	Ge (111)	NaCl (200)	LiF (200)	LiF (220)
Fe	$K\beta_{1,3}$		1	20	26	6	7.057	1.75661	5	a	2.00	2.88	7.09	7.56	7.71	10.16	13.28	19.01	23.01	23.18	30.36	31.20	36.29	51.73	76.16
Fe	$K\beta_5$	B	1	.03	26	5	7.107	1.74420	4	a	1.99	2.86	7.04	7.51	7.66	10.09	13.19	18.87	22.88	23.02	30.18	30.97	36.02	51.38	75.53
Fe	AK		1		26	6	7.110	1.74346	5		1.98	2.85	7.04	7.51	7.65	10.08	13.18	18.86	22.83	23.01	30.13	30.96	36.01	51.31	75.49
Co	Ll		1	9	27	6	0.678	18.29199	3		20.93	30.30	80.22	86.76	88.90	134.43									
Co	Lη		1	2	27	6	0.694	17.87000	2		20.44	29.59	78.01	84.29	86.33	128.50									
Co	$L\alpha_{1,2}$		1	100	27	6	0.776	15.97200	3		18.25	26.38	68.47	73.70	75.39	107.23									
Co	AL_3		1		27	6	0.779	15.91500	3		18.19	26.29	68.19	73.39	75.07	106.68									
Co	$L\beta_1$		1	19	27	6	0.791	15.66600	3		17.90	25.96	66.98	72.06	73.70	104.30									
Co	AL_2		1		27	6	0.794	15.51800	3		17.84	25.79	66.75	71.81	73.44	103.95									
Co	$L\beta_{3,4}$		1	.5	27	6	0.866	14.31000	2		16.34	23.59	60.53	65.00	66.44	92.32	140.80								
Co	$K\alpha_2$		1	50	27	5	6.914	1.79285	6		2.04	2.94	7.24	7.72	7.87	10.37	13.56	19.40	23.49	23.67	31.00	31.86	37.06	52.88	78.03
Co	$K\alpha_{1,2}$		1	150	27	5	6.924	1.79026	5		2.04	2.93	7.23	7.71	7.86	10.35	13.54	19.37	23.45	23.63	30.96	31.81	37.01	52.80	77.89
Co	$K\alpha_1$		1	100	27	5	6.929	1.78896	5		2.04	2.93	7.23	7.70	7.85	10.35	13.53	19.36	23.44	23.62	30.93	31.79	36.98	52.76	77.83
Co	$K\beta_{1,3}$		1	20	27	5	7.648	1.62079	5	a	1.84	2.65	6.55	6.98	7.11	9.37	12.25	17.52	21.21	21.37	27.96	28.73	33.40	47.47	69.37
Co	$K\beta_5$	B	1	.03	27	6	7.705	1.60891	5	a	1.83	2.63	6.50	6.93	7.06	9.30	12.16	17.39	21.05	21.21	27.75	28.52	33.14	47.10	68.79
Co	AK		1		27	6	7.708	1.60815	5		1.83	2.63	6.49	6.92	7.06	9.30	12.15	17.39	21.04	21.20	27.74	28.51	33.13	47.08	68.76
Ni	Ll		1	9	28	6	0.743	16.69299	3		19.08	27.59	72.32	77.63	79.44	114.57									
Ni	Lη		1	3	28	6	0.762	16.26999	2		18.60	26.99	69.93	75.31	77.05	110.18									
Ni	$L\alpha_{1,2}$		1	100	28	5	0.851	14.56100	3		16.53	24.01	61.71	66.29	67.76	94.43	146.91								
Ni	AL_3		1		28	6	0.853	14.52500	3		16.50	23.95	61.58	66.10	67.57	94.13	145.97								
Ni	$L\beta_1$		1	21	28	5	0.869	14.27100	3		16.29	23.53	60.35	64.80	66.23	91.99	139.94								
Ni	AL_2		1		28	6	0.870	14.24200	3		16.25	23.49	60.21	64.66	66.08	91.75	139.30								
Ni	$L\beta_{3,4}$		1	.9	28	6	0.941	13.18000	2		15.04	21.71	55.32	59.33	60.51	83.26	120.38								
Ni	$K\alpha_2$		1	50	28	5	7.460	1.66175	6		1.89	2.72	6.71	7.15	7.29	9.61	12.56	17.97	21.75	21.92	28.69	29.48	34.27	48.75	71.39
Ni	$K\alpha_{1,2}$		1	150	28	5	7.471	1.65919	6		1.89	2.72	6.70	7.14	7.29	9.59	12.54	17.94	21.72	21.88	28.64	29.43	34.21	48.67	71.26
Ni	$K\alpha_1$		1	100	28	5	7.477	1.65791	5		1.89	2.71	6.70	7.14	7.28	9.59	12.53	17.93	21.70	21.86	28.62	29.41	34.18	48.63	71.20
Ni	$K\beta_{1,3}$		1	20	28	6	8.263	1.50013	5		1.71	2.46	6.05	6.45	6.59	8.67	11.34	16.21	19.61	19.76	25.85	26.55	30.84	43.75	63.57
Ni	$K\beta_5$	B	1	.03	28	5	8.327	1.48862	5	a	1.69	2.44	6.01	6.41	6.53	8.61	11.25	16.09	19.46	19.61	25.64	26.35	30.60	43.39	63.03
Ni	AK		1		28	5	8.330	1.48807	5		1.69	2.44	6.01	6.41	6.53	8.60	11.24	16.08	19.45	19.60	25.63	26.34	30.59	43.38	63.00
Cu	AM_2		1		29	6	0.078	158.50000	1																
Cu	AM_1		1		29	6	0.112	110.59999	1																
Cu	Ll		1	5	29	6	0.811	15.28600	3		17.46	25.23	65.15	70.06	71.63	100.79									
Cu	Lη		1	1	29	5	0.832	14.90000	2		17.02	24.58	63.31	68.08	69.56	97.36	157.57								
Cu	$L\alpha_{1,2}$		1	100	29	6	0.930	13.33600	3		15.22	21.97	56.03	60.10	61.40	85.37	122.79								
Cu	AL_3		1		29	5	0.933	13.28800	3		15.17	21.89	55.81	59.86	61.16	85.10	122.24								
Cu	$L\beta_1$		1	20	29	5	0.950	13.05300	3		14.90	21.49	54.74	58.70	59.96	82.28	118.49								
Cu	AL_2		1		29	5	0.953	13.01400	3		14.85	21.43	54.55	58.51	59.76	81.99	117.31								
Cu	$L\beta_{3,4}$		1	1	29	5	1.023	12.12200	3		13.83	19.94	50.55	54.15	55.30	75.32	105.89								
Cu	$K\alpha_2$		1	50	29	6	8.026	1.54439	5		1.76	2.53	6.24	6.65	6.79	8.93	11.57	15.59	20.20	20.35	26.62	27.35	31.79	45.11	65.68
Cu	$K\alpha_{1,2}$		1	150	29	5	8.040	1.54184	5		1.75	2.52	6.23	6.63	6.77	8.91	11.55	15.56	20.16	20.32	26.59	27.31	31.72	45.03	65.55
Cu	$K\alpha_1$		1	100	29	5	8.046	1.54056	5		1.75	2.52	6.22	6.63	6.76	8.91	11.54	15.55	20.15	20.30	26.55	27.29	31.70	44.99	65.49
Cu	$K\beta_3$		1	5	29	5	8.901	1.39263	4		1.58	2.29	5.62	5.99	6.11	8.05	10.52	15.04	18.19	18.33	23.95	24.62	28.59	40.47	58.55
Cu	$K\beta_{1,3}$		1	20	29	5	8.904	1.39222	5		1.58	2.29	5.62	5.99	6.11	8.05	10.52	15.04	18.19	18.33	23.95	24.61	28.58	40.46	58.53
Cu	$K\beta_5$	B	1	.03	29	5	8.975	1.38109	5		1.57	2.26	5.58	5.95	6.06	7.99	10.43	14.92	18.08	18.19	23.75	24.41	28.38	40.12	58.02
Cu	AK		1		29	5	8.979	1.38059	5		1.57	2.26	5.57	5.94	6.06	7.98	10.43	14.91	18.08	18.17	23.75	24.40	28.33	40.10	57.99

FIGURE 7.3. Portion of a page of a line-to-2θ x-ray spectrometer table giving data for all crystals together. [E. W. White and G. G. Johnson, Jr., "X-Ray Emission and Absorption Wavelengths and Two-Theta Tables," 2nd ed., *ASTM Data Series* **DS-37A**, 293 pp., American Society for Testing and Materials (1970); courtesy of the authors and ASTM.]

A portion of a page of the line-to-2θ part of this table is shown in Figure 7.3. Elements are listed in order of increasing atomic number, and, for each element, spectral lines and absorption edges are listed in order of increasing energy (decreasing wavelength). Satellite line symbols are prefixed by S, for example, Al $SK\alpha_3$. Lines to which no symbol has been assigned are designated by the initial and final atom states of the electron transition resulting in that line, for example, Br M_3–N_1. Absorption edges are designated AK, AL_1, AL_2, AL_3, AM_1, etc. The column headings signify, in order, the element (El), line (Line), order (N), relative intensity (I), atomic number (Z), energy (Kev), and wavelength (Lambda); columns C, R, P, and Δ need not concern us here. The relative intensity values in column I are corrected for variation with atomic number; this is a very useful feature of this table. The next 15 columns give the 2θ angles at which the lines are diffracted, or at which absorption edges fall, for each of the 15 indicated crystals arranged in order of decreasing $2d$ spacing. Incidentally, the column for the graphite crystal (Graph) applies also to quartz(101). The first edition of this table gave data for orders 1, 2, and 3 for all lines, the current edition lists only the first order. Although there is certainly no need for higher-order data for very weak lines, it is most unfortunate that such data was not retained in the current edition for the strong lines.

A portion of a page of the 2θ-to-line part of the ASTM table is shown in Figure 7.4. The lines and absorption edges are listed in order of increasing 2θ or increasing product of order and wavelength $n\lambda$ up to $n\lambda$ ~160 Å. Lines having relative intensity <1 are entered in orders up to 3, the strong lines in orders up to 10. 2θ angles are given for 23 crystals—actually 24 because data for graphite applies also to quartz(101). Data for all 24 crystals is not given for all lines, but the various crystals are phased in and out of the table over their useful regions. The format is the same as for the line-to-2θ part of the table described above, except that here the *Lambda* column gives values of $n\lambda$.

Most 2θ-to-line tables have substantially the format shown in Figure 7.4. However, line-to-2θ tables may have different formats. The format represented by Figure 7.3 lists for each line 2θ values for all crystals covered in the table together. The formats represented by Figures 7.5 and 7.6 give data for each crystal separately. Figure 7.5 shows a portion of a page of a table for the LiF(200) crystal in which each spectral line has its own column and each of the first five orders its own row. Figure 7.6 shows a portion of a page of a LiF(200) table in which each line has its own row and each order its own column.

El	Line	C	N	I	Z	R	KeV	Lambda	P Δ 10⁻	Gyps (020)	ADP (101)	EDDT (020)	PET (002)	Qtz (100)	ADP (200)	Graph (002)	Ge (111)	Si (111)	ADP (112)	NaCl (200)	LiF (200)	LiF (220)	Qtz (203)	Topaz (303)
Gd	$L\gamma_4$		1	.1	64	6	7.553	1.64123	4	12.41	17.75	21.48	21.64	22.21	25.28	28.32	29.10	30.34	31.01	33.83	48.11	70.38	73.31	74.48
Ir	$K\beta_2''$		10	10	77	6	75.516	1.64150	5	12.41	17.75	21.48	21.65	22.22	25.28	28.33	29.11	30.35	31.01	33.84	48.12	70.39	73.33	74.50
Pb	$L\gamma_2$		2	1	82	6	15.099	1.64200	4	12.41	17.76	21.49	21.65	22.22	25.29	28.34			31.02	33.85	48.13	70.42	73.35	74.52
Sm	$K\beta_3$		6	10	62	6	45.281	1.64256	5	12.42	17.76	21.50	21.66	22.23	25.30	28.35	29.13	30.37	31.03	33.86	48.15	70.44	73.38	74.55
La	$K\beta_3$		5	9	57	6	37.714	1.64343	6	12.42	17.77	21.51	21.67	22.24	25.32	28.36	29.14	30.39	31.05	33.88	48.18	70.49	73.43	74.60
Rb	$K\beta_4$		2	.06	37	6	15.082	1.64383	4	12.42	17.77	21.51	21.68	22.25	25.32	28.37			31.06	33.88	48.19	70.90	73.45	74.62
Bk	$L\alpha_2$		2	10	97	3	15.080	1.64400		12.43	17.78	21.51	21.68	22.25	25.32	28.37			31.06	33.89	48.19	70.51	73.46	74.63
Pt	$K\beta_3$		10	12	78	6	75.355	1.64501	6	12.43	17.79	21.53	21.69	22.26	25.34	28.39	29.17	30.42	31.08	33.91	48.23	70.56	73.51	74.68
Dy	$K\alpha_2$		6	50	66	6	45.200	1.64548	6	12.44	17.79	21.53	21.70	22.27	25.35	28.40	29.18	30.42	31.09	33.92	48.24	70.59	73.54	74.71
Pb	Lv	B	2	.01	82	6	15.057	1.64654	5	12.45	17.80	21.55	21.71	22.29	25.36	28.42			31.11	33.94	48.27	70.64	73.59	74.76
Pb	$L\gamma_8$		2	.1	82	6	15.050	1.64730	5	12.45	17.81	21.56	21.72	22.30	25.38	28.43			31.13	33.96	48.30	70.68	73.63	74.81
Ho	$L\beta_1$		1	50	67	6	7.524	1.64750	4	12.45	17.82	21.56	21.73	22.30	25.38	28.43	29.22	30.46	31.13	33.96	48.30	70.69	73.64	74.82
Bi	L_2-N_3	B	2	.01	83	6	15.029	1.64960	4	12.47	17.84	21.59	21.75	22.33	25.41	28.47			31.17	34.01	48.37	70.79	73.75	74.93
Tb	AL_3		1		65	6	7.514	1.64970	4	12.47	17.84	21.59	21.75	22.33	25.41	28.47	29.26	30.50	31.17	34.01	48.37	70.80	73.76	74.93
Tb	$L\beta_5$		1	.1	65	6	7.508	1.65100	4	12.48	17.85	21.61	21.77	22.35	25.43	28.50	29.28	30.53	31.20	34.04	48.41	70.86	73.82	75.00
Pb	$K\alpha_1$		10	100	82	6	74.956	1.65376	6	12.50	17.88	21.64	21.81	22.38	25.48	28.55	29.33	30.58	31.25	34.10	48.50	71.00	73.97	75.15
Pm	$K\beta_2$	M	6	8	61	6	44.929	1.65540	4	12.51	17.90	21.67	21.83	22.41	25.50	28.57	29.36	30.61	31.28	34.13	48.55	71.08	74.05	75.24
Pr	$L\beta_3$	M	2	10	87	6	14.973	1.65578	5	12.52	17.91	21.67	21.84	22.41	25.51	28.58			31.29	34.14	48.56	71.10	74.07	75.26
Th	$L\beta_6$		2	.1	90	6	14.973	1.65580	5	12.52	17.91	21.67	21.84	22.41	25.51	28.58			31.29	34.14	48.56	71.10	74.07	75.26
Sm	$L\gamma_3$		1	1	62	6	7.485	1.65601	5	12.52	17.91	21.67	21.84	22.42	25.51	28.58	29.37	30.62	31.29	34.14	48.57	71.11	74.08	75.27
Yb	$K\alpha_1$		7	100	70	6	52.380	1.65658	6	12.52	17.91	21.68	21.85	22.42	25.52	28.59	29.38	30.64	31.30	34.16	48.59	71.14	74.11	75.30
Bi	$K\alpha_2$		10	50	83	6	74.802	1.65717	6	12.53	17.92	21.69	21.85	22.43	25.53	28.61	29.39	30.65	31.32	34.17	48.61	71.16	74.15	75.33
Pb	L_1-N_1	B	2	.01	82	6	14.960	1.65718	5	12.53	17.92	21.69	21.85	22.43	25.53	28.61			31.32	34.17	48.61	71.17	74.15	75.33
Rb	$K\beta_1$		2	16	37	6	14.959	1.65736	5	12.53	17.92	21.69	21.86	22.43	25.53	28.61			31.32	34.17	48.61	71.17	74.16	75.34
Eu	$L\gamma_1$		1	5	63	6	7.479	1.65740	4	12.53	17.92	21.69	21.86	22.43	25.53	28.61	29.40	30.65	31.32	34.17	48.61	71.18	74.16	75.34
Tl	$L\gamma_{11}$	B	2	.01	81	6	14.957	1.65758	5	12.53	17.92	21.69	21.86	22.44	25.54	28.61			31.32	34.18	48.62	71.18	74.17	75.35
Y	$K\alpha_1$		2	100	39	6	14.956	1.65768	5	12.53	17.93	21.70	21.86	22.44	25.54	28.61			31.33	34.18	48.62	71.19	74.17	75.36
Ni	$K\alpha_1$		1	100	28	6	7.477	1.65791	6	12.53	17.93	21.70	21.86	22.44	25.54	28.62	29.41	30.66	31.33	34.18	48.63	71.20	74.18	75.37
Cm	$L\alpha_1$		2	100	96	3	14.953	1.65800		12.53	17.93	21.70	21.87	22.44	25.54	28.62			31.33	34.19	48.63	71.21	74.19	75.38
Rb	$K\beta_3$		2	8	37	6	14.949	1.65842	5	12.54	17.93	21.71	21.87	22.45	25.55	28.63			31.34	34.19	48.64	71.23	74.21	75.40
Tb	$L\beta_7$		1	.1	65	5	7.474	1.65850	4	12.54	17.93	21.71	21.87	22.45	25.55	28.63	29.42	30.67	31.34	34.20	48.65	71.23	74.21	75.40
Pa	Ln		2	1	91	6	14.944	1.65900	4	12.54	17.94	21.71	21.88	22.46	25.56	28.64			31.35	34.21	48.66	71.26	74.24	75.43
Ni	$K\alpha_{1,2}$		1	150	28	6	7.471	1.65919	6	12.54	17.94	21.72	21.88	22.46	25.56	28.64	29.43	30.68	31.36	34.21	48.67	71.26	74.25	75.44
Nd	$K\alpha_1$		5	100	60	6	37.355	1.65923	6	12.54	17.94	21.72	21.89	22.46	25.56	28.64	29.43	30.69	31.36	34.21	48.67	71.27	74.25	75.44
W	$K\beta_1$		9	25	74	6	67.233	1.65937	6	12.54	17.94	21.72	21.88	22.46	25.56	28.64	29.43	30.69	31.36	34.21	48.67	71.27	74.26	75.45
Ho	$L\beta_4$		1	5	67	6	7.470	1.65950	4	12.54	17.95	21.72	21.89	22.46	25.57	28.65	29.44	30.69	31.36	34.22	48.69	71.28	74.27	75.46
Tl	L_1-N_4	B	2	.01	81	6	14.935	1.66002	5	12.55	17.95	21.73	21.89	22.47	25.58	28.66			31.37	34.23	48.69	71.31	74.29	75.48
Sm	$L\gamma_2$		1	1	62	6	7.465	1.66044	5	12.55	17.96	21.73	21.90	22.48	25.58	28.66	29.45	30.71	31.38	34.24	48.71	71.33	74.32	75.51
Y	$K\alpha_{1,2}$		2	150	39	6	14.931	1.66049	5	12.55	17.96	21.73	21.90	22.48	25.58	28.66			31.38	34.24	48.71	71.33	74.32	75.51
Re	$K\alpha_2$		9	50	75	6	59.708	1.66089	6	12.55	17.96	21.74	21.90	22.48	25.59	28.67	29.46	30.72	31.39	34.25	48.72	71.35	74.34	75.53
Ni	$K\alpha_2$		1	50	28	6	7.460	1.66175	5	12.56	17.97	21.75	21.92	22.49	25.60	28.69	29.48	30.73	31.41	34.27	48.75	71.39	74.38	75.58
Am	$L\gamma_2$		3	1	95	6	22.359	1.66320	4	12.57	17.99	21.77	21.94	22.51	25.62	28.71	29.50	30.76	31.43	34.30	48.79	71.46	74.46	75.65
Hf	Ls	B	1	.01	72	6	7.452	1.66346	5	12.57	17.99	21.77	21.94	22.52	25.63	28.72	29.51	30.77	31.44	34.30	48.80	71.49	74.47	75.67
Ba	$K\beta_2$		5	7	56	6	37.251	1.66385	5	12.58	17.99	21.78	21.94	22.52	25.64	28.72	29.51	30.77	31.45	34.31	48.81	71.50	74.49	75.69
Ta	$K\beta_2'$		9	9	73	6	67.001	1.66510	6	12.59	18.01	21.79	21.95	22.54	25.65	28.75	29.54	30.80	31.47	34.34	48.85	71.56	74.56	75.75

FIGURE 7.4. Portion of a page of a 2θ-to-line x-ray spectrometer table [E. W. White and G. G. Johnson, Jr., "X-Ray Emission and Absorption Wavelengths and Two-Theta Tables," 2nd ed., *ASTM Data Series* **DS-37A**, 293 pp., American Society for Testing and Materials (1970); courtesy of the authors and ASTM.]

ASCENDING ATOMIC NUMBER DATA FOR LIF   2D   4.028

ELEM	ORD	KA1	KA2	KB1	KB3	LA1	LB1	LB2	LY1	LB3	LB5	LL	LB4	LY6
81 TL	1	4.84	4.98	4.27	4.30	34.87	29.19	29.04	24.89	28.78	28.19	40.22	29.90	24.22
	2	9.68	9.97	8.54	8.60	73.64	60.53	60.20	51.06	59.61	58.30	86.89	62.11	49.61
	3	14.55	14.98	12.83	12.91	128.04	98.22	97.57	80.55	96.41	93.88		101.40	78.00
	4	19.44	20.02	17.13	17.25				119.08	167.48	153.90			114.10
	5	24.36	25.09	21.46	21.61									
82 PB	1	4.70	4.84	4.15		33.92	28.22	28.25	24.07	27.84	27.37	39.16	28.96	23.41
	2	9.40	9.68	8.31		71.38	58.36	58.43	49.30	57.52	56.48	84.18		74.96
	3	14.12	14.55	12.49		122.12	94.00	94.13	77.46	92.39	90.43		97.18	74.96
	4	18.86	19.44	16.67			154.41	154.93	113.06	148.42	142.30		180.00	108.45
	5	23.64	24.36	20.88										
83 BI	1	4.58	4.70	4.04	4.07	33.00	27.34	27.43	23.32	26.96	26.58	38.17	28.07	22.65
	2	9.17	9.40	8.09	8.14	69.23	56.42	56.61	47.68	55.58	54.75	81.68	58.04	46.25
	3	13.77	14.12	12.14	12.23	116.87	90.31	90.68	74.64	89.75	87.21	157.56	93.38	72.19
	4	18.40	18.86	16.21	16.33		141.95	143.01	107.87	137.65	133.72		151.96	103.53
	5	23.06	23.64	20.30	20.45									158.15

FIGURE 7.5. Portion of a page of a line-to-2θ x-ray spectrometer table giving data for each crystal separately and having the spectral lines in columns and the diffraction orders in rows. [M. C. Powers, *X-Ray Fluorescent Spectrometer Conversion Tables for Topaz, LiF, NaCl, EDDT, and ADP Crystals*, Philips Electronic Instruments, Inc. (1957); courtesy of the author and Philips Electronic Instruments.]

Tables for energy-dispersive x-ray spectrometry are of *line-to-energy* and *energy-to-line* forms, analogous to line-to-2θ and 2θ-to-line tables, respectively. Portions of pages of such tables are shown in Figures 7.7 and 7.8 respectively. These tables list all lines in the ASTM wavelength tables described above up to wavelength 50 Å (0.25 keV). The column

LIF   2D   4.0267

ELEMENT	LAMBDA	ORDER	1	2	3	4	5
82 PB	.14077	K	ABS	4.01			
82 PB	.14155	KB4	4.03	8.06	12.11	16.17	20.25
82 PB	.14191	KB2	4.04	8.08	12.14	16.21	20.30
82 PB	.14494	KB5	4.13	8.26	12.40	16.56	20.74
82 PB	.14596	KB1	4.15	8.31	12.49	16.67	20.88
82 PB	.14680	KB3	4.18	8.36	12.56	16.77	21.01
82 PB	.16536	KA1	4.71	9.42	14.15	18.91	23.70
82 PB	.16701	KA	4.75	9.52	14.29	19.10	23.94
82 PB	.17029	KA2	4.85	9.70	14.58	19.48	24.41
82 PB	.78152	L I	ABS	22.38			
82 PB	.78588	LY4	22.51	45.95	71.68	102.64	154.76
82 PB	.78710	LYP4	22.54	46.03	71.81	102.87	155.56
82 PB	.80231	LY11	22.99	46.97	73.42	105.69	170.05
82 PB	.81484	LY3	23.35	47.75	74.76	108.08	
82 PB	.81508	L II	ABS	23.36			
82 PB	.81686	LY6	23.41	47.87	74.97	108.47	
82 PB	.82127	LY2	23.54	48.15	75.45	109.34	
82 PB	.82327	LYP5	23.59	48.27	75.67	109.73	
82 PB	.82368	LY8	23.61	48.30	75.71	109.81	
82 PB	.83971	LY1	24.07	49.30	77.45	113.05	
82 PB	.86645	LY5	24.85	50.98	80.41	118.79	
82 PB	.92677	LB9	26.61	54.81	87.33	134.03	
82 PB	.93383	LB10	26.82	55.27	88.17	136.14	
82 PB	.95029	LIII	ABS	27.30			

FIGURE 7.6. Portion of a page of a line-to-2θ x-ray spectrometer table giving data for each crystal separately and having the spectral lines in rows and the diffraction orders in columns. [*X-Ray Wavelengths for Spectrometer*, 5th ed., Diano Corp., Industrial X-Ray Division; courtesy of the publisher.]

Page 2.

El	Line	I	Z	R	KeV	Lambda	El	Line	I	Z	R	KeV	Lambda
Cr	Kα₁	100	24	6	5.414	2.290	Ga	Ll	1	31	6	0.957	12.953
Cr	Kβ₁,₃	18	24	6	5.946	2.085	Ga	Ln	1	31	6	0.984	12.597
Cr	Kβ₅	.03	24	6	5.986	2.071	Ga	Lα₁,₂	100	31	6	1.098	11.292
Mn	Ll	2	25	6	0.556	22.290	Ga	Lβ₁	35	31	6	1.125	11.023
Mn	Ln	1	25	6	0.567	21.850	Ga	Lβ₃,₄	2	31	6	1.197	10.359
Mn	Lα₁,₂	100	25	6	0.637	19.450	Ga	Kα₂	50	31	6	9.223	1.344
Mn	Lβ₁	30	25	6	0.649	19.110	Ga	Kα₁,₂	150	31	6	9.241	1.341
Mn	Lβ₃,₄	.1	25	6	0.721	17.190	Ga	Kα₁	100	31	6	9.250	1.340
Mn	Kα₂	50	25	6	5.887	2.106	Ga	Kβ₃	7	31	6	10.259	1.208
Mn	Kα₁,₂	150	25	6	5.894	2.103	Ga	Kβ₁	14	31	6	10.263	1.208
Mn	Kα₁	100	25	6	5.898	2.102	Ga	Kβ₅	.04	31	6	10.346	1.198
Mn	Kβ₁,₃	20	25	6	6.489	1.910	Ga	Kβ₂	.3	31	6	10.365	1.196
Mn	Kβ₅	.03	25	6	6.534	1.897	Ge	Ll	1	32	6	1.036	11.965
Fe	Ll	8	26	6	0.615	20.150	Ge	Ln	1	32	6	1.068	11.609
Fe	Ln	2	26	6	0.628	19.750	Ge	Lα₁,₂	100	32	6	1.188	10.436
Fe	Lα₁,₂	100	26	6	0.705	17.590	Ge	Lβ₁	35	32	6	1.218	10.175
Fe	Lβ₁	20	26	6	0.718	17.260	Ge	Lβ₄	1	32	6	1.286	9.640
Fe	Lβ₃,₄	.5	26	6	0.792	15.650	Ge	Lβ₃	1	32	6	1.294	9.581
Fe	Kα₂	50	26	6	6.390	1.940	Ge	Kα₂	50	32	6	9.854	1.258
Fe	Kα₁,₂	150	26	6	6.398	1.937	Ge	Kα₁,₂	150	32	6	9.874	1.255
Fe	Kα₁	100	26	6	6.403	1.936	Ge	Kα₁	100	32	6	9.885	1.254
Fe	Kβ₁,₃	20	26	6	7.057	1.757	Ge	Kβ₃	7	32	6	10.976	1.129
Fe	Kβ₅	.03	26	6	7.107	1.744	Ge	Kβ₁	14	32	6	10.980	1.129
Co	Ll	9	27	6	0.678	18.292	Ge	Kβ₅	.05	32	6	11.073	1.119
Co	Ln	2	27	6	0.694	17.870	Ge	Kβ₂	.5	32	6	11.099	1.117

FIGURE 7.7. Portion of a page of a line-to-energy x-ray spectrometer table. [G. G. Johnson, Jr., and E. W. White, "X-Ray Emission Wavelengths and keV Tables for Nondiffractive Analysis," *ASTM Data Series* **DS-46**, 38 pp., American Society for Testing and Materials (1970); courtesy of the authors and ASTM.]

headings are the same as for Figures 7.3 and 7.4, insofar as applicable. Inasmuch as most 2θ tables also list photon and excitation energies (keV), the principal difference in the energy tables—aside from the absence of crystal data—is that no higher-order lines are listed; of course, orders have no significance in nondiffractive systems.

Tables for electron-probe microanalysis have still another form. X-ray spectrometers for use on these instruments usually have 2θ-drive drums (odometers) calibrated to indicate wavelength in angstroms directly when used with a specified crystal, usually LiF(200) or ADP(101). If a crystal other than the one for which it is calibrated is used on such an instrument, the odometer no longer indicates wavelength, but a reading related to wavelength by

$$R_X = (d_C/d_X)\lambda_L \qquad (7.1)$$

where R_X is the odometer reading for the interchanged crystal X; d_C and d_X are the interplanar spacings (Å) for the diffracting planes of, respectively, the crystal for which the odometer was calibrated C and the interchanged crystal X; and λ_L is the wavelength of the line, or the odometer reading for that line with the original crystal C. Interchange tables are available that give odometer settings for various crystals used on a spectrometer calibrated for a specified crystal. These tables have line-to-reading and reading-to-line forms analogous to the forms of conventional x-ray spectrometer tables. Figure 7.9 shows a portion of a page of the reading-to-line form of such a table. The first seven column headings have the same significance as in Figure 7.3. The six columns ADP/LSD, ADP/KAP, etc., give odometer readings for LSD, KAP, etc., used on a spectrometer calibrated for ADP. The six columns LIF/LSD, LIF/KAP, etc., have the same significance for a spectrometer calibrated for LiF. The line-to-reading part of this table has the format shown in Figure 7.3 except that the crystal columns are replaced by the crystal interchange columns, ADP/LSD, etc., and LIF/LSD, etc. A later edition of this table gives data only for spectrometers calibrated for LiF(200) crystals. The line-to-reading table gives data for 10 interchange crystals. The reading-to-line table gives data for 19 interchange crystals, which are phased in and out of the table over their useful wavelength regions, as is done in the table shown in Figure 7.4.

Page 24.

El	Line	I	Z	R	KeV	Lambda	El	Line	I	Z	R	KeV	Lambda
Th	$L\gamma_2$	1	90	6	19.302	0.642	U	$L\beta_3$	6	92	6	17.452	0.710
Ru	$K\alpha_1$	100	44	6	19.276	0.643	Mo	$K\alpha_{1,2}$	150	42	6	17.441	0.711
Ru	$K\alpha_{1,2}$	150	44	6	19.233	0.645	Mo	$K\alpha_2$	50	42	6	17.371	0.714
Ra	$L\gamma_{13}$	.01	88	6	19.215	0.645	Pr	$L\gamma_1$	10	87	6	17.300	0.717
Ra	$L_1-O_{4,5}$	.01	88	6	19.165	0.647	Ra	$L\gamma_5$	.1	88	6	17.271	0.718
Ru	$K\alpha_2$	50	44	6	19.147	0.547	Pu	$L\beta_2$	20	94	6	17.252	0.719
Th	L_1-N_1	.01	90	6	19.143	0.648	U	$L\beta_1$	50	92	6	17.217	0.720
Pu	$L\beta_{10}$	.01	94	6	19.124	0.649	Pu	$L\beta_{15}$	1	94	6	17.205	0.720
Am	$L\beta_3$	6	95	6	19.103	0.649	U	$L_3-P_{4,5}$	.01	92	6	17.159	0.722
Ra	$L\gamma_4$	.1	88	6	19.081	0.650	Th	$L\beta_6$	.01	90	6	17.135	0.723
Ra	$L\gamma_4 P$	.1	88	6	19.032	0.651	U	$L_3-P_{2,3}$	.01	92	6	17.115	0.724
Th	L_2-N_5	.01	90	6	19.009	0.652	U	L_3-P_1	.01	92	6	17.093	0.725
Cf	$L\beta_2$	20	98	3	18.983	0.653	U	$L\beta_5$	1	92	6	17.067	0.726
Th	$L\gamma_1$	10	90	6	18.979	0.653	Np	$L\beta_4$	4	93	6	17.058	0.727
Nb	$K\beta_4$	.01	41	6	18.978	0.653	Y	$K\beta_4$	.01	39	6	17.033	0.728
Nb	$K\beta_2$	4	41	6	18.949	0.654	Y	$K\beta_2$	4	39	6	17.013	0.729
Pa	$L\gamma_5$	.1	91	6	18.925	0.655	Th	$L\beta_{10}$	.01	90	6	16.978	0.730
Am	$L\beta_1$	50	95	6	18.949	0.653	U	L_3-O_3	.01	92	6	16.960	0.731
Th	L_2-N_3	.01	90	6	18.725	0.662	Pa	$L\beta_6$	6	91	6	16.927	0.732
Ra	$L\gamma_{11}$	.01	88	6	18.629	0.665	U	L_3-O_2	.01	92	6	16.904	0.733
Nb	$K\beta_1$	.6	41	6	18.619	0.666	Am	$L\beta_4$	.1	95	6	16.884	0.734
Nb	$K\beta_3$	7	41	5	18.603	0.666	Y	$K\beta_5$	.07	39	6	16.877	0.734
Ra	L_1-N_4	.01	88	6	18.596	0.667	U	$L\beta_7$	.1	92	6	16.842	0.736
Pu	$L\beta_3$	6	94	6	18.537	0.669	Np	$L\beta_2$	20	93	6	16.837	0.736
Bk	$L\beta_2$	20	97	3	18.529	0.669	U	Lu	.01	92	6	16.783	0.739

FIGURE 7.8. Portion of a page of an energy-to-line x-ray spectrometer table. [G. G. Johnson, Jr., and E. W. White, "X-Ray Emission Wavelengths and keV Tables for Nondiffractive Analysis," *ASTM Data Series* **DS-46**, 38 pp., American Society for Testing and Materials (1970); courtesy of the authors and ASTM.]

PAGE 84.0

EL	LINE	N	I	Z	KEV	LAMBDA	ACP LSD	ACP KAP	ALP EDDT	ALP SIO2	ADP NACL	ADP LIF	LIF LSD	LIF KAP	LIF ADP	LIF EDDT	LIF SIO2	LIF NACL
TL	L3-M3	3	1	81	12.05	3.0860		1.2324	3.7778	4.9101	5.8712	8.1533		0.4666	1.1680	1.4109	1.8584	2.2013
W	LG4+	1	1	74	12.05	3.0860		1.2329	3.7278	4.9101	5.8712	8.1533		0.4666	1.1680	1.4109	1.8584	2.2013
CA	SKB5	1	20	20	4.02	3.0864		1.2330	3.7283	4.9108	5.8712	8.1283		0.4667	1.1681	1.4111	1.8587	2.2016
EU	LG4,	1	1	63	8.03	3.0876		1.2335	3.7298	4.9127	5.8642	8.1576		0.4669	1.1686	1.4117	1.8594	2.2044
RH	KA2	5	50	45	20.07	3.0884		1.2337	3.7705	4.9136	5.8251	8.1591		0.4669	1.1688	1.4120	1.8598	2.2048
CU	KA2	2	50	29	8.03	3.0886		1.2339	3.7110	4.9143	5.8261	8.1602		0.4670	1.1690	1.4121	1.8600	2.2051
TM	LG4	2	4	69	8.03	3.0886		1.2341	3.7110	4.9143	5.8261	8.1602		0.4670	1.1690	1.4121	1.8600	2.2051
RE	SLB2+	10	1	75	12.04	3.0890		1.2342	3.7317	4.9152	5.8269	8.1613		0.4671	1.1691	1.4123	1.8604	2.2054
SM	KA1	10	100	62	40.13	3.0892		1.2343	3.7322	4.9159	5.8272	8.1618		0.4671	1.1691	1.4123	1.8604	2.2054
CA	KB1	1	15	20	4.01	3.0896		1.2343	3.7322	4.9159	5.8420	8.1629		0.4672	1.1694	1.4126	1.8606	2.2058
CF	LG1	6	10	98	24.07	3.0900		1.2345	3.7326	4.9164	5.8287	8.1638		0.4672	1.1695	1.4128	1.8608	2.2061
PR	LL1	8	100	87	12.03	3.0902		1.2346	3.7320	4.9168	5.8291	8.1644		0.4672	1.1696	1.4140	1.8611	2.2063
BA	KA1,2	8	150	56	32.07	3.0926		1.2355	3.7358	4.9206	5.8337	8.1708		0.4676	1.1705	1.4140	1.8624	2.2080
PA	LI-01	4	20	91	16.02	3.0946		1.2363	3.7382	4.9238	5.8374	8.1761		0.4679	1.1714	1.4149	1.8636	2.2094
W	LI-01	3	1	74	12.02	3.0951		1.2366	3.7388	4.9246	5.8383	8.1773		0.4680	1.1718	1.4151	1.8644	2.2097
AU	SLB2+	1	75	79	12.01	3.0955		1.2372	3.7409	4.9259	5.8399	8.1795		0.4681	1.1721	1.4155	1.8649	2.2103
RE	LG2	9	100	75	12.01	3.0960		1.2372	3.7409	4.9273	5.8416	8.1819		0.4682	1.1721	1.4159	1.8649	2.2110
RE	KA1	1	50	75	19.96	3.0969		1.2372	3.7409	4.9273	5.8417	8.1821		0.4682	1.1721	1.4159	1.8650	2.2110
PB	LI-M1	10	150	82	39.02	3.0969		1.2373	3.7411	4.9277	5.8420	8.1824		0.4683	1.1722	1.4160	1.8651	2.2111
MO	LB10	2	1	53	12.01	3.0970		1.2373	3.7411	4.9277	5.8420	8.1824		0.4683	1.1722	1.4160	1.8651	2.2111
IN	KA2	6	50	49	24.00	3.0992		1.2382	3.7438	4.9312	5.8461	8.1883		0.4686	1.1730	1.4170	1.8664	2.2127
MG	LB3	1	80	12	11.96	3.1004		1.2388	3.7453	4.9331	5.8484	8.1914		0.4688	1.1735	1.4175	1.8671	2.2136
CA	SKB+	1	20	20	4.00	3.1011		1.2394	3.7461	4.9342	5.8497	8.1933		0.4689	1.1737	1.4179	1.8675	2.2141
AU	SLB5+	3	1	79	11.98	3.1022		1.2400	3.7474	4.9359	5.8518	8.1962		0.4691	1.1741	1.4184	1.8682	2.2149
U	LI-M2	1	5	92	11.98	3.1037		1.2400	3.7493	4.9383	5.8547	8.2002		0.4693	1.1747	1.4191	1.8691	2.2159
BI	LB2-M2	3	42	83	11.96	3.1038		1.2404	3.7493	4.9384	5.8547	8.2003		0.4693	1.1751	1.4191	1.8691	2.2160
RO	LG2	9	50		19.96	3.1047		1.2405	3.7508	4.9399	5.8565	8.2028		0.4694	1.1752	1.4195	1.8697	2.2166
BK	LB1	1	150		39.02	3.1050		1.2406	3.7508	4.9403	5.8570	8.2035		0.4695	1.1753	1.4194	1.8699	2.2168
BK	KA1,2	10	62			3.1052		1.2406	3.7511	4.9407	5.8575	8.2042		0.4695	1.1753	1.4197	1.8700	2.2170
L	SLA9	2	1	53	11.95	3.1055		1.2407	3.7511	4.9412	5.8580	8.2049		0.4696	1.1754	1.4199	1.8702	2.2172
BR	SKA4	3	1	35	11.97	3.1058		1.2420	3.7518	4.9417	5.8586	8.2057		0.4696	1.1755	1.4200	1.8704	2.2174
BR	SKA3,	3	1	35	11.96	3.1091		1.2421	3.7558	4.9469	5.8648	8.2145		0.4701	1.1768	1.4215	1.8724	2.2198
RE	LG6	1	1	75	35.06	3.1109		1.2428	3.7580	4.9498	5.8683	8.2192		0.4704	1.1774	1.4224	1.8735	2.2211
PR	KA1,2		150	59	35.88	3.1112		1.2430	3.7583	4.9503	5.8688	8.2200		0.4704	1.1774	1.4225	1.8736	2.2213
AC	SLB2+5	4	70	89	11.95	3.1124		1.2434	3.7598	4.9522	5.8711	8.2232		0.4708	1.1780	1.4230	1.8746	2.2222
CS	KB2	5	55	55	35.02	3.1128		1.2436	3.7602	4.9527	5.8717	8.2241		0.4707	1.1781	1.4232	1.8744	2.2224
CS	KB2	7	55		37.02	3.1140		1.2444	3.7627	4.9560	5.8756	8.2296		0.4710	1.1790	1.4241	1.8758	2.2239
SM	LB9	1	1	62	27.06	3.1149		1.2444	3.7628	4.9563	5.8758	8.2301		0.4710	1.1790	1.4242	1.8759	2.2239
L	SLA8	1	3	53	11.96	3.1150		1.2445	3.7629	4.9564	5.8760	8.2301		0.4710	1.1790	1.4242	1.8759	2.2240

FIGURE 7.9. Portion of a page of a reading-to-line x-ray spectrometer crystal-interchange table for use when certain other crystals are used on spectrometers calibrated directly in angstroms for LiF and ADP crystals. [E. W. White, G. V. Gibbs, G. G. Johnson, Jr., and G. R. Zechman, Jr., "X-Ray Wavelengths and Crystal Interchange Settings for Wavelength-Geared, Curved-Crystal Spectrometers," *Special Publication 3-64*, 195 pp., Mineral Industries Experiment Station, Pennsylvania State University (1965); courtesy of the authors and Pennsylvania State University.]

7.1.2.2. Identification of Peaks

The product of a qualitative analysis on a wavelength-dispersive spectrometer is a wavelength spectrum—a ratemeter–recorder chart of peaks on a grid of intensity *versus* 2θ, or wavelength. The product on an energy-dispersive spectrometer is an energy spectrum—an X–Y recorder chart or a photograph of a cathode-ray tube display of peaks on a grid of intensity *versus* photon energy. There is a systematic procedure for identification of the peaks on these charts or photographs by use of tables. The following discussion is in terms of wavelength spectra, but is applicable as well to energy spectra on X–Y recorder charts. The last three paragraphs of this section give special techniques applicable to cathode-ray tube displays of energy spectra.

Peaks may be as much as $0.5°$ 2θ away from their tabulated values because of slight misalignment of the goniometer, or lag in ratemeter–recorder response, and this displacement may vary progressively from start to end of the spectrum. Moreover, it may be difficult to read the chart more precisely than $\pm 0.2°$ 2θ. Peaks in energy-dispersed spectra may also be displaced from their tabulated values due to miscalibration of the energy scale, and it may be difficult to read this scale accurately (Figure 7.1).

First, by use of a line-to-2θ (or line-to-energy) table, all peaks known to be present are identified. These include peaks of elements known to be present in the specimen and coherently and incoherently scattered lines of the target and its contaminants. The incoherently scattered peaks appear as relatively broad peaks on the high-2θ (long-wavelength) sides of the coherently scattered peaks. All scattered peaks are more intense the lower the mean atomic number of the specimen. The incoherent-to-coherent intensity ratio increases as wavelength decreases (photon energy increases) and mean atomic number of the specimen decreases.

Then, by use of a 2θ-to-line (or energy-to-line) table, all remaining peaks are identified, starting with the most intense. This peak should be a $K\alpha$, or $L\alpha_1$ line. Once the element associated with the strongest line is identified, the other lines in this series are identified with a line-to-2θ table. For example, if the strongest line is Au $L\alpha_1$, then Au $L\beta_1$, $L\beta_2$, $L\beta_3$, $L\gamma_1$, etc., are found. The remaining lines are then identified in about the same way, again starting with the most intense.

One soon learns to recognize groups of associated lines. The K x-ray spectrum of an element consists of the strong $K\alpha$ line and the weaker ($\frac{1}{5}$ or less) $K\beta$ line at shorter wavelength (lower 2θ). In the second and higher orders, the $K\alpha$ peak splits into the $K\alpha_1\alpha_2$ doublet in 2 : 1 intensity ratio, the more intense $K\alpha_1$ component occurring at lower 2θ. The L x-ray spectrum of an element usually consists of at least the following three lines more or less equally spaced at progressively shorter wavelengths (lower 2θ): the strongest (usually) $L\alpha_1$, the somewhat weaker (usually) $L\beta_1$, and the still weaker ($\sim$0.2 or less) $L\gamma_1$. For elements of atomic number $\sim$82 (lead), the relatively intense $L\beta_2$ line lies very close to the $L\beta_1$ line and may be unresolved. At lower atomic number, $L\beta_2$ lies at shorter wavelength (lower 2θ) than $L\beta_1$, and at higher atomic number, at longer wavelength. If the element is present in high concentration, the many other L lines appear, most of them between $L\alpha_1$ and $L\gamma_1$, although Ll lies at longer wavelength (higher 2θ) than $L\alpha_1$, and $L\gamma_2$ lies at shorter wavelength (lower 2θ) than $L\gamma_1$.

Insofar as possible, more than one of its lines should be identified for an element to be regarded as present. If a peak is believed to be Fe $K\alpha$, the weaker ($\sim\frac{1}{5}$) Fe $K\beta$ should be present. If a peak is believed to be Pb $L\beta_1$, the slightly stronger Pb $L\alpha_1$, and weaker Pb $L\gamma_1$ peaks should be present. If relatively strong Mo $K\alpha$–$K\beta$ peaks are present, their second- and perhaps third-order peaks should be present. If the associated peaks cannot be found, their absence must be accounted for. Some examples follow: If a $K\alpha$ peak is very weak, the $K\beta$ peak may not be distinguishable from the background; $K\beta$ lines of elements 9–17 (fluorine to chlorine) are very weak relative to the $K\alpha$ lines; a peak may be lost in an interfering peak; and second-order peaks may undergo extinction by the crystal.

Relative intensities of peaks must be considered and gross anomalies accounted for. For example, suppose that a strong peak is believed to be Cu $K\alpha$ but the Cu $K\beta$ peak is very weak, when it should be $\sim\frac{1}{5}$ as intense as Cu $K\alpha$. Perhaps another line is superimposed on Cu $K\alpha$, giving it an anomalously high apparent intensity, or perhaps a matrix element is preferentially absorbing the Cu $K\beta$—nickel, for example. Then again, perhaps the strong line is not Cu $K\alpha$ after all! In considering relative intensities, one can be guided by certain general relationships: The relative intensities of the lines within a series (K, L, etc.) are given in the tables, especially *ASTM Data Series* **37-A** (see the reading list at the end of this chapter), and approximately in Figure 1.10; in the same 2θ region and for the same concentrations, first-order $K\alpha$ lines are 7–10 times as intense as first-order $L\alpha_1$ lines; and for LiF crystals, second-order lines are $\sim\frac{1}{10}$ as intense as first-order lines.

A peak may appear anomalously intense if: (1) another peak—an emission line from a sample element or a coherently or incoherently scattered target line—is superimposed on it; (2) it lies on the short-wavelength side of an absorption edge of the detector gas or of the iodine in a scintillation counter phosphor; and/or (3) it is preferentially excited by a high-concentration matrix-element line or a strong x-ray tube target line. A peak may appear anomalously weak if: (1) it lies on the short-wavelength side of an absorption edge of a high-concentration matrix element; and/or (2) it lies on the long-wavelength side of a detector absorption edge.

Energy-dispersed x-ray spectra on cathode-ray tube displays are subject to special techniques that aid in peak identification. Even instruments not having computer-control units provide two features that facilitate identification by use of tables. An energy scale along the bottom of the cathode-ray display (Figure 7.10) permits estimation of the photon energy corresponding to each peak. Most instruments also provide a manually movable electronic bright-line cursor that can be placed at the centroid of each peak in turn, and the instrument then displays numerically the photon energy corresponding to that peak. (The peak *centroid* is the maximum or "top" of the peak.) In Figure 7.10A, a bright-line cursor is shown at the centroid of the Ta $L\alpha_1$ peak, and the Ta $L\alpha_1$ photon energy is displayed at the top (8143 EV). However, if the energy-dispersive spectrometer has a computer-control unit, extremely rapid, convenient, all-electronic means of peak identification are possible. Two examples follow.

Most instruments provide for electronically marking the cathode-ray tube display with bright-line ticks showing the positions of the $K\alpha$, $K\beta$, $L\alpha_1$, $L\beta_1$, $L\beta_2$, $L\gamma_1$, $M\alpha$, and $M\beta$ lines of any selected element and with an alphanumeric display of the atomic number and symbol of that element. All of the eight lines that lie within the displayed spectral region are thus marked simply by entering the atomic number of the element on a telephone-type dial or digital keyboard. The α, β, and γ ticks are long, medium, and short, respectively, suggestive of their relative intensities. The spectrum in Figure 7.10B shows this mode of peak identification.

In another instrument, an electronic bright-line cursor can be moved over the cathode-ray tube display by means of a knob on the control unit. When the cursor is set at the centroid of a peak of interest, an alphanumeric list is displayed of the atomic numbers and symbols of any elements having a major line within a preset energy interval of the cursor. A maximum of eight such elements can be displayed. The major lines are the same eight lines cited above. The preset energy interval can be up to ± 0.250 keV. Of course, the bright-line tick method (above) can then be applied to each of the prospective elements—at least, to each that cannot be disregarded on the basis of prior knowledge. Figure 7.10A shows application of the

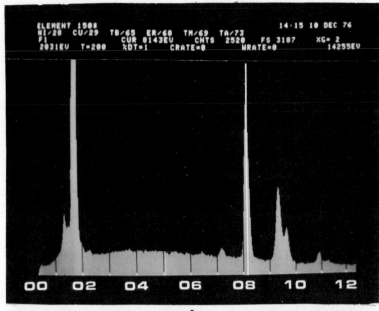

A

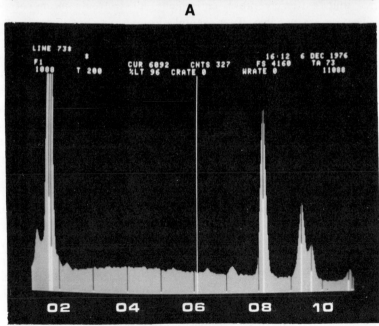

B

moving-cursor method to the strongest peak in a spectrum. Note that one of the six prospective elements is tantalum (TA/73). Figure 7.10B shows the L- and M-line marker ticks for tantalum, confirming it to be the analyte.

7.2. SEMIQUANTITATIVE ANALYSIS

7.2.1. General

In quantitative x-ray spectrometric analysis (Chapter 9), analyte concentration is derived from precisely measured and carefully corrected intensity data and from standards very similar to the samples in analyte concentration, matrix composition, surface texture, and particle size. The result is analyte concentration within relatively narrow known limits of precision and accuracy.

Semiquantitative x-ray spectrometric analysis differs from quantitative analysis in two basic features: The intensities are measured with speed and convenience, rather than precision, as primary objectives; and the standards used are applicable to a variety of more or less related sample types that may have a relatively wide range of analyte concentration, matrix composition, surface texture, and particle size. When the analytical samples happen to be very similar to the standards, a semiquantitative analysis may be very accurate, but the more they differ, the poorer the accuracy is.

If two specimens A and B are identical except that A has twice the,

---

FIGURE 7.10. Electronic peak identification in energy-dispersed x-ray spectra.

A. X-ray energy spectrum in the 0–12-keV region. The movable bright-line cursor is placed at the centroid of the peak to be identified. The instrument displays alphanumerically the photon energy corresponding to the position of the cursor (CUR 8143EV) and the atomic numbers and symbols of elements having strong lines ($K\alpha$, $K\beta$, $L\alpha_1$, $L\beta_1$, $L\beta_2$, $L\gamma_1$, $M\alpha$, $M\beta$) within ± 150 eV (preselected and displayed, ELEMENT 150*) of the cursor; six such elements are indicated (NI/28, CU/29, TB/65, ER/68, TM/69, TA/73). This mode of peak identification is exclusive to Princeton Gamma Tech instruments.

B. The atomic number of a prospective element (tantalum) is entered into the instrument. The instrument displays alphanumerically the symbol and atomic number of this element (TA 73). The spectrum is automatically marked with bright-line ticks showing the positions of any of the same eight lines (see above) of that element lying in the energy region displayed. Six Ta lines are marked (from extreme left) $M\alpha$, $M\beta$, $L\alpha_1$, $L\beta_1$, $L\beta_2$, $L\gamma_1$. The cursor lies at an arbitrary setting of 6092 eV (CUR 6092). This mode of peak identification is available on substantially all manufacturers' instruments.

(*Note*: On these displays, any alphanumeric data that is not referred to specifically above need not concern the reader.)

say, iron concentration of B, the Fe $K\alpha$ net intensity from A will be about twice that from B. However, this simple relationship is not generally true for different elements in the *same* specimen for many reasons, principally the following. (1) *Excitation potential*—each element has its own excitation potential, and the amount by which this potential is exceeded by the x-ray tube operating potential affects the excitation efficiency. (2) *Atomic weight* —if a specimen is, say, 50 wt% iron (at. wt. 56) and 50 wt% silicon (at. wt. 28), there are about twice as many silicon atoms as iron atoms. (3) *Surface texture* and/or *particle size* of the specimen affect analyte lines having different wavelengths to different degrees. (4) *Absorption–enhancement effects* do the same. (5) Finally, *crystal "reflectivity"* and *detector quantum efficiency* also vary with wavelength. Thus, simple comparison of peak heights on a ratemeter–recorder chart, or of scaled intensities, does not necessarily permit semiquantitative analysis (estimation) of relative concentrations, although it does permit designation of the elements present as major, minor, or trace constituents.

Intensity data for semiquantitative analysis is often derived directly from ratemeter–recorder charts (Section 5.3.1), but may also be obtained from digital scalers. Standards may be obtained by independent chemical analysis of selected samples of the same or related types, or from any of the sources given in Section 10.3.2. The samples and standards are subjected to the same measurement procedure to obtain intensity data from which to prepare calibration curves or equations.

Tables 7.1 and 7.2 are examples of semiquantitative analysis by application of simple corrections to ratemeter–recorder chart peak heights. Table 7.1 shows data for analysis of a mixture of oxides of selenium, zirconium, and antimony in concentrations 25, 50, and 25 wt%, respectively, containing a trace iodate impurity. This is a very favorable case: The three major heavy elements do not differ widely in atomic number ($Z = 34$, 40, and 51 for selenium, zirconium, and antimony, respectively); the same line ($K\beta$) is measured for each; and the only other major element is oxygen. The table shows that the relative intensities (chart peak heights above background) give a very passable semiquantitative analysis. Correction for oxide form improves the results only slightly, although presumably this refinement would be more beneficial with less favorable systems.

Table 7.2 shows data for analysis of a 50–50-wt% tin–lead solder containing trace antimony and bismuth impurities. This is a more complex case: The two major elements differ widely in atomic number ($Z = 50$ and 82 for tin and lead, respectively); lines of different series are measured for each (Sn $K\alpha$, Pb $L\alpha_1$); and the excitation potentials differ substantially

TABLE 7.1. Semiquantitative Analysis of an Oxide Mixture[a]

Element and line	Oxide	Oxide factor[b]	Relative intensity,[c] I_{rel}	Composition (wt%)	
				Estimated	True
Se $K\beta$	SeO_2	1.40 $\times$ 28 = 39.2		39.2/133 = 29	25
Zr $K\beta$	ZrO_2	1.35 $\times$ 48 = 64.8		64.8/133 = 49	50
Sb $K\beta$	Sb_2O_5	1.33 $\times$ 20 = 26.6		26.6/133 = 20	25
I $K\beta$	IO_3^-	1.38 $\times$ 2 = 2.8		2.8/133 = 2	Trace
			133.4	100	

[a] From R. Jenkins, *An Introduction to X-Ray Spectrometry*, Heyden & Son Ltd., London, p. 101 (1974).
[b] M/nA, where M is the molecular weight of the oxide, A is the atomic weight of the heavy element, and n is the number of atoms of heavy element in a molecule of oxide: $SeO_2/Se = 111/79$; $ZrO_2/Zr = 123/91$; $Sb_2O_5/2Sb = 324/244$; $IO_3^-/I = 175/127$.
[c] Ratemeter–recorder chart divisions.

($V_{SnK} = 29$ kV, $V_{PbLIII} = 13$ kV). The table shows that the uncorrected relative intensities are unsuitable, and how correction is made for line series by use of "weighted fluorescent yields" and for excitation potential by use of Equation (1.22). The corrected semiquantitative analysis is in excellent agreement with the known composition.

7.2.2. Comparison Analysis

An energy-dispersive x-ray spectrometer having cathode-ray tube display, computer, and storage units makes possible a new type of x-ray spectrometric analysis known variously as *comparison*, "signature," or "fingerprint" analysis. This technique consists in direct visual or electronic comparison of the x-ray energy spectra of the unknown sample and prospective known materials. The spectra may be compared side by side or superimposed. For example, in a 400-channel analyzer, channels 1–200 may display the sample spectrum while channels 201–400 display simultaneously, one at a time, on the same energy scale (eV/channel), various reference spectra for direct visual comparison. This technique is shown in Figures 7.1A and C. Alternatively, the sample and comparison spectra may be superimposed, both as bar graphs, or one as bars, the other as dots. Either of the two superimposed spectra can be displaced vertically relative to the other to any desired degree of overlap by means of a manual offset

TABLE 7.2. Semiquantitative Analysis of a Tin-Lead Solder[a]

Element and line	Relative intensity,[b] I_{rel}	Weighted fluorescent yield,[c] ω'	Excitation potential, V_{exc} (kV)	Excitation factor,[d] F	$\omega'F$	Corrected intensity,[e] $I_{rel,cor}$	Composition (wt%) Estimated,[f] C_{est}	Composition (wt%) True
Sn $K\alpha$	250	0.7	29	130	91	2.75	49.2	50
Sb $K\alpha$	2.5	0.7	30	121	85	0.03	0.5	Trace
Pb $L\alpha_1$	180	0.2	13	323	64.6	2.79	49.9	50
Bi $L\alpha_1$	1.5	0.2	13	323	64.6	0.02	0.4	Trace
						Σ 5.59		Σ 100.0

[a] From R. Jenkins, *An Introduction to X-Ray Spectrometry*, Heyden & Son Ltd., London, p. 102 (1974).

[b] Ratemeter-recorder chart divisions.

[c] The weighted fluorescent yield $\omega' = \omega g$, where ω and g are the Sn K fluorescent yield and probability factor [Equation (4.2)], respectively. For example, the weighted fluorescent yield for Sn $K\alpha$ is the Sn K fluorescent yield ($\omega_{SnK} = 0.845$) multiplied by $g_{SnK\alpha}$, the fraction of all Sn K-line photons that are Sn $K\alpha$ (Sn $K\alpha_1$ or Sn $K\alpha_2$) photons. This fraction is the relative intensity of Sn $K\alpha$ divided by the sum of the relative intensities of all the principal Sn K lines ($I_{rel} \geqslant 1$):

Sn $K\alpha_1$:Sn $K\alpha_2$:Sn $K\beta_1$:Sn $K\beta_2$:Sn $K\beta_3 = 100$:50:19:5:9

$(100 + 50)/(100 + 50 + 19 + 5 + 9) = 150/183 = 0.82$

$0.845 \times 0.82 = 0.7$

[d] $F = (V - V_{exc})^{1.7}$ [Equation (1.22)], where V is x-ray tube operating potential, 50 kV in this case.

[e] $I_{rel,cor} = I_{rel}/(\omega'F)$.

[f] $C_{est} = (I_{rel,cor}/\Sigma I_{rel,cor}) \times 100$.

The best source of relative intensities is *ASTM Data Series* DS-37A (see the reading list at the end of this chapter).

control. This technique is shown in Figures 7.1A and B. The peak intensities in the sample and reference spectra can be compared electronically, eliminating subjectivity from the comparison. The reference spectra can be accumulated "on the spot" from the reference materials, entered from magnetic or punched tape, or stored in the computer memory, if it has sufficient capacity.

It is evident that comparison analysis combines features of qualitative, semiquantitative, and quantitative analysis. Like qualitative, but unlike semiquantitative and quantitative analysis, comparison analysis does not require actual measurement and correction of intensities; this is because when the correct comparison spectrum is found, the sample and reference material are in fact the same—the same analytes in substantially the same concentrations in the same matrix. The terms "signature" and "fingerprint" analysis arise from the fact that the x-ray energy spectrum of a material actually constitutes a kind of "signature" or "fingerprint" from which that material can be identified unequivocally.

The comparison method is extremely valuable for rapid, convenient identification and sorting of materials when there is a limited number of possibilities. For example, suppose that a laboratory is required to identify incoming metal samples as one of, say, 30 alloys. The x-ray energy spectra ("fingerprints" or "signatures") of the 30 alloys are recorded on magnetic or punched tape or stored in the computer memory. The spectrum of each sample is rapidly compared with the taped or stored spectra one by one and thereby identified. The comparison can be done visually on the cathode-ray tube display or electronically by the computer. Other examples are found in the field of forensic investigation: Counterfeit currency is readily identified by comparison of the x-ray spectra of the metallic elements in the inks on the suspected and genuine currency. The manufacturer and model of an automobile used in a crime is established by comparison of the spectra of paint chips found at the crime scene and taken from test panels provided by the automobile manufacturers.

SUGGESTED READING

Adler, I., *X-Ray Emission Spectrography in Geology*, Elsevier, Amsterdam (1966); pp. 79–88.

Bertin, E. P., *Principles and Practice of X-Ray Spectrometric Analysis*, 2nd ed., Plenum Press, New York; Chap. 10, pp. 435–457.

Jenkins, R., *Introduction to X-Ray Spectrometry*, Heyden, London (1974); Chap. 5, pp. 99–106.

X-Ray Spectrometer Tables

ANONYMOUS, *X-Ray Wavelengths for Spectrometer*, 5th ed., Diano Corp., Industrial X-Ray Division, Woburn, Mass. (1969).

JOHNSON, G. G., JR., and E. W. WHITE, "X-Ray Emission Wavelengths and keV Tables for Nondiffractive Analysis," *ASTM Data Series* **DS-46**, American Society for Testing and Materials, Philadelphia (1970).

POWERS, M. C., *X-Ray Fluorescent Spectrometer Conversion Tables for Topaz, LiF, NaCl, EDDT, and ADP Crystals*, Philips Electronic Instruments, Inc., Mount Vernon, N.Y. (1957).

WHITE, E. W., and G. G. JOHNSON, JR., "X-Ray Emission and Absorption-Edge Wavelengths and Interchange Settings for LiF-Geared, Curved-Crystal Spectrometers," *Special Publication* **1-70**, Earth and Mineral Sciences Experiment Station, Pennsylvania State University, University Park, Pa. (1970).

WHITE, E. W., and G. G. JOHNSON, JR., "X-Ray Emission and Absorption Wavelengths and Two-Theta Tables," 2nd ed., *ASTM Data Series* **DS-37A**, American Society for Testing and Materials, Philadelphia (1970).

WHITE, E. W., G. V. GIBBS, G. G. JOHNSON, JR., and G. R. ZECHMAN, JR., "X-Ray Emission and Absorption Wavelengths and Two-Theta Tables," *ASTM Data Series* **DS-37**, American Society for Testing and Materials, Philadelphia (1965).

Chapter 8

Problems
and Performance Criteria

This chapter considers the principal problems that beset a quantitative x-ray spectrochemical analysis (especially absorption–enhancement, surface-texture, particle-size, and heterogeneity effects and spectral interference) and analytical performance criteria (precision and accuracy, sensitivity, and resolution).

8.1. SOURCES OF ERROR

The errors besetting an x-ray spectrometric analysis may be classified by source as follows:

1. *Statistical counting error* (Section 8.4.3) constitutes the best possible attainable precision and depends only on the total accumulated count.

2. *Instrumental errors* consist of short-term and long-term variation, instability, and drift in instrumental components, conditions, and parameters, principally the following: (a) x-ray tube potential (kV) and current (mA); (b) intensity and distribution (over the specimen plane) of the primary x-ray beam (caused by changes in dimension and position of internal components of the tube); (c) crystal interplanar spacing (caused by changes in temperature; such changes cause slight displacement in 2θ for the spectral line); (d) gas amplification of the proportional counter (caused by drift in detector-tube potential, changes in ambient temperature, and, in flow counters, changes in ambient pressure); (e) secondary-emission ratio of the scintillation counter (caused by drift in detector-tube potential and changes in ambient temperature); (f) coincidence (dead-time) losses in the detector and electronic components; (g) shift and distortion of pulse-

height distributions (only when pulse-height selection is used); and (h) electronic components.

3. *Operational errors* (manipulative and resetting errors) consist in slight nonreproducibility in settings of instrument conditions, principally x-ray tube potential (kV) and current (mA) and goniometer angle (2θ). Operational errors are most severe in manual instruments, minimal in semiautomatic and automatic instruments, where preset values are set mechanically.

4. *Specimen errors* arise in the specimen itself. However, specimen errors do not include *sampling errors* arising from failure of the submitted sample to be representative of the bulk of the material to be analyzed. Specimen errors may be classified as follows: (a) absorption–enhancement effects; (b) physical attributes, including surface texture, particle size (average and distribution), heterogeneity of composition, heterogeneity of density (porosity, voids, cracks, etc.), packing density (of loose and briqueted powders), thickness (if less than infinite), and problems with liquids (volatility, expansion, bubble formation, radiolysis, etc.); (c) specimen insertion and position effects, including variations in specimen plane, takeoff angle, position, orientation, and flatness; and (d) chemical effects on spectral-line wavelength. Incidentally, the effect of variation in specimen plane is more serious in modern instruments, which have short x-ray tube target-to-specimen distances; as this distance decreases, variations in specimen plane assume an increasing proportion of the mean distance.

5. *Error in estimation of concentration from the calibration curve.*

6. *Spectral-line interference.* Inasmuch as a nearby spectral line can contribute to an intensity measurement of an analyte line, spectral interference may constitute a source of error.

Only the statistical counting error is easily and accurately calculated. Instrumental and operational errors are minimal in modern automatic instruments, and are compensated by ratio analytical methods. Absorption–enhancement effects are dealt with by appropriate choice of analytical method, most other specimen errors by appropriate choice of specimen preparation and presentation. Spectral interference is discussed in Section 8.3.

The chemical effect consists in small shifts in analyte-line wavelength caused by differences in oxidation number (valence) of analyte atoms and in the identity and number of atoms or radicals neighboring the analyte atoms. Such wavelength shifts occur when the analyte line arises from an electron transition from the outermost (valence) shell of the analyte atom.

This is the case for K lines of elements having atomic number $\lesssim 18$ (argon). The magnitude of the shift may be as much as 0.03 Å. For example, in the determination of magnesium, if the goniometer 2θ angle for Mg $K\alpha$ is peaked for magnesium, the line may fall at a slightly different angle for standards prepared from magnesium oxide; moreover, for complex mineral samples, the peak profile may become asymmetric and show minor peaks. In wavelength dispersion, simple peaks can be dealt with by peaking the goniometer for each specimen, but complex profiles can be measured only by integrated counting. In energy dispersion, the window can be set to admit all analyte pulses, unless another distribution is too close. Incidentally, the chemical shift provides a valuable method for investigation of chemical bonding.

8.2. MATRIX EFFECTS

8.2.1. Introduction

Consider a thick specimen free from all positional and chemical specimen error in an x-ray spectrometer free from all instrumental and operational error. One would expect the intensity $I_{A,M}$ of a spectral line of element A in matrix M (see below) to be given by

$$I_{A,M} = W_{A,M}I_{A,A} \quad \text{or} \quad I_{A,M}/I_{A,A} = W_{A,M} \qquad (8.1)$$

where $W_{A,M}$ is weight fraction of analyte A in matrix M, and $I_{A,A}$ is analyte-line intensity from pure A, and both the sample and pure analyte are infinitely thick. Unfortunately, even under these ideal—and unattainable— conditions, this simple relationship rarely applies. Matrix effects account for this disparity between the observed analyte-line intensity and that predicted by Equation (8.1).

The *matrix* consists of the entire specimen except the particular analyte under consideration. Thus, in a multielement system, the matrix of the same specimen is different for each analyte in the specimen, and each analyte constitutes a part of the matrix of every other analyte. The term *matrix* applies to the specimen as measured in the spectrometer. For example, if an alloy is to be dissolved and measured as a solution, one may consider the alloy itself as consisting of certain analytes in a certain basic matrix, but it is the solution matrix that is of analytical significance. In such cases, the alloy and solution matrixes may be referred to as the *original* and *specimen* matrixes, respectively. The same applies if an internal standard is added; the treated matrix is the significant one.

The effects of the matrix on analyte-line intensity may be classified in two categories: those arising from the *chemical* composition of the matrix (absorption–enhancement effects), and those arising from the *physical* features of the specimen (surface-texture, particle-size, and heterogeneity effects). These two classes of matrix effects are discussed in Sections 8.2.2 and 8.2.3, respectively.

In general, absorption–enhancement effects become more severe and particle-size and surface-texture effects less severe the deeper the active layer thickness is, that is, the deeper the excitation radiation penetrates the specimen and, more important, the greater the depth from which analyte-line radiation can emerge.

In general, matrix effects are substantially the same for wavelength- and energy-dispersive and nondispersive x-ray spectrometry, provided that the excitation is by x-radiation. However, if the excitation is by electrons, ions, or α or β radioisotopes, the exciting radiation does not penetrate as deeply into the specimen surface. In these cases, absorption–enhancement effects become less severe, and surface-texture and particle-size effects more severe.

Absorption–enhancement effects are dealt with principally by the choice of analytical method (Chapter 9), particle-size, surface-texture, and heterogeneity effects by the specimen preparation and presentation techniques (Chapter 10).

8.2.2. Absorption–Enhancement Effects

8.2.2.1. General

Absorption–enhancement effects arise from the following phenomena:

1. The matrix absorbs primary x-rays (*primary-absorption effect*); it may have a larger or smaller absorption coefficient than the analyte for primary x-rays, and it may preferentially absorb or transmit those wavelengths that excite the analyte line most efficiently, that is, those near the short-wavelength side of the analyte absorption edge.

2. The matrix absorbs the secondary analyte-line radiation (*secondary-absorption effect*); it may have a larger or smaller absorption coefficient than the analyte for the analyte-line radiation, and it may preferentially absorb or transmit this wavelength.

3. The matrix elements emit their own characteristic lines, which may lie on the short-wavelength side of the analyte absorption edge, and thereby

excite the analyte to emit line radiation in addition to that excited by the primary x-rays (*enhancement*).

Because analyte-line radiation is a discrete wavelength, whereas the primary radiation is usually a continuum, the secondary-absorption effect is usually more severe than the primary-absorption effect, but is also more easily predicted, evaluated, and corrected.

The absorption–enhancement effects may be classified in two ways. On the basis of their effect on the analyte-line intensity, they may be *positive* or *negative* absorption effects, or *true* or *apparent* enhancement effects. On the basis of their origin or general nature, they may be *nonspecific* (*general*), *specific*, *secondary* (*second order*), or *unusual* (*special*).

In the *positive absorption effect*, the matrix has a smaller absorption coefficient than the analyte for the primary and analyte-line radiation, and the analyte-line intensity for a specified analyte concentration is higher than would be predicted from Equation (8.1). In the *negative absorption effect*, the matrix has a larger absorption coefficient than the analyte, and the analyte-line intensity is lower than would be predicted. In the *true enhancement effect*, one or more spectral lines of matrix elements have wavelengths shorter than the analyte absorption edge. Thus, the matrix actually excites analyte-line radiation in addition to that excited by the primary source, and the intensity is higher than that predicted by Equation (8.1). The *apparent enhancement effect* is simply the positive absorption effect. The analyte-line intensity is higher, but only because matrix absorption is lower; the matrix excites no additional analyte-line radiation.

True enhancement may take either or both of two forms: *direct enhancement* and the *third-element effect*. Consider a ternary system A–B–C in which: (1) A is the analyte; (2) the respective strongest lines λ_A, λ_B, λ_C have progressively shorter wavelength; and (3) λ_C can excite λ_B and λ_A, and λ_B can excite λ_A. In direct enhancement, λ_B and λ_C excite λ_A directly. In the third-element effect, λ_C excites λ_B, which in turn excites λ_A. For example, in the chromium–iron–nickel system, the wavelengths of the respective $K\alpha$ lines are 2.29, 1.94, and 1.66 Å, and the respective K absorption edges 2.07, 1.74, and 1.49 Å. Thus, Ni $K\alpha$ excites iron and chromium, and Fe $K\alpha$ excites chromium. The contributions to the emitted Cr $K\alpha$ intensity from primary-beam excitation, direct enhancement by Fe $K\alpha$, direct enhancement by Ni $K\alpha$, and third-element enhancement by Fe $K\alpha$ excited by Ni $K\alpha$ are 72.5, 23.5, 2.5, and 1.5%, respectively.

Nonspecific or *general* absorption–enhancement effects result simply from differences in the absorption coefficients of analyte and matrix elements for the primary and, especially, analyte-line radiation. Specific absorption edges are not involved. Nonspecific effects involve only absorption—unless one considers apparent enhancement. *Specific* absorption–enhancement effects result from interaction of analyte and matrix spectral lines and absorption edges in close proximity. *Secondary* or *second-order* absorption–enhancement effects arise from the influence of the overall matrix on a nonspecific or specific effect on a certain analyte–matrix element pair. They take the form of departures from the effects predicted from the absorption coefficients and the wavelengths of spectral lines and absorption edges of the specified analyte–matrix pair. *Unusual* or *special* absorption–enhancement effects include cases in which analyte-line intensity remains substantially constant, or even decreases, as analyte concentration increases.

Whatever absorption–enhancement effects a specified analyte–matrix system may be subject to, they are most severe at and above infinite thickness, decrease in severity as thickness decreases below infinite thickness, and substantially disappear in thin films. Also, whatever absorption–enhancement effects a specified matrix element may cause, they decrease in severity as the concentration of that element decreases. Quantitative x-ray spectrometric analysis is applied most successfully to specimens that are either very thin (absorption–enhancement effects negligible) or infinitely thick (absorption–enhancement effects maximum, but constant) for the analyte lines to be measured.

8.2.2.2. Nonspecific Absorption Effects

The relative intensities I_1 and I_2 measured under identical conditions from two samples having concentrations C_1 and C_2 of the same analyte in the same matrix are given by

$$\frac{I_1}{I_2} \approx \frac{C_1}{C_2} \frac{(\mu/\varrho)_2'}{(\mu/\varrho)_1'} \tag{8.2}$$

where $(\mu/\varrho)'$ is mass-absorption coefficient of the sample for both primary and secondary x-rays. The ratio of the slopes m_1 and m_2 of the calibration curve (Figure 9.1) at C_1 and C_2 is

$$\frac{m_1}{m_2} = \frac{(\mu/\varrho)_2'}{(\mu/\varrho)_1'} \tag{8.3}$$

Equations (8.2) and (8.3) show that as the ratio C_1/C_2 changes, I_1/I_2 also changes, but not necessarily at the same rate, as determined by

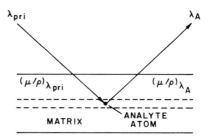

FIGURE 8.1. Origin of absorption effects for thick (solid lines) and thin (dashed lines) specimens.

$(\mu/\varrho)_2'/(\mu/\varrho)_1'$. This difference in rate may lead to positive and negative deviations from linearity of the calibration function.

It follows from Equation (8.2) that for the same analyte concentration in two samples 1 and 2 having different matrixes with mass-absorption coefficients $(\mu/\varrho)_1'$ and $(\mu/\varrho)_2'$ for primary and secondary x-rays, the relative analyte-line intensities I_1 and I_2 are

$$\frac{I_1}{I_2} = \frac{(\mu/\varrho)_2'}{(\mu/\varrho)_1'} \tag{8.4}$$

In Equations (8.2)–(8.4), for practical purposes, $(\mu/\varrho)'$ may be replaced with the corresponding mass-absorption coefficient for the analyte line. Then, to a first approximation, analyte-line intensity may be corrected for matrix absorption simply by use of mass-absorption coefficients of the samples.

Figure 8.1 illustrates the origin of nonspecific absorption and apparent enhancement effects. A ray of primary x-rays λ_{pri} must pass through a certain distance in matrix M having mass-absorption coefficient $(\mu/\varrho)_{M,\lambda_{pri}}$ to reach a specified analyte atom. If this atom emits analyte-line x-rays λ_A, they must emerge through a certain distance in matrix having mass-absorption coefficient $(\mu/\varrho)_{M,\lambda_A}$. The sum of $(\mu/\varrho)_{M,\lambda_{pri}}$ and $(\mu/\varrho)_{M,\lambda_A}$ is the $(\mu/\varrho)'$ of Equations (8.2)–(8.4). Figure 8.2 shows the effect of non-

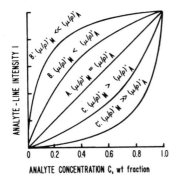

ANALYTE-LINE INTENSITY I

ANALYTE CONCENTRATION C, wt fraction

FIGURE 8.2. Nonspecific absorption effects; $(\mu/\varrho)'$ is mass-absorption coefficient for both primary and analyte-line x-rays.

specific absorption and apparent enhancement effects on the form of the calibration curve, as discussed in the following paragraphs.

In a *neutral matrix* (curve *A*), the absorption coefficient of the matrix $(\mu/\varrho)_M'$ and analyte $(\mu/\varrho)_A'$ are substantially the same for both the analyte-line and the primary radiation that excites it most efficiently. The primary x-rays reach each incremental specimen volume, and the excited analyte-line radiation emerges from the specimen with about the same attenuation in the matrix as in pure analyte. Thus, analyte-line intensity I_A increases at about the same rate as C_A, the I_A–C_A curve is substantially linear, and the absorption effect is very small.

Elements within one or two atomic numbers of the analyte usually constitute neutral matrixes for that analyte, at least down to atomic number 22 (titanium). For these medium-Z and high-Z elements, absorption–enhancement properties of adjacent elements are so similar that they can serve as internal standards for one another. However, at lower atomic numbers, the differences in wavelength (Figure 1.11) and absorption coefficient for adjacent elements have increased to a degree such that the Z $K\alpha$ line is very strongly absorbed by element $Z - 1$.

In a *light matrix* (curves *B* and *B'* in Figure 8.2A), $(\mu/\varrho)_M'$ is less than $(\mu/\varrho)_A'$ for the analyte-line and primary x-rays. In such a matrix, the primary x-rays undergo less attenuation in reaching any incremental specimen volume and analyte-line radiation undergoes less attenuation in emerging from the specimen in the matrix than in pure analyte. Thus, I_A increases at a faster rate than C_A. At higher concentrations, the specimen composition approaches pure analyte, and the difference in rates of increase for I_A and C_A decreases. The I_A–C_A curves are nonlinear and show the positive absorption effect; the greater the difference between $(\mu/\varrho)_M'$ and $(\mu/\varrho)_A'$, the more distorted is the curve, as is evident from a comparison of curves *B* and *B'*. This positive absorption effect constitutes an apparent or pseudo enhancement effect. Examples of light matrixes are usually provided by analytes having relatively short-wavelength lines in relatively low-Z matrixes. Lead as tetraethyllead in gasoline is an example of an extreme case (curve *B'*).

In a *heavy matrix* (curves *C* and *C'*), $(\mu/\varrho)_M'$ is greater than $(\mu/\varrho)_A'$, primary and analyte-line radiation undergo greater attenuation in the matrix than in pure analyte, I_A increases at a slower rate than C_A, and the curves show the negative absorption effect. Again, at higher analyte concentration, the difference in the two rates decreases, and the greater the difference in $(\mu/\varrho)_M'$ and $(\mu/\varrho)_A'$, the more distorted is the curve. Examples of heavy matrixes are usually provided by analytes having relatively long-wavelength lines in relatively high-Z matrixes. Determinations of magnesium, aluminum, and silicon in molybdenum alloys provide examples of extreme cases (curve C').

As specimen thickness decreases from infinite thickness, the paths of the primary and analyte-line x-rays in the specimen decrease, and so do the absorption effects. For thin-film specimens, represented by the dashed lines in Figure 8.1, these effects substantially disappear.

It must be emphasized that whether a matrix is "heavy," "light," or "neutral" for a specified analyte line is determined by the mass-absorption coefficient of that matrix for that line, not by the effective atomic number of the matrix. Admittedly, it is usually true that a high-Z element constitutes a heavy matrix for a long-wavelength line, and a low-Z element constitutes a light matrix for a short-wavelength line. It is also usually true that a high-Z element constitutes a heavy matrix for the spectral lines of low-Z elements, a low-Z element constitutes a light matrix for the lines of high-Z elements, and any element constitutes a more or less neutral matrix for lines of adjacent and neighboring elements. However, these simple guidelines are far from generally applicable for reasons given in paragraphs 1 and 2 of Section 2.1.3 and must be used with great caution. For example, lead (Z 82) is a heavy matrix (μ/ϱ 5968), as expected, for Na $K\alpha$ (Z 11, λ 11.9 Å), but is a relatively light matrix (μ/ϱ 85) for Br $K\alpha$ (Z 35, λ 1.04 Å). Similarly, aluminum (Z 13) is a light matrix (μ/ϱ 34 to 6), as expected, for the $K\alpha$ lines of elements 31–42 (gallium to molybdenum), but is a heavy matrix for Si $K\alpha$ (Z 14, λ 7.13 Å, μ/ϱ 3493) and for Zr $L\alpha_1$ (Z 40, λ 6.07 Å, μ/ϱ 2236), for which the "rules" above would predict it to be neutral and light, respectively.

8.2.2.3. Specific Absorption–Enhancement Effects

If the analyte line occurs at a wavelength just less than that of an absorption edge of a particular matrix element B, the A line is highly absorbed by B, and A-line intensity is reduced in proportion to B concentration. This is the case for Fe $K\alpha$ (1.94 Å) in the presence of chromium ($\lambda_{K_{ab}}$ 2.07 Å), and chromium has a similar, but progressively weaker, negative absorption effect on the $K\alpha$ lines of elements of successively higher atomic number—cobalt, nickel, copper, etc.

Conversely, if the analyte absorption edge occurs at a wavelength just greater than that of a line of a particular matrix element B, the B line is absorbed by A, and A-line intensity is enhanced in proportion to B concentration. This is the case for Fe K_{ab} (1.74 Å) and Ni $K\alpha$ (1.66 Å), and nickel has a similar, but progressively weaker, enhancement effect on the

$K\alpha$ lines of elements of successively lower atomic number—manganese, chromium, vanadium, etc.

The effects just described are specific, in that they depend on the proximity of spectral lines and absorption edges of analyte and matrix elements. These specific absorption–enhancement effects are now considered in detail, with Fe $K\alpha$ serving as an example of a typical analyte line. Ideally, in the determination of iron in various matrixes, Fe $K\alpha$ intensity should be given by Equation (8.1):

$$I_{\mathrm{Fe}K\alpha,\mathrm{M}} = W_{\mathrm{Fe},\mathrm{M}} I_{\mathrm{Fe}K\alpha,\mathrm{Fe}} \qquad (8.5)$$

That is, Fe $K\alpha$ intensity in matrix M should be given by the product of the weight fraction of iron in matrix M and the Fe $K\alpha$ intensity from pure iron. Let us consider specific absorption–enhancement effects on the Fe $K\alpha$ intensity measured from binary alloys or mixtures consisting of iron (atomic number 26) in matrixes of aluminum (13), chromium (24), manganese (25), cobalt (27), nickel (28), cerium (58), and lead (82). Figure 8.3 shows the spectral lines, absorption edge, and mass-absorption curve for each of these elements in the spectral region of the iron K absorption edge and lines.

The effect of each matrix element on Fe $K\alpha$ intensity must be considered in terms of three phenomena: (1) absorption of that portion of the primary beam that excites Fe $K\alpha$ most efficiently, that is, having wavelength near the short-wavelength side of the iron K absorption edge; (2) absorption of the Fe $K\alpha$ line; and (3) enhancement of the Fe $K\alpha$ line by matrix lines having wavelengths on the short side of the iron K edge.

With respect to absorption of the primary x-rays, at all wavelengths on the short side of the iron K edge, manganese absorption is about the same as that of iron, cerium absorption is substantially higher, chromium and lead absorption are substantially lower, and aluminum absorption is very low. Cobalt and nickel absorption are very low in the extremely critical region within, respectively, ∼0.15 Å and ∼0.25 Å of the iron K edge, but they rise to about the same value as iron at shorter wavelengths.

With respect to the much more significant absorption of Fe $K\alpha$ radiation, aluminum, manganese, iron, cobalt, and nickel all have about the same very low absorption. Lead, chromium, and, especially, cerium have much higher absorption.

With respect to enhancement, aluminum, chromium, manganese, cerium, and lead have no spectral lines at wavelengths anywhere near the short side of the iron K edge, and therefore do not enhance Fe $K\alpha$. The Co $K\beta$ and Co $K\alpha$ lines bracket the iron K edge, with Co $K\beta$ on the short side; thus, cobalt mildly enhances Fe $K\alpha$. Both Ni $K\alpha$ and Ni $K\beta$ lie on the short side of the iron K edge, and nickel strongly enhances Fe $K\alpha$.

The net effect of each of these matrix elements on Fe $K\alpha$ intensity may be presented in terms of its predicted effect based on the foregoing discussion as compared with the intensity calculated from Equation (8.5); that is,

$$R = \frac{I_{FeK\alpha,M}(\text{measd.})}{W_{Fe,M}I_{FeK\alpha,Fe}} = \frac{I_{\text{measured}}}{I_{\text{predicted}}} \tag{8.6}$$

These effects are summarized below and in Table 8.1.

Aluminum. $R \gg 1$ because of the very low absorption of primary and Fe $K\alpha$ radiation and despite the absence of any enhancing spectral lines (strong positive absorption or apparent enhancement effect).

Chromium. $R \ll 1$ because the high absorption of Fe $K\alpha$ outweighs the somewhat lower absorption of the primary x-rays (strong negative absorption effect).

Manganese. $R \approx 1$ because manganese has substantially the same absorption coefficient as iron for primary and Fe $K\alpha$ radiation and causes no enhancement (absorption–enhancement effects largely absent; neutral matrix).

Iron. $R = 1$.

Cobalt. R is somewhat greater than 1. The absorption of primary radiation within ~ 0.15 Å of the short side of the iron K edge is very low, and Co $K\beta$ enhances Fe $K\alpha$ (combination of positive absorption and mild enhancement).

Nickel. $R \gg 1$ because the absorption of primary radiation within ~ 0.25 Å of the short side of the iron K edge is very low, and both Ni $K\alpha$ and Ni $K\beta$ enhance Fe $K\alpha$ (combination of positive absorption and strong enhancement).

Cerium. $R \ll 1$ because of the very high absorption of both primary and Fe $K\alpha$ radiation and the absence of enhancement (strong negative absorption).

Lead. $R < 1$. The effect of lead is similar to that of cerium, except that the absorption coefficient of lead is not as high in this spectral region (relatively strong negative absorption).

An analysis of the possible absorption–enhancement effects should be made whenever an analytical method is to be developed for a new combination of analyte(s) and matrix. Often, it is sufficient to plot on a wavelength scale the principal spectral lines and absorption edges of all elements present, in the manner shown at the top of Figure 8.3.

TABLE 8.1. Specific Absorption–Enhancement Effects of Several Matrix Elements on Fe $K\alpha$ Intensity (Refer to Figure 8.3)

Matrix element and atomic number, M	Absorption of primary x-rays at $\lambda < \lambda_{FeK_{ab}}$	Absorption of $\lambda_{FeK\alpha}$	Enhancement of $\lambda_{FeK\alpha}$	$R = \dfrac{I_{FeK\alpha,M}(\text{measd.})}{C_{Fe,M}I_{FeK\alpha,Fe}}$
$_{13}$Al	Very low	Very low	None	$\gg 1$
$_{24}$Cr	Low	High	None	$\ll 1$
$_{25}$Mn	Similar to Fe	$\sim$Fe	None	~ 1
$_{26}$Fe	—	—	—	1
$_{27}$Co	Very low at λ just $< \lambda_{FeK_{ab}}$; $\sim$Fe at shorter wavelengths	Very low	By Co $K\beta$ only	> 1
$_{28}$Ni	Same as for $_{27}$Co	Very low	By Ni $K\alpha$ and Ni $K\beta$	$\gg 1$
$_{58}$Ce	High	Very high	None	$\ll 1$
$_{82}$Pb	Low	High	None	< 1

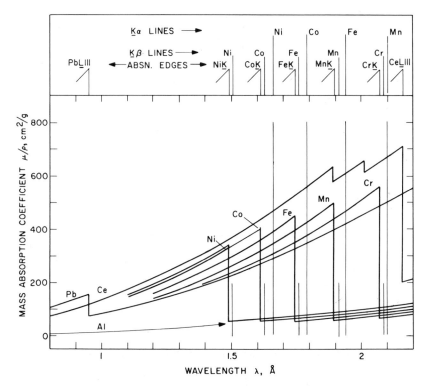

FIGURE 8.3. Specific absorption–enhancement effects of several matrix elements on Fe $K\alpha$ intensity.

8.2.2.4. Secondary Absorption–Enhancement Effects

Mitchell and Kellam have advanced the characterization of absorption–enhancement effects a significant step farther by consideration of the effect of the total matrix composition on the absorption–enhancement effect of a particular matrix element on a particular analyte line. They describe secondary effects that may markedly influence the effect predicted for an analyte-matrix element pair on the basis of the nonspecific and specific effects described above. A consideration of these effects is beyond the scope of this introductory text, but two examples follow.

Consider the effect on Fe $K\alpha$ intensity from a constant iron concentration, say 10 wt%, in a matrix consisting of chromium and one other element. Chromium has a much higher mass-absorption coefficient for Fe $K\alpha$ (490 cm²/g) than does iron itself (73), so as chromium concentration increases, Fe $K\alpha$ intensity should decrease. However, if the other matrix element has a substantially higher absorption for Fe $K\alpha$ than does chromium, the expected decrease in Fe $K\alpha$ intensity on increase in chromium concentration may actually become an apparent enhancement. For example, if the other matrix element is cerium [$(\mu/\varrho)_{Ce,FeK\alpha}$ 636], as chromium replaces cerium, the total matrix absorption for Fe $K\alpha$ decreases slightly and Fe $K\alpha$ intensity increases slightly. This is an example of a secondary absorption effect.

Now consider the effect on Fe $K\alpha$ intensity from a constant 10-wt% iron concentration in a matrix consisting of nickel and one other element. Nickel strongly enhances Fe $K\alpha$ because both Ni $K\alpha$ and Ni $K\beta$ lie close to the short-wavelength side of the iron K-absorption edge, so as nickel concentration increases, Fe $K\alpha$ intensity should increase. However, if the other matrix element has a substantially lower absorption for Fe $K\alpha$ than does nickel (90), the expected increase in Fe $K\alpha$ intensity on increase in nickel concentration may actually become an apparent absorption. For example, if the other matrix element is carbon [$(\mu/\varrho)_{C,FeK\alpha}$ 11], as nickel replaces carbon, the total matrix absorption for Fe $K\alpha$ actually increases and more than compensates the enhancement, and Fe $K\alpha$ intensity decreases markedly. This is an example of a secondary enhancement effect.

8.2.2.5. Unusual Absorption–Enhancement Effects

This term is applied to cases in which analyte-line intensity remains substantially constant or even decreases with increasing analyte concentration. Two examples follow.

The x-ray fluorescence spectrometer is usually unable to distinguish various concentrations of an analyte having a relatively short-wavelength line in a very light matrix. For example, in general, it is not possible to distinguish the various oxides of an element of medium or high atomic number simply by comparing the intensities of its spectral line measured from the untreated oxides. Table 8.2 gives data for iron and its oxides.

TABLE 8.2. Unusual Absorption Effect for Oxides of Iron[a]

	Fe	FeO	Fe_3O_4	Fe_2O_3
Fe concentration (wt%)	100	77.8	72.5	70.0
Fe $K\alpha$ intensity, relative	100	95.7	95.5	95.0
Critical thickness for Fe $K\alpha$ (cm $\times 10^{-3}$)	1.36	2.31	2.72	2.77

[a] From H. A. Liebhafsky, H. G. Pfeiffer, E. H. Winslow, and P. D. Zemany, *X-Ray Absorption and Emission in Analytical Chemistry*, John Wiley & Sons, New York, p. 184 (1960).

Assuming an effective primary wavelength of 1.39 Å, the oxygen is highly transparent to Fe $K\alpha$ (μ/ϱ 22) and to the primary x-rays (μ/ϱ 8), so that iron (μ/ϱ 73 and 250, respectively) largely determines the absorption coefficient of the oxide and thereby the critical thickness or depth from which Fe $K\alpha$ can emerge from the specimen. Thus, substantially the same number of iron atoms contribute to the measured Fe $K\alpha$ intensity for each oxide, and the relative intensity increases only from 95.0% to 95.7% while iron concentration increases from 70.0% to 77.8%. The difficulty is easily remedied in several ways: (1) elimination of the absorption effect by dispersion as a thin film; (2) leveling the matrix to a constant absorption by low-absorption dilution or (3) high-absorption dilution; (4) for some elements, by measurement of an L or M line instead of a K line; or (5) by use of an internal standard.

The basis of this phenomenon is explained by reference to Equations (4.7) and (4.12), the fundamental excitation equations. For an analyte having a relatively high atomic number and short-wavelength analyte line, in a light matrix,

$$(\mu/\varrho)_{A,\lambda_{\mathrm{pri}}} \gg (\mu/\varrho)_{i,\lambda_{\mathrm{pri}}}$$

$$(\mu/\varrho)_{A,\lambda_L} \gg (\mu/\varrho)_{i,\lambda_L}$$

where i is a matrix element excluding the analyte A, and λ_L and λ_{pri} are the wavelengths of the analyte-line and primary x-rays, respectively. Then $\sum [C_i(\mu/\varrho)_{i,\lambda_{\mathrm{pri}}}]$, again excluding the analyte, can be disregarded as compared with $C_A(\mu/\varrho)_{A,\lambda_{\mathrm{pri}}}$, assuming only that C_A is not too small. The same applies to $\sum [C_i(\mu/\varrho)_{i,\lambda_L}]$. Then I_L, or I_L/I_0, becomes substantially independent of C_A, a calibration curve of I_L versus C_A becomes substantially horizontal, and the analysis is not feasible.

It is also possible for analyte-line intensity to actually decrease as analyte concentration increases. This phenomenon occurs when the following two conditions occur simultaneously: (1) a specific matrix element B has a high absorption coefficient for the analyte line, and (2) the concentration of element B increases at a faster rate than analyte concentration. The phenomenon was first noted in mixtures of aluminum-containing chromites [FeO·(Cr,Al)$_2$O$_3$] and olivines [2(Mg,Fe)O·SiO$_2$] in the composition range 5–25% iron as shown in Figure 8.4. Chromium has a high mass-absorption coefficient for Fe $K\alpha$ (490 cm²/g) compared with that of iron (73 cm²/g). Pure chromite (FeO·Cr$_2$O$_3$) and olivine (FeO·MgO·SiO$_2$) contain 25 and 10% iron, respectively, and have mass-absorption coefficients for Fe $K\alpha$ of 231 and 68 cm²/g, respectively; chromite contains 46% chromium. Thus, as chromite minerals are added to olivines, chromium con-

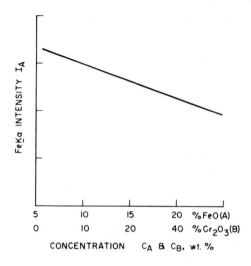

FIGURE 8.4. Unusual matrix effect of chromium on Fe $K\alpha$ intensity. [B. J. Mitchell and J. E. Kellam, *Appl. Spectrosc.* **22**, 742 (1968).]

centration increases at a faster rate than iron concentration. The high absorption of chromium and disproportionate rates of increase of chromium and iron concentrations result in the negative absorption effect of chromium outweighing the increased Fe $K\alpha$ emission from the iron, and the iron calibration curve has a negative slope, as shown.

A simple calculation proves the point [Equation (8.4)]. Assume that the Fe $K\alpha$ intensity from pure olivine is 100 counts/s per 1% iron. Then, for pure olivine (10% Fe), Fe $K\alpha$ intensity is $10 \times 100 = 1000$ counts/s; for pure chromite (25% Fe), Fe $K\alpha$ intensity is $(25 \times 100)(68/231) = 735$ counts/s, where 68 and 231 are the mass-absorption coefficients of olivine and chromite, respectively.

8.2.3. Surface-Texture, Particle-Size, and Heterogeneity Effects

The derivations of the basic excitation Equations (4.7) and (4.12) (Section 4.1.4.2) assume a homogeneous, smooth specimen. This means that the relatively thin surface layer that emits the measured x-ray spectral-line intensities contains all elements present in the specimen homogeneously distributed in their true concentration ratios.

A powder or polycrystalline solid is said to be homogeneous or heterogeneous according to whether all particles (grains, crystallites) have the same chemical composition or the material is a mixture of particles having two or more compositions. A powder or solid is said to have uniform or nonuniform particle size according to whether all particles have the same

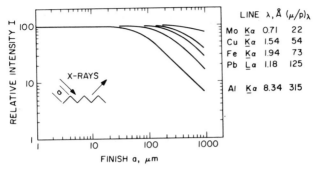

FIGURE 8.5. Effect of roughness of parallel grind marks on intensity for analyte lines of various wavelengths measured from plane solid surfaces. The grind marks were perpendicular to the plane defined by the central rays of the primary and secondary x-ray beams as shown. [R. O. Müller, *Spektrochemische Analysen mit Roentgenfluoreszenz*, R. Oldenbourg, Munich, Germany (1961); courtesy of the author and publisher.]

size or the material is a mixture of particles of different sizes. The particle-size distribution is a curve showing the relative numbers of particles of each size present—rather analogous to a pulse-height distribution.

Measured analyte-line intensity may be affected by surface texture of a massive solid specimen even if both composition and particle size are uniform. The intensity is affected not only by the roughness of the surface finish, but by the orientation of the grind or polish marks with respect to the directions of the primary and secondary x-ray beams. These effects are shown in Figures 8.5 and 8.6, respectively. Surface texture may have

FIGURE 8.6. Effect of roughness and direction of parallel grind marks on intensity of analyte lines of long (Si Kα) and short (Sn Kα) wavelengths measured from plane solid surfaces. Grind marks for curves *A* were parallel, for curves *B* perpendicular to the plane defined by the central rays of the primary and secondary x-ray beams. [K. Togel, in *Zerstörungsfreie Materialprüfung*, R. Oldenbourg, Munich, Germany (1961); courtesy of the author and publisher.]

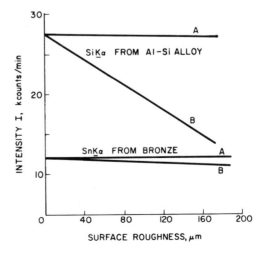

one or more of three principal effects: (1) The path lengths of the primary and analyte-line x-rays within the specimen may vary from point to point. (2) There may be shielding and shadowing effects; that is, surface topography may obstruct primary x-rays from reaching, and analyte-line x-rays from leaving certain points on the surface. (3) Finally, extremely coarse surface topography may actually influence the effective distance between the x-ray tube target and specimen. These effects become progressively more severe as: (1) the wavelength of the analyte line increases; (2) the mass-absorption coefficient of the specimen for primary and, more important, analyte-line x-rays increases; and (3) the wavelength of the x-ray tube target line increases. The third condition applies to Cr $K\alpha$ and rhodium, palladium, and silver L lines, but only when the target lines make the predominant contribution to the excitation. In general, surface finish should be finer than the path length of the longest-wavelength analyte line to be measured; $\sim$20-μm finishes are usually sufficient, even for Mg $K\alpha$.

Measured analyte-line intensity may be affected by the particle size and particle-size distribution of a powder, briquet, or polycrystalline solid even if composition is uniform. The intensity may be affected by heterogeneity of composition even if the particle size is uniform. In powders having nonuniform particle size, segregation may occur by size; in powders having nonuniform composition, segregation may occur by density. Both conditions are worsened by sifting. If particles have size greater than the path length of the analyte-line x-rays, analyte-line intensity may be reduced. The effect is more severe the longer the analyte-line wavelength and the higher the absorption of the specimen for the line and the primary radiation that excites it. In general, particle size should be smaller than the path length of the longest-wavelength analyte line to be measured. It is evident that particle size and surface texture are analogous effects pertaining to polycrystalline and massive-solid specimens, respectively.

Figure 8.7 shows portions of the two outermost layers of particles in powder (or briquet or polycrystalline solid) specimens having large and small particles, respectively. The entire shaded area in each particle represents the volume penetrated by the primary beam—that is, the volume in which analyte-line x-rays are generated. The crosshatched area represents the volume from which analyte-line x-rays can emerge—that is, the effective volume. For the larger particles in the figure, only $\sim\frac{1}{3}$ of the volume is effective, for the smaller particles, $\sim\frac{2}{3}$. For very small particles, the entire volume in the outermost layer and perhaps several deeper layers would be effective. The higher the x-ray tube operating potential, the larger will be the excited volume. The shorter the analyte-line wavelength, the more

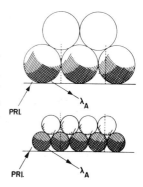

FIGURE 8.7. Portions of the two outermost layers of particles in powder specimens having large and small particles, respectively, showing the volume penetrated by the primary x-rays (total shaded area) and the effective volume (crosshatched area) from which analyte-line x-rays λ_A can emerge.

nearly will the effective volume equal the excited volume, and the longer this wavelength, the smaller will be the effective volume. In general, the particle size should be no larger than—and should be substantially smaller than—the effective volume for the longest wavelength to be measured. A similar guideline applies to surface roughness of massive solid specimens.

De Jongh describes three distinct particle-size effects for a two-phase heterogeneous powder: In the *particle-size effect*, the analyte is present in only one phase, and both phases have substantially the same mass-absorption coefficient for the analyte line. In such systems, the analyte-line intensity is influenced only by the relative particle sizes of the two phases, that is, by the actual fraction of the analyte phase in the effective specimen layer. The effect is most severe when the analyte-line path length in the specimen is less than the average particle size. In the *intermineral effect*, the analyte is present in only one phase, but the two phases have substantially different absorption coefficients for the analyte line. Analyte-line intensity depends not only on particle size, but also on the relative absorption coefficients. In the *mineralogical effect*, the analyte is present in both phases, which have different absorption coefficients for the analyte line. The effect is similar to that in the intermineral effect, but more complex. The nomenclature was suggested by De Jongh. However, inasmuch as all three effects are particle-size effects and occur in all kinds of powders, more appropriate terms are to be preferred.

Because of surface-texture, particle-size, and heterogeneity effects, in a multielement system, the spectral-line intensities of the elements may all increase or all decrease, or one or more may increase while the others decrease. Clearly, if these effects vary among samples and standards, the measured intensities are likely to be difficult to correlate with each other and with analyte concentration. If these effects are not severe, specimen rotation during measurement may be sufficient to correct them. Otherwise, they are dealt with by appropriate specimen preparation and presentation (Chapter 10). Mathematical methods have also been developed for correcting these effects, notably by Berry and Criss.

8.3. SPECTRAL INTERFERENCE

Spectral interference may be defined as that condition whereby (1) photons of a line other than the measured line enter the detector at the 2θ angle of the measured line, or (2) pulses arising from a line other than the measured line enter the pulse-height selector window set to pass the measured distribution. These definitions represent wavelength and energy spectral interference, respectively. In wavelength interference, the measured and interfering lines may have the same or nearly the same wavelength (λ) or product of order and wavelength ($n\lambda$). In energy interference, the interfering pulses may be from a photo, escape, or sum peak. In general, spectral interference is not a serious problem in x-ray spectrochemical analysis because x-ray spectra are simple and consist of relatively few lines.

Interfering lines may arise in the specimen itself or its support, if any (Mylar, filter paper, ion-exchange resin, etc.), or its mask, or they may be scattered x-ray tube target or other source lines. In energy dispersion, crystal emission is another possible source of interference.

The majority of cases of wavelength spectral interference may be classified in one of the following four categories, regardless of the source of the interfering lines.

1. Superposition of first-order lines of the same series (K, L, M) from adjacent or neighboring elements in the periodic table. Some examples follow. The $K\alpha$ lines of adjacent elements above technetium (Z 43) lie within 0.03 Å of one another, as do the $L\alpha_1$ lines of adjacent elements above mercury (Z 80). For titanium through cobalt (Z 22–27), the Z $K\alpha$ line lies within 0.03 Å of the ($Z - 1$) $K\beta$ line, and for rhodium through indium (Z 45–49), the Z $L\alpha_1$ line lies within 0.03 Å of the ($Z - 1$) $L\beta_1$ line.

2. Superposition of first-order lines of different series; for example, As $K\alpha$ (λ 1.177 Å) and Pb $L\alpha_1$ (λ 1.175 Å).

3. Superposition of first-order K lines and higher-order lines from elements of higher atomic number; for example, Ni $K\alpha$ (λ 1.659 Å) and Y $K\alpha$ (2λ 1.660 Å), P $K\alpha$ (λ 6.155 Å) and Cu $K\alpha$ (4λ 6.168 Å), Al $K\alpha$ (λ 8.337 Å) and Cr $K\beta$ (4λ 8.340 Å) or Ba $L\alpha_1$ (3λ 8.325 Å). This type of interference is particularly troublesome in the determination of low-Z

elements in multicomponent samples having many medium-Z and high-Z elements. Higher orders of scattered target lines also cause this type of interference with low-Z elements.

4. Superposition of first-order L lines and higher-order K lines from elements of lower atomic number. The classic examples are Hf $L\alpha_1$ (λ 1.570 Å) and Zr $K\alpha$ (2λ 1.574 Å), and Ta $L\alpha_1$ (λ 1.522 Å) and Nb $K\alpha$ (2λ 1.494 Å). Similar interferences also exist between $L\beta_1$ and second-order $K\beta$ lines of these elements.

The higher the intensity of the measured line, and the lower the intensity and greater the separation of the interfering line, the less serious is the interference. Orders higher than the fifth, satellite lines, and M and N lines are seldom troublesome in x-ray spectrochemical analysis except for weak measured lines.

The most practical methods for reducing spectral interference are the following, selected as appropriate for the type and source of interference: (1) pulse-height selection (if measured and interfering lines have substantially different wavelengths); (2) alternative measured-element line, free of interference; (3) higher (better dispersed) order of the measured line; (4) finer (longer and/or closer spaced) collimation; (5) analyzer crystal having higher resolution (smaller $2d$ spacing) or no even orders; (6) different x-ray tube target or other excitation source; (7) reduced x-ray tube current (if the interfering line has low intensity relative to the measured line); (8) selective excitation (if the interfering line has much higher excitation potential than the measured line); and (9) filter having an absorption edge at wavelength just longer than that of the interfering line.

8.4. PRECISION

8.4.1. Definitions

The *error* in a measurement (of intensity, for example) or in an overall analytical result is the difference between the measured value and the "true" value. However, since the true value must itself be determined by measurement, it might seem impossible to evaluate the error. Strictly speaking, this may be true, but in practice, error can be evaluated satisfactorily in terms of precision and accuracy, which, although often used synonymously, are quite different.

The *precision* of a measurement or analysis is the degree of agreement among replicate determinations made under conditions as nearly identical as possible. Quantitatively, precision p_i is the difference between the individual measurement m_i or analysis and the mean $\bar{m}$ of a large number or *set* of independent replicate measurements or analyses, usually expressed relative to the mean and as percent, that is,

$$p_i = \frac{m_i - \bar{m}}{\bar{m}} \times 100 \qquad (8.7)$$

Thus, the greater ("better") the precision, the smaller is its numerical value. The mean $\bar{m}$ is regarded as the "best known" value, that is, the value most likely to be true. Precision can be evaluated experimentally or calculated by summation of the individual contributing errors [Equation (8.17)].

Incidentally, the difference between the highest and lowest individual values m_i in a related set of measurements is known as the *range* or *spread*.

The *accuracy* of a measurement or analysis is the degree of agreement with the "true," accepted, or most reliably known value. Quantitatively, accuracy a_i is the difference between the individual measurement m_i or analysis and the true value t, usually expressed relative to the true value and as percent, that is,

$$a_i = \frac{m_i - t}{t} \times 100 \qquad (8.8)$$

Thus again, the greater ("better") the accuracy, the smaller is its numerical value.

Incidentally, the difference between the individual and mean values, $m_i - \bar{m}$ in Equation (8.7), and between the individual and true values, $m_i - t$ in Equation (8.8), is known as the *deviation*.

In short, precision is a measure of the reproducibility of the measurement or analysis, and accuracy is a measure of its correctness. Accuracy is unattainable without precision, but precision does not necessarily guarantee accuracy. A measurement or analysis may be very precise, that is, reproducible, but very inaccurate. The *reliability* of a measurement or analysis is the degree to which it possesses precision and accuracy.

Error may be expressed in absolute or relative terms. *Absolute error* is the degree by which the measured and true values may be expected to differ, expressed in the physical unit of the measured quantity. *Relative error* is the degree by which the measured and true values may be expected to differ, expressed as a fraction or percent of the measured value. Thus, in x-ray spectrometric analysis, absolute error is expressed in terms of analyte concentration in weight percent, micrograms per square centimeter, or other unit, relative error in fraction or percent "of the amount present." In either case, the degree of certainty of the indicated error must be stated, as discussed in Section 8.4.2. For example, if an analyst states that the analyte concentration is $20\% \pm 1\%$ absolute (with 95% certainty), he means that he is 95% sure that the true concentration lies between 19 and 21%, which is $20\% \pm 5\%$ *relative*. If he states that the analyte concentration is $20\% \pm 1\%$ relative (with 95% certainty), he means that he is 95% sure that the true concentration lies between 19.8 and 20.2%, which is $20\% \pm 0.2\%$ *absolute*.

The error is said to be *high* or *low*, or *positive* or *negative*, depending on whether the measured value is, respectively, greater or less than the true value.

The errors affecting an x-ray spectrometric analysis may be classified as random or systematic. *Random errors* consist of small differences in successive values of a measurement made repetitively with great care by the same competent person under conditions as nearly constant as possible. The magnitude of random errors can be evaluated, and the errors can be minimized, but not corrected for. *Systematic errors* are those that can be avoided, or at least evaluated and corrected for; they may be constant, in which case they account for the deviation or *bias* of the experimental result from the true value; or they may fluctuate about a mean value, in which case they contribute to the precision. To a first approximation, random and systematic errors limit the precision and accuracy, respectively. Some statisticians recognize a third type of error, *wild errors*, occasional values that diverge very widely from the distribution of the other values in the set.

8.4.2. Elementary Statistics

Each measurement or overall analytical determination results in a numerical value x. Replicate measurements or determinations result in a series of values $x_1, x_2, \ldots, x_n$, which constitute a population of n members. If the measurements are subject only to random errors, a plot of frequency of occurrence of individual values x_i *versus* the value of x will have a *Gaussian* or *normal* distribution having the form shown in Figure 8.8 and given by the equation

$$P(x) = \frac{1}{(2\pi\bar{x})^{1/2}} \exp\left[-\frac{(x - \bar{x})^2}{2\bar{x}}\right] \qquad (8.9)$$

where $P(x)$ is the probability of occurrence of a specified value x, $\bar{x}$ is the mean value of x in the population (see below), and $(x - \bar{x})$ is the deviation of an individual value from the mean.

A series of replicate measurements is characterized by its population

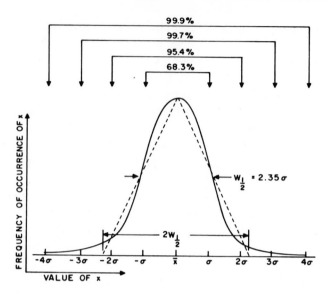

FIGURE 8.8. Gaussian (normal) distribution. The full width at half maximum height (FWHM) is 2.35 σ intervals, as shown. The Gaussian distribution may be represented approximately by an isosceles triangle having base twice the width at half-height (dashed lines); such a triangle contains ~92% (~2σ) of the total population of the distribution.

(the number of measurements or determinations), range or spread (the difference between the highest and lowest values), mean, and scatter or divergence. The Gaussian distribution of the series is characterized simply by the mean and variance. The *mean* $\bar{x}$ is the best approximation of the true value and is given by

$$\bar{x} = (\textstyle\sum x_i)/n \tag{8.10}$$

The scatter in the individual values is a measure of the precision of the measurement or analysis and is evaluated by the variance and standard deviation. The *variance v* is the sum of the squares of the deviations of the individual values x_i from their mean $\bar{x}$, divided by the number of degrees of freedom, which is 1 less than the number in the population:

$$v = \frac{\sum (x_i - \bar{x})^2}{n - 1} \tag{8.11}$$

The *standard deviation* σ is the square root of the sum of the squares of the individual deviations divided by the degrees of freedom:

$$\sigma = \left[\frac{\sum (x_i - \bar{x})^2}{n - 1} \right]^{1/2} \tag{8.12}$$

Variance and standard deviation are related as follows:

$$v = \sigma^2; \qquad \sigma = v^{1/2} \tag{8.13}$$

Actually, in statistics, the *standard deviation* σ is defined as the square root of the mean of the squares (rms) of the deviations of an infinite set of measurements, or at least of all possible measurements v, from their arithmetic mean μ. In practice, the *estimate of the standard deviation s* is the rms deviation of a limited set of measurements n from their mean $\bar{x}$, which is itself the *estimate of the mean.* However, the practice of using the symbol σ instead of s in the x-ray literature is very common, as is the practice of referring to precision in terms of one, two, and three "sigma." Therefore, the symbol σ is used in this book.

The practical significance of the standard deviation is shown in Figure 8.8 and Table 8.3 and may be stated in various ways, including the following:

1. The probability is 68.3% ($\sim$7/10) that any individual value of x will deviate from the average $\bar{x}$ of a very large number of values by $\leq \sigma$.

2. The probability is 68.3% that any x will have a value between $\bar{x} \pm \sigma$.

3. Of a very large number of measurements of x, 68.3% will have values of x lying between $\bar{x} \pm \sigma$.

4. It follows that 68.3% of the area under the Gaussian distribution curve lies between $\bar{x} \pm \sigma$.

TABLE 8.3. Statistical Precision

Symbol	Term	Confidence level (%)	Probability
0.67σ	Probable error	50.0	1/2
σ	Standard deviation, sigma	68.3	7/10
2σ	Two sigma	95.4	19/20
3σ	Three sigma	99.7	997/1000
4σ	Four sigma	99.9	999/1000

Similarly, the probability is 95.4% (19/20) that an individual x will lie between $\bar{x} \pm 2\sigma$, 99.7% (997/1000) that it will lie between $\bar{x} \pm 3\sigma$, and 99.9% (999/1000) that it will lie between $\bar{x} \pm 4\sigma$. Finally, the probability is 50% (1/2) that an individual x will lie between $\bar{x} \pm 0.67\sigma$; 0.67σ is known as the *probable error* or *probable fractional error*.

Figure 9.1 shows a typical calibration curve on which the associated error is indicated on each calibration point in the conventional manner. The ranges may indicate $\pm 1\sigma$ or some other specified precision. For example, if $\pm 1\sigma$ is indicated, the probability is 68.3% that the true intensity for each point is neither greater nor less than the limits shown. If the fixed-count method is used, the range is the same for each point, as shown, but if different numbers of counts are accumulated for each point, the range is also different. In addition to the random error shown for all the points, the point for the standard having second highest concentration also shows a substantial systematic error.

The larger σ is, the flatter is the Gaussian distribution curve; the width of the distribution at half its maximum height is $2(2 \log_e 2)^{1/2}\ \sigma$ intervals, or 2.35σ (Figure 8.8). The flatter the curve, the greater is the spread in replicate intensity measurements or analytical results, and the lower is the precision in the measurements or results. The larger the number n of replicate measurements, the more nearly their distribution corresponds to the Gaussian curve and the more reliable is the calculated value of σ.

The standard deviation expressed relative to the mean is the *relative standard deviation* or *coefficient of variation* ε and is often expressed as a percent, although in this chapter, for simplicity, ε is expressed as a fraction:

$$\varepsilon = \sigma/\bar{x} \qquad \text{or} \qquad \varepsilon = 100\sigma/\bar{x} \tag{8.14}$$

If n replicate determinations are made and averaged of a quantity having standard deviation σ and relative standard deviation ε, the terms for the mean are given by

$$\sigma_n = (1/n^{1/2})\sigma \tag{8.15}$$

$$\varepsilon_n = (1/n^{1/2})\varepsilon \tag{8.16}$$

These equations are used to calculate the standard counting error (Section 8.4.3) for replicate measurements of an intensity or accumulated count.

If it is required to calculate the standard deviation or relative standard deviation of a combination of measurements or determinations, each having its own standard deviation, the following equations apply.

For the sum or difference of two values, $x_1 + x_2$ or $x_1 - x_2$, each having its own σ and ε,

$$\sigma = (\textstyle\sum \sigma_i^2)^{1/2} \tag{8.17}$$

$$\varepsilon = \frac{[(x_1\varepsilon_1)^2 + (x_2\varepsilon_2)^2]^{1/2}}{x_1 \pm x_2} \tag{8.18}$$

Note that Equation (8.17) applies to any number of individual values added and/or subtracted. In Equation (8.18), the denominator is $x_1 + x_2$ or $x_1 - x_2$ according to whether the calculation is for a sum or difference, respectively. These equations are used to calculate the standard counting error (Section 8.4.3) of a net intensity ($I_P - I_B$) or net count ($N_P - N_B$), where P and B refer to peak and background, respectively.

For the product of two measurements $x_1 x_2$,

$$\sigma = x_1 x_2 [(\sigma_1/x_1)^2 + (\sigma_2/x_2)^2]^{1/2} \tag{8.19}$$

$$\varepsilon = (\varepsilon_1^2 + \varepsilon_2^2)^{1/2} \tag{8.20}$$

For the quotient of two measurements x_1/x_2,

$$\sigma = (x_1/x_2)[(\sigma_1/x_1)^2 + (\sigma_2/x_2)^2]^{1/2} \tag{8.21}$$

Equation (8.20) applies also for the quotient.

One final statistical term: The highest and lowest values of x between which the actual measurement is likely to fall with a specified probability are known as the *confidence limits* or *confidence levels*. For example, one may predict that a measured x will lie between $\bar{x} \pm \sigma$ at the 68.3% confidence level, and $\bar{x} + \sigma$ and $\bar{x} - \sigma$ are the 68.3% confidence limits. The most frequently cited confidence limits are given in Table 8.3.

8.4.3. Counting Error

Suppose that a perfectly stable homogeneous specimen is placed in an x-ray spectrometer, all components of which, from the x-ray generator to the readout components, are perfectly stable. Suppose further that a series of preset-time intensity measurements 1, 2, 3, ..., n is made of an x-ray spectral line of an element in the specimen without disturbing the specimen or any instrument setting during the series. Even under these ideal—and unattainable—conditions, the numbers of counts accumulated $N_1, N_2, N_3, \ldots, N_n$ during successive time intervals will differ. Since the x-ray photons have a random time distribution, the n accumulated counts

N_i have a Gaussian distribution about a mean value $\bar{N}$ having a standard deviation or *standard counting error* equal to the square root of the mean count.

Following are equations for calculation of the *standard counting error* σ_N and *relative counting error* ε_N in terms of accumulated count N measured by various counting techniques.

For a *single measurement* of N counts accumulated in time T s at intensity I counts/s,

$$\sigma_N = N^{1/2} \tag{8.22}$$

$$\varepsilon_N = \frac{\sigma_N}{N} = \frac{N^{1/2}}{N} = \frac{1}{N^{1/2}} = \frac{1}{(IT)^{1/2}} \tag{8.23}$$

For a *net count* of N_P counts at the analyte peak corrected for a background of N_B counts, both accumulated for the same time, from Equation (8.17),

$$\sigma_N = (\sigma_P{}^2 + \sigma_B{}^2)^{1/2} \tag{8.24}$$

$$= (N_P + N_B)^{1/2} \tag{8.25}$$

$$\varepsilon_N = \frac{(N_P + N_B)^{1/2}}{N_P - N_B} \tag{8.26}$$

For a *ratio count uncorrected for background*, where the time to accumulate N_S counts from a standard S is also used to accumulate N_X counts from the sample X,

$$\varepsilon_N = \left(\frac{1}{N_X} + \frac{1}{N_S} \right)^{1/2} \tag{8.27}$$

For a *ratio count corrected for background*, with the corresponding backgrounds $(N_B)_S$ and $(N_B)_X$ accumulated in the same time as N_S and N_X,

$$\varepsilon_N = \left[\frac{(N_P + N_B)_X}{(N_P - N_B)_X{}^2} + \frac{(N_P + N_B)_S}{(N_P - N_B)_S{}^2} \right]^{1/2} \tag{8.28}$$

In Equations (8.27) and (8.28), N_X is measured at the analyte line; N_S may be measured at the analyte line from a standard, or it may be measured from the sample at an internal-standard line, or a scattered x-ray tube target line, or a wavelength in the scattered continuum. The method is widely used with automatic spectrometers.

The foregoing equations apply to single measurements of N, $N_P - N_B$, or N_X/N_S, as the case may be. If n replicate determinations are made of each of these quantities, the standard counting error and relative fractional counting error are given by Equations (8.15) and (8.16):

$$(\sigma_N)_n = \frac{1}{n^{1/2}}\, \sigma_N \qquad (8.29)$$

$$(\varepsilon_N)_n = \frac{1}{n^{1/2}}\, \varepsilon_N \qquad (8.30)$$

Equations of this form are also used to calculate the standard deviation of the overall average of a set of n averages, each derived from a series of measurements and each having its own σ and ε.

It is evident that standard counting error decreases as N increases, and, in principle, may be made as small as required if a sufficiently long counting time is permissible. However, accumulation of a large N at low intensity requires a very long counting time; this may be disadvantageous for several reasons. Analysis time is increased. Extremely high stability is required in the x-ray tube potential and current and in the electronic detector and readout components. Finally, difficulties arise on prolonged exposure of liquid specimens to the x-ray beam.

Table 8.4 gives values of σ, 2σ, and 3σ and ε, 2ε, and 3ε for a range of values of N. Figure 8.9 shows the correlation of ε for $N = 10^2$–10^7 counts at 0.67σ, σ, 2σ, and 3σ.

Following are equations similar to those above, but in terms of intensity I rather than accumulated count N.

For a *single measurement* of I counts/s for time T s,

$$\sigma_I = \frac{I}{(IT)^{1/2}} = \left(\frac{I}{T}\right)^{1/2} \qquad (8.31)$$

$$\varepsilon_I = \frac{(I/T)^{1/2}}{I} = \frac{1}{(IT)^{1/2}} \qquad (8.32)$$

For a *net intensity* of I_P counts/s at the analyte peak corrected for I_B counts/s at the background counted for times T_P and T_B s, respectively,

$$(\sigma_I)_P = \left(\frac{I_P}{T_P}\right)^{1/2}; \qquad (\sigma_I)_B = \left(\frac{I_B}{T_B}\right)^{1/2} \qquad (8.33)$$

From Equation (8.17),

$$\sigma_I = (\sigma_P{}^2 + \sigma_B{}^2)^{1/2} \qquad (8.34)$$

$$= \left(\frac{I_P}{T_P} + \frac{I_B}{T_B}\right)^{1/2} \qquad (8.35)$$

$$\varepsilon_I = \frac{[(I_P/T_P) + (I_B/T_B)]^{1/2}}{I_P - I_B} \qquad (8.36)$$

TABLE 8.4. Standard and Relative Counting Error

Number of counts, N	σ^a (counts/s)	$\varepsilon_{68.3}{}^b$ (%)	$2\sigma^a$ (counts/s)	$\varepsilon_{95.4}{}^b$ (%)	$3\sigma^a$ (counts/s)	$\varepsilon_{99.7}{}^b$ (%)
100	10	10.0	20	20.0	30	30.0
200	14	7.1	28	14.1	42	21.2
500	22	4.5	44	9.0	66	13.4
1,000	32	3.2	64	6.3	96	9.5
2,000	45	2.2	90	4.5	135	6.7
5,000	71	1.4	142	2.8	213	4.2
10,000	100	1.0	200	2.0	300	3.0
20,000	141	0.70	282	1.4	423	2.1
50,000	224	0.45	448	0.90	672	1.3
100,000	316	0.32	632	0.63	948	0.95
200,000	447	0.22	894	0.45	1341	0.67
500,000	707	0.14	1414	0.28	2121	0.42
1,000,000	1000	0.10	2000	0.20	3000	0.30
2,000,000	1414	0.07	2828	0.14	4242	0.21
5,000,000	2236	0.05	4472	0.09	6708	0.13
10,000,000	3162	0.03	6324	0.06	9486	0.10

[a] $\sigma = N^{1/2}$.
[b] $\varepsilon = 100(n\sigma/N)$; the subscript of ε is the confidence level; n is 1, 2, or 3.

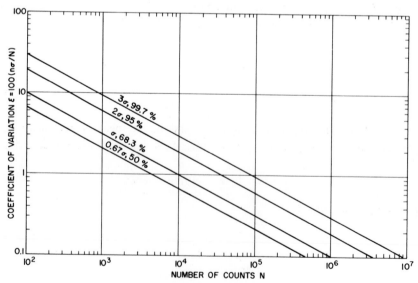

FIGURE 8.9. Correlation of coefficient of variation and accumulated count for various standard deviations and confidence levels.

For a *ratio measurement uncorrected for background*, where the time T_S to measure intensity I_S from a standard S is also used to measure intensity I_X from the sample X,

$$\varepsilon_I = \frac{1}{T_S^{1/2}} \left(\frac{1}{I_X} + \frac{1}{I_S} \right)^{1/2} \qquad (8.37)$$

There are three basic methods for measurement of the peak and background intensities: *fixed-count*, in which the same total count is accumulated for both; *fixed-time*, in which the same counting time is used for both; and *optimal fixed-time*, in which the total counting time is apportioned between the two measurements so as to give minimal error in the net intensity. Equations for calculation of counting error and relative counting error for these three methods follow.

For the *fixed-count method*, characterized by the conditions $(IT)_P = (IT)_B$, $T_P/T_B = I_B/I_P$, and $T_P + T_B = T$ (total counting time),

$$\sigma_{\mathrm{FC}} = \left(\frac{1}{T^{1/2}} \right)(I_P + I_B)^{1/2}\left(\frac{I_P}{I_B} + \frac{I_B}{I_P} \right)^{1/2} \qquad (8.38)$$

For the *fixed-time method*, characterized by $T_P = T_B = T/2$ and $T_P + T_B = T$,

$$\sigma_{\mathrm{FT}} = \left(\frac{2}{T} \right)^{1/2} (I_P + I_B)^{1/2} \qquad (8.39)$$

$$\varepsilon_{\mathrm{FT}} = \left(\frac{2}{T} \right)^{1/2} \left(\frac{I_P + I_B}{I_P - I_B} \right)^{1/2} \qquad (8.40)$$

The *optimal-fixed-time method* is essentially the same as the fixed-time method except that the total counting time T is apportioned between the peak and background to give minimal error in the measured net intensity. This condition exists when $T_P + T_B = T$ and

$$\frac{T_P}{T_B} = \left(\frac{I_P}{I_B} \right)^{1/2} \qquad (8.41)$$

Then,

$$\sigma_{\mathrm{FTO}} = \frac{1}{T^{1/2}} (I_P^{1/2} + I_B^{1/2}) \qquad (8.42)$$

$$\varepsilon_{\mathrm{FTO}} = \frac{1}{T^{1/2}} \left(\frac{I_P^{1/2} + I_B^{1/2}}{I_P - I_B} \right) \qquad (8.43)$$

$$= \frac{1}{T^{1/2}} \left(\frac{1}{I_P^{1/2} - I_B^{1/2}} \right) \qquad (8.44)$$

Incidentally, an equation similar to Equation (8.41) applies when an optimal time apportionment is required for the measurement of the *ratio* of two intensities I_1/I_2:

$$\frac{T_1}{T_2} = \left(\frac{I_1}{I_2}\right)^{1/2} \tag{8.45}$$

Equations (8.29) and (8.30) for replicate measurements having σ_N and ε_N apply to replicate measurements having σ_I and ε_I as well.

The precision of *ratemeter–recorder* intensity measurements is also subject to statistical evaluation. Measurement of an intensity with a rate-meter having a time constant of RC seconds is equivalent to accumulation of counts during a time interval of $2RC$ seconds. Then the relative fractional counting error is

$$\varepsilon_I = 1/[2I(RC)]^{1/2} \tag{8.46}$$

Actually, R signifies resistance in ohms (Ω), and C signifies capacitance in farads (F). Dimensionally, $1\,\Omega = 1$ s/esu-cm, and $1\,\text{F} = 1$ esu cm. Then $RC\ [=]\,\Omega\text{F}\ [=]$ (s/esu-cm)(esu cm) $[=]$ s.

8.4.4. Counting Strategy and Figure of Merit

Counting strategy is the selection of counting technique, counting time, and/or accumulated count to make a specified intensity or count measurement or series of measurements with a specified precision in the shortest time. The choice of counting strategy for the measurement of a net intensity consists in choice of the fixed-count, fixed-time, or optimal-fixed-time method and apportionment of the total counting time between measurement of peak and background intensities.

Incidentally, one must distinguish the terms *preset* time and count (Section 5.3.1) and *fixed* time and count as defined in Section 8.4.3 and below. The former refer simply to the mechanics of making the measurement, the latter to the strategy of the measurement.

In the fixed-count method, the same preset count N is accumulated at the peak and background; then,

$$N_P = N_B, \qquad (IT)_P = (IT)_B, \qquad T_P + T_B = T \tag{8.47}$$

where all the notation has the usual significance except T (without subscript), which is total counting time for peak and background. The fixed-count method has the advantages that only 2θ need be changed between peak and background measurements and that, since $N_B = N_P$, the net counting error is only slightly greater than that for the peak alone [Equation (8.24)]. However, accumulation of the same number of counts at the background as at the peak may require a long counting time, and the lower I_B is, the longer the counting time must be.

The fixed-time method has three variations, one of which is not really a fixed-time method at all, and each of which has a different counting error.

In case I, the counting time at the peak T_P is selected to accumulate about the same number of counts N_P as would be selected if the fixed-count method were to be used; then T_B is the same as T_P.

In case II, the counting times at the peak and background are also equal, but each is about half the *total* counting time used to accumulate the peak and background counts by the fixed-count method. For example, if the fixed-count method requires peak and background counting times of, say, 30 and 70 s, respectively, the total counting time is 100 s, and in the fixed-time method, T_P and T_B would each be 50 s.

In case III, the fixed times for peak and background are different and are apportioned as specified by Equation (8.41). This is the *optimal fixed-time method*. This is a preselected-time method (Section 5.3.1), but not truly a fixed-time method.

The true fixed-time method, then, is characterized by equality of counting times for peak and background (cases I and II above). Like the fixed-count method, it also requires only a change of 2θ between peak and background measurements, but does not require background counting times as long as the fixed-count method. The net counting error for fixed time is greater or less than that for fixed count, depending on whether one considers case I or case II above.

The *optimal fixed-time method* has the advantages that net counting error is minimal and that long background counting times are not required. However, in addition to 2θ, the counting time must also be changed between peak and background measurements. A more often cited disadvantage is that it is necessary to make preliminary measurements of I_P and I_B and to calculate the optimal apportionment of the counting times. These calculations are somewhat elaborate, and because of this, some workers reject the optimal fixed-time method. Other workers use the fixed-time method for small numbers of specimens, but take advantage of the economy of counting time of the optimal method for analyses involving large numbers of specimens and for trace analyses where the counting times are very long.

For case I, the net counting error is the same or greater for fixed time than for fixed count, but never more than 10% greater, and this increase is justified by the economy of background counting time.

For case II—*total* counting time the same for fixed-time and fixed-count methods—the net counting error is the same or less for fixed-time than for fixed-count. Moreover, for a fair comparison, the precision of the two methods should be compared for a specified total counting time.

Gaylor has shown that: (1) for a specified total counting time for peak and background, the variance of the fixed-time method is always less than or equal to that of the fixed-count method; and (2) for a specified precision, the total counting time required for the fixed-time method is always less than or equal to that for the fixed-count method. It is also demonstrable that when the ratio of two intensities is required, the fixed-count and fixed-time methods give the same counting error, but when the difference of two intensities or a net intensity is required, the fixed-time method always gives a lower counting error in a specified total measuring time with peak and background counting times the same.

The counting errors of all three methods may be compared by solution of Equations (8.38) for σ_{FC}, (8.39) for σ_{FT}, and (8.42) for σ_{FTO} for the *same* total counting time with the following typical conditions: $I_P = 10,000$ counts/s, $I_B = 100$ counts/s, and T (*total* counting time) $= 100$ s. The result is $(\sigma_{FC} = 100) > (\sigma_{FT} = 14) > (\sigma_{FTO} = 11)$, where the numbers are counts.

In practice, Equations (8.38), (8.39), and (8.42) can be used to calculate the total counting time T required to realize a specified standard counting error for the peak and background intensities actually encountered from the specimens. Then the optimal apportionment of T between peak I_P and background I_B intensities can be calculated from the following equations:

$$T_P + T_B = T \tag{8.48}$$

$$T_P/T_B = (I_P/I_B)^{1/2} \tag{8.49}$$

Of course, the instrument is not likely to provide the exact preset times given by these calculations, and in such cases, the closest preset times provided are used.

The *figure of merit* is a guide to the selection of excitation and other operating parameters to realize measurements having highest attainable precision. Figures of merit can also be used to compare the precision of intensity measurements on different instruments at substantially the same conditions and on the same instrument at different conditions.

Peak-to-background ratio (P/B, I_P/I_B) or the product of peak-to-background ratio and analyte-line intensity $[(I_P/I_B) \times I_L]$ provide commonly recognized figures of merit to be attained in setting instrument operating conditions. Peak-to-background ratios with secondary excitation may be as high as 10,000 : 1, whereas with primary excitation, they are at best a few thousand to 1. Counting precision decreases rapidly as peak-to-background ratio decreases below $\sim$3 : 1.

Jenkins and de Vries propose other figures of merit based on Equation (8.44). For major and minor constituents (relatively high analyte-line intensity) and a properly apportioned total counting time T, if ε is to be minimum, $I_P^{1/2} - I_B^{1/2}$ should be maximum. For constituents having very low concentrations (low analyte-line intensity), I_P is not much greater than I_B and, in fact, approaches it at or near the detection limit. Then $(I_P I_B)^{1/2} \approx I_P$. In such cases, if ε is to be minimum, $(I_P - I_B)/I_B^{1/2}$ should be maximum.

If $I_P - I_B$ (counts/s) is replaced by the slope m (counts/s per % analyte) of the intensity *versus* concentration calibration curve, the figure of merit is $m/I_B^{1/2}$. This is the figure of merit proposed by Spielberg and Brandenstein. It is most useful at low concentrations where I_P approaches I_B, and, like $(I_P - I_B)/I_B^{1/2}$, should be as large as possible.

8.4.5. Analytical Precision

Table 8.5 gives data for 12 (n) determinations of germanium made on the same sample by timing (T) 12 preset counts of 51,200 (N) of the Ge $K\beta$ line, calculating the intensities ($I = N/T$), and applying them to a calibration curve. The curve was prepared previously from measurements on standards regarded as accurate. The specimen was reloaded without regard to orientation for each measurement, but no instrument setting was disturbed. The standard deviation σ of the set of n results x from their mean $\bar{x}$ is calculated from Equation (8.12),

$$\sigma = \left(\frac{\sum d^2}{n-1} \right)^{1/2}$$

where d is the deviation of an individual determination x_i from the average $\bar{x}$ of the n replicate determinations. The coefficient of variation ε is then calculated from Equation (8.14):

$$\varepsilon = 100(\sigma/\bar{x})$$

The standard deviation is a measure of absolute precision; the coefficient of variation is a measure of relative precision, that is, precision relative to the "amount" (concentration) of analyte present.

TABLE 8.5. Precision of an X-Ray Spectrometric Analysis

i	Counting time T (s)	Ge $K\beta$ net intensity I (counts/s)	Ge concentration C (at %) x_i	Deviation, $d = x_i - \bar{x}$	Deviation squared, d^2
1	26.1	1962	34.7	+0.3	0.09
2	26.0	1969	35.1	+0.7	0.49
3	26.3	1947	33.9	−0.5	0.25
4	26.2	1954	34.3	−0.1	0.01
5	26.1	1962	34.7	+0.3	0.09
6	26.2	1954	34.3	−0.1	0.01
7	26.2	1954	34.3	−0.1	0.01
8	26.2	1954	34.3	−0.1	0.01
9	26.4	1939	33.4	−1.0	1.00
10	26.0	1969	35.1	+0.7	0.49
11	26.2	1954	34.3	−0.1	0.01
12 = n	26.2	1954	34.3	−0.1	0.01

$$\Sigma I = 23472 \qquad \Sigma x = 412.7 \qquad\qquad \Sigma d^2 = 2.47$$

$$\Sigma I/n = 1956 \qquad \Sigma x/n = 34.4 = \bar{x} \qquad \Sigma d^2/(n-1) = 0.22$$

$$\sigma = [\Sigma d^2/(n-1)]^{1/2} = 0.47 \text{ at } \% \text{ Ge}$$

$$\varepsilon = 100(\sigma/\bar{x}) = 1.4 \%$$

Standard counting error for $N = 51{,}200$: $\sigma_N = N^{1/2} = 226$

$$\varepsilon_N = 100(\sigma_N/N) = 0.44\%$$

In Table 8.5, the analytical result of, say, the first determination ($i = 1$) may be reported by any combination of the following statements: 34.7% Ge $\pm$ 0.5% absolute, or 34.7% Ge $\pm$ 1.4% relative, or 34.7% Ge $\pm$ 1.4% of the amount present; followed by: for one standard deviation, or for one sigma (1σ), or at the 68.3% confidence level. This means that the probability is 68.3% (7/10) that the "true" germanium concentration, that is, the mean of a very large number of determinations, lies within $\pm 0.5\%$ absolute of 34.7%, or that it lies within (0.014×34.7) of 34.7%. Similarly, the result may be reported as 34.7% Ge $\pm 0.9\%$ absolute or $\pm 2.8\%$ relative for 2σ or at the 95.4% confidence level. Table 8.3 gives other degrees of precision that are used in reporting analytical results.

Table 8.5 also gives the standard counting error σ_N and relative counting error ε_N for the preset count N of 51,200. It is evident that the coefficient of variation of a single determination of germanium (1.4%) is nearly three times the relative counting error for a count of 51,200 (0.44%). This discrepancy shows that serious errors other than the counting error are impairing the precision of the determination. Standard counting error is—or is derived from—the square root of the mean. Standard deviation is a measure of the dispersion of a group of individual results about their mean. The two are identical in an x-ray spectrometric analysis only if the analytical process is truly random and subject to no error other than that arising from the random time distribution of the x-ray photons.

In the preceding paragraph, it is implied that in the absence of any source of error other than the counting error, the relative standard deviation of the analytical concentration should be the same as the relative counting error of the intensity or combination of intensities (difference, ratio, etc.) from which it is derived. This is substantially true only for linear calibration curves that pass through the origin, but is not generally true, as shown in Figure 8.10, where σ_I is the standard counting error.

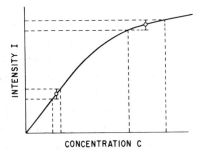

FIGURE 8.10. Relationship of counting error and that portion of the analytical concentration error contributed by the counting error. The indicated spread may represent, say, $\pm \sigma$.

The standard counting error σ_N has twofold significance in x-ray spectrometric analysis: (1) It represents the maximum precision attainable, and cannot be eliminated from a process based on counting photons; and (2) it can be accurately calculated and used as a standard against which to evaluate the precision actually attained.

The precision of an x-ray spectrometric analysis is obtained by combining the variances for the individual sources of error [Equation (8.17)], so that the precision may be represented by

$$\sigma_T{}^2 = \sigma_N{}^2 + \sigma_I{}^2 + \sigma_O{}^2 + \sigma_S{}^2 + \sigma_M{}^2 \qquad (8.50)$$

where the subscripts represent, in order, total, counting, instrumental, operational, specimen, and miscellaneous errors. A similar equation can be written for each of the individual errors, expressing the variance for each as the sum of the variances of several specific sources of error.

The counting error is calculated by use of the appropriate equation from Section 8.4.3. Long- and short-term errors are evaluated from repeated measurements on a smooth, homogeneous, stable specimen without disturbing the specimen or the instrument settings. Operational error is evaluated from repeated measurements on the same type of specimen without disturbing the specimen, but resetting, say, x-ray tube potential (kV) (but always to the same value) for each measurement. Similar series of measurements evaluate the effect of resetting x-ray tube current (mA), goniometer angle (2θ), etc. Specimen error is evaluated from repeated measurements on a specimen of the type to be analyzed, without disturbing instrument settings, but reorienting, rotating, reloading, etc., the specimen for each measurement. Specimen-preparation error is evaluated from measurements on several portions ("aliquots") of a carefully homogenized sample, each portion subjected to the same specimen preparation process.

8.5. SENSITIVITY

Sensitivity in x-ray spectrometric analysis may be defined in either of two ways—in terms of minimum detectable amount of analyte, or rate of change of analyte-line intensity with change in amount of analyte.

The *minimum detectable amount, minimum detection limit* MDL, *lower limit of detectability* LLD, or *concentration at the detection limit* C_{DL} may be defined in several ways, but the definition given by Birks is probably the most widely accepted: that amount of analyte that gives a net line intensity equal to three times the square root of the background intensity for a speci-

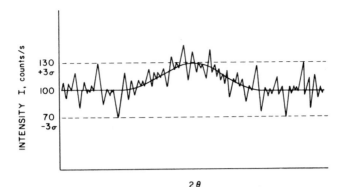

FIGURE 8.11. Detection limit.

fied counting time, or, in statistical terms, that amount that gives a net intensity equal to three times the standard counting error of the background intensity. "Amount" signifies concentration (%, mg/ml, etc.) for infinitely thick specimens, area density (mg/cm²) for specimens less than infinitely thick, or mass (μg) for extremely small particulate or filamentary specimens. Figure 8.11 shows a ratemeter–recorder 2θ scan of a weak peak on a uniform background. Suppose that in a specified preset time, the accumulated count for the background is 100 counts. The standard counting error for the background count σ_N [Equation (8.22)] is then $100^{1/2}$, or 10 counts. The background count remains within $\pm 3\sigma$ (± 30 counts) of its mean (100 counts) 99.7% of the time. Then, if there is to be 95% certainty (2σ) of detection of the peak, its accumulated count in the same preset time must exceed the background by $3\sigma_N$, or 30 counts; that is, the accumulated peak count must be 130. Conversely, if the accumulated count is 130, then there is 95% probability that there is really a peak present. Whatever concentration, area density, or isolated mass is required to give analyte-line intensity corresponding to this count is the minimum detectable amount, or limit of detectability.

The value $3(I_B)^{1/2}$ is obtained as follows. From Equation (8.34), the standard counting error for a net count is calculated from the counting errors of the individual peak σ_P and background σ_B measurements:

$$\sigma = (\sigma_P^2 + \sigma_B^2)^{1/2} \tag{8.51}$$

For trace concentrations, I_P approaches I_B, and $\sigma_P \approx \sigma_B$; then,

$$\sigma = (2\sigma_B^2)^{1/2} = 2^{1/2}\sigma_B \tag{8.52}$$

For 95% confidence,

$$2\sigma = 2(2^{1/2}\sigma_B) \approx 3\sigma_B = 3(I_B)^{1/2} \tag{8.53}$$

The counting time should also be specified and is usually 100 s; that is, $T_P = T_B = 50$ s. From Equation (8.33), $\sigma_B = (I_B/T_B)^{1/2}$, so that the minimum detectable count is $3(I_B/T_B)^{1/2}$. If now one includes the slope m (counts per 1% analyte) of the count *versus* concentration calibration curve, the minimum detectable *amount* becomes

$$C_{DL} = (3/m)(I_B/T_B)^{1/2} \tag{8.54}$$

It is evident that the minimum detectable amount is smaller the lower the background intensity, the greater the slope, and the longer the counting time.

To return to our example (Figure 8.11), suppose that application of a net analyte-line count of 30 to a previously prepared calibration curve represents 0.01% analyte; this is the *concentration at the detection limit* or *minimum detection limit*. The precision of the analysis is given by Equation (8.26): $(130 + 100)^{1/2}/(130 - 100) = (230)^{1/2}/30 = 15.2/30 \approx 0.5$, that is, $\pm 50\%$ of the amount present or ± 0.005 wt% for 1σ.

For two analytes 1 and 2 in the same matrix, the detection limits $C_{DL,1}$ and $C_{DL,2}$ are related to the two respective net intensities I, accumulated counts N (for the same counting time), and counting times T (for the same count) as follows:

$$C_{DL,1}/C_{DL,2} = (I_1/I_2)^{1/2} = (N_1/N_2)^{1/2} = (T_2/T_1)^{1/2} \tag{8.55}$$

If the detection limit of an analyte is known in a specified matrix, it may be calculated approximately for another matrix by multiplying the known limit by the ratio of the mass-absorption coefficients of the two matrixes for the analyte line. For example, suppose that the detection limit of titanium in iron is 0.0001%, and the detection limit in silver is required. The mass-absorption coefficients of iron and silver for Ti $K\alpha$ are 186 and 1042 cm²/g, respectively. The detection limit of titanium in silver is then $(1042/186)(0.0001) \approx 0.00056\%$. Unfortunately, this simple estimation is less successful for analyte lines of short wavelength because they occur in a spectral region having high and varying background.

With energy-dispersive analysis using a 10-W x-ray tube or the equivalent amount of radioisotope for excitation, in favorable cases, sensitivity as low as 10–20 ng/cm² may be realized for specimens on Millipore substrates.

With proton or other ion excitation, continuous background is extremely low. In favorable cases, sensitivity as low as 1–2 ng/cm^2 may be realized for specimens on film substrates of evaporated carbon or ultrathin plastic having 10–20 μg/cm^2 mass thickness. Filter paper, Millipore filters, and even 3-μm (0.00012-inch) Mylar substrates are destroyed rapidly by ion beams and cannot be used.

The lower limit of detectability or detection limit applies to *detection* of the analyte, that is, to qualitative analysis. The lower limit of *determination* or determination limit for a quantitative analysis is substantially greater and is, in fact, defined as three times the detection limit.

Alternatively, sensitivity may be stated in terms of intensity per unit analyte concentration (counts/s per %, counts/s per mg/ml, etc.) for infinitely thick specimens; intensity per unit area density (counts/s per mg/cm^2) for thin specimens; or intensity per unit mass (counts/s per μg) for very small specimens. Sometimes, sensitivity is given per unit x-ray tube current—for example, counts/s per % per mA. With radioisotope and ion-bombardment excitation, sensitivity may be stated in terms of amount of isotope in millicuries (mCi) and amount of incident ions in microcoulombs (μC)—counts/s per % per mCi and counts/s per % per μC, respectively. Actually, all these definitions constitute the slope of the calibration function—analyte-line intensity *versus* amount of analyte, and when the calibration function is nonlinear, the sensitivity varies with the concentration or amount of analyte. Although this may not be the case with thin-film and very small specimens, it is often the case with the more common solid, briquet, powder, and liquid specimens. In such cases, it is necessary to express sensitivity at a specified concentration level, for example, counts/s per % "at the 10% concentration level."

Sometimes, the true x-ray spectrometric sensitivity stated in terms of the measured sample is more conveniently stated in terms of the original sample from which the sample specimen was prepared. This is most often the case when the analytes have been preconcentrated. For example, suppose that a certain analyte, preconcentrated and distributed on a filter-paper disk, has a detection limit of 1 μg. If the material on the filter disk was separated from 10 g of solid, 10 l of water, or 10 m^3 of air, the detection limit of the analyte may be stated as 0.1 μg/g, 0.1 μg/l, or 0.1 μg/m^3, respectively.

In general, for a standard 50-kV instrument, a sensitivity curve of counts/s per 1% analyte *versus* atomic number has the general appearance of a lopsided M, and a curve of minimum detectable concentration C_{DL} *versus* atomic number has the general appearance of a lopsided W. The left half of the M or W represents elements of atomic number 9–55 (fluorine to cesium), for which $K\alpha$ lines are used; the right half represents the heavier elements, for which $L\alpha_1$ lines are used. The reasons for the shapes of the curves follow.

Sensitivity is highest (C_{DL} lowest) for elements of atomic number 20 (calcium) to 40 (zirconium) because of favorable combination of efficient excitation, high fluorescent yield, low analyte-line absorption, high-reflectivity crystals, and high-efficiency detectors. Incidentally, possibly the most sensitive element of all is titanium (Z 22).

As atomic number decreases from 20 to 9 (fluorine), the $K\alpha$ wavelength increases from 3.4 to 18 Å and sensitivity decreases (C_{DL} increases) because of: (1) poor excitation due to low transmission of the long-wavelength primary x-rays through the x-ray tube window; (2) decreasing fluorescent yield; and (3) increasing analyte-line absorption in the matrix, path, and cell and detector windows.

As atomic number increases from 40 to ∼55 (cesium), the $K\alpha$ wavelength decreases from 0.8 to 0.4 Å and sensitivity decreases (C_{DL} increases) because of: (1) poor excitation due to increasing excitation potential; (2) failure of the analyzer crystal to intercept the entire secondary beam as 2θ decreases; (3) high background consisting mostly of first-order scattered radiation; and (4) decreased spectrometer resolution as 2θ decreases [Equation (8.61)].

For elements 55 to ∼80 (mercury), $L\alpha_1$ lines are used and, for these elements, have wavelengths 2.9–1.2 Å, about the same as the $K\alpha$ lines of elements 20–40 above. Sensitivity increases (C_{DL} decreases) for substantially the same reasons as given for elements 20–40. However, sensitivity does not attain its magnitude for the most sensitive $K\alpha$-line elements because of the lower intensity of $L\alpha_1$ lines relative to $K\alpha$.

As atomic number increases beyond 80, sensitivity decreases again for substantially the same reasons as given for $K\alpha$-line elements 40–55 above.

Because of the multiplicity of factors that affect sensitivity, it is difficult to cite "typical" performance. For extremely small particulate, filamentary, or residue specimens, a flat-crystal spectrometer can measure as little as 10^{-6}–10^{-7} g, a curved-crystal spectrometer as little as 10^{-8}–10^{-9} g. For comparison, the electron-probe microanalyzer can detect as little as 10^{-12}–10^{-15} g. For thin-film specimens having area 1 cm² or more, a flat-crystal instrument can measure a monolayer of any element having atomic number 12 (magnesium) or more, and perhaps as little as 0.01 monolayer of the more sensitive elements.

A good, modern, commercial x-ray spectrometer, using optimal accessories (target, crystal, collimator, detector) and maximum excitation (50 kV, 50 mA, CP), should give typical $K\alpha$ net intensities (counts/s) for "pure element" specimens as follows: $_9F$ (in NaF), 300; $_{11}Na$ (in NaF), 3000; $_{13}Al$, 250,000; $_{42}Mo$, 10^6; and $_{56}Ba$ [in $Ba(NO_3)_2$], 320,000.

In Equation (8.54), it is evident that there are three ways to improve sensitivity: increase analyte-line intensity and thereby the slope m of the calibration curve, reduce background intensity, and increase counting time. For a specified model of spectrometer, the most practical means for improving sensitivity are the following.

The x-ray tube power should have constant-potential, rather than peak, wave form, especially for analyte lines having high excitation potentials. Analyte-line excitation is favored by x-ray tube target (or other source) lines and/or primary continuum hump near the short-wavelength side of the associated absorption edge. Excitation conditions and other instrument operating components and conditions should be chosen to give high peak-to-background ratio or other favorable figure of merit (Section 8.4.4).

The specimen surface texture and/or particle size should be smaller than the path length of the longest wavelength to be measured. Surface grind marks should be oriented parallel to the plane defined by the directions of the primary and secondary x-rays.

The optimal analyte line should be used: For atomic number less than 55 (cesium), the $K\alpha$ line is most intense; for Z 55–65 (terbium), the $L\beta_1$ line is most intense; for $Z > 65$, the $L\alpha_1$ line is most intense. However, sometimes an L line, although less intense than a K line, may give a more favorable peak-to-background ratio if it occurs in a less intense region of the scattered continuum. In practice, with a 50-kV generator, K lines are almost always used for elements up to atomic number 40 (zirconium) and L lines are almost always used above atomic number 60 (neodymium); K or L lines may be used for the intermediate elements. First-order lines have about 10 times the intensity of second-order lines with a LiF(200) crystal.

All practical means should be applied to reduce background (Section 5.3.2.3). Collimation should be the minimum—as short and/or widely spaced as possible—for the resolution required. The crystal should have high "reflectivity" and diffract the analyte line at a 2θ angle high enough so that the entire secondary beam is intercepted. Helium or vacuum path should be used for atomic number 22 (titanium) and lower, vacuum for 11 (sodium) and lower. The detector should have high quantum efficiency for the analyte line. The pulse-height selector should be omitted if feasible; otherwise, the window should be set to admit as many analyte-line pulses as possible without also admitting interfering pulses.

8.6. RESOLUTION

Resolution, or resolving power, is a measure of the ability of the spectrometer to distinguish, or recognize as separate, two closely spaced spectral lines, and is a function of their dispersion and divergence—that is, their separation and breadth. Depending on whether wavelength (2θ) or energy dispersion is used, x-ray spectrometer resolution is expressed in terms of *wavelength resolution*—separation of the 2θ peaks of the two lines —or *energy resolution*—separation of their pulse-height distributions. Wavelength resolution or resolving power may be expressed in terms of the difference in wavelength $\Delta\lambda$ of two spectral lines of average wavelength λ that are just resolved, that is, $\lambda/\Delta\lambda$; $\Delta\lambda$ is placed in the denominator so that the higher the resolution, the higher is the number expressing it. Energy resolution may be expressed analogously in terms of the difference in photon energy ΔE of two spectral lines of average photon energy E that are just resolved, that is, $E/\Delta E$.

Perhaps more practically, x-ray spectrometric wavelength resolution is usually expressed as the quotient of the (angular) dispersion D of two peaks that are just resolved, by their divergence or breadth B, that is, D/B, where both terms are in degrees 2θ. Dispersion and divergence are discussed below.

Energy resolution may be expressed analogously as the quotient of the separation $\bar{V}_1 - \bar{V}_2$ of the peaks of two pulse-height distributions that are just resolved, by their half-widths $W_{1/2}$ or FWHM (Section 5.1.6.4), that is, $(\bar{V}_1 - \bar{V}_2)/W_{1/2}$, where all terms are in volts. However, energy resolution of an x-ray detector or spectrometer is usually expressed as the quotient of the half-width of an *individual* pulse-height distribution by the *average* pulse height in that distribution $\bar{V}$, that is, $W_{1/2}/\bar{V}$, usually as percent. Again, both terms are in volts. It is evident that by this definition, the higher the resolution, the lower is the number expressing it.

Detector resolution is discussed in more detail in Section 5.1.6.4.

Dispersion is the separation of the two lines in θ or of their pulse-height distributions in average pulse height, that is, $\Delta\theta/\Delta\lambda$ (degrees θ per angstrom) or $\Delta V/\Delta E$ (volts per electron volt). The angular dispersion of an analyzer crystal for two lines $\lambda_1 < \lambda_2$ diffracted at Bragg angles θ_1 and θ_2 is given by differentiation of the Bragg law [Equation (8.56)] and taking the reciprocal of the result.

$$\lambda = (2d/n) \sin \theta \qquad (8.56)$$

$$\frac{\theta_2 - \theta_1}{\lambda_2 - \lambda_1} = \frac{d\theta}{d\lambda} = \frac{n}{2d \cos \theta} \qquad (8.57)$$

where θ is in radians. It is evident that $d\theta/d\lambda$ increases (improves) with increasing diffraction order n, decreasing crystal spacing $2d$, and increasing Bragg angle θ. In practice, with adequate collimation and crystals having small rocking angle, the dispersion is likely to largely constitute the resolution.

Divergence is a measure of the intensity distribution of the diffracted peak in 2θ (wavelength divergence), or of its pulse-height distribution in baseline potential (energy divergence). It is expressed as the half-width $W_{1/2}$ or breadth B of the diffracted peak in degrees 2θ, or of the pulse-height distribution in volts, measured midway between the peak maximum and the background. The choice of symbol, $W_{1/2}$ or B, is largely a matter of personal choice. Still another term having the same significance is *full width at half-maximum*, or FWHM.

The intensity distribution of the x-ray beam diffracted by the analyzer crystal is known as its rocking curve. The distribution is Gaussian and has divergence B_D, as shown in Figure 8.12A. The divergence B_D is smaller the more nearly perfect is the mosaic structure of the crystal. The intensity distribution of the x-ray beam transmitted by a single collimator C_1 is triangular, as shown in Figure 8.12B. If the collimator has length L and foil spacing S, the divergence is given by

$$B_{C_1} = \tan^{-1}(S/L) \tag{8.58}$$

When a crystal is used with a single collimator, the resultant intensity distribution, shown in Figure 8.12D, is otained from the convolution of the crystal rocking curve (Figure 8.12A) and the intensity distribution of the collimator (Figure 8.12B). The total divergence B_T is calculated by

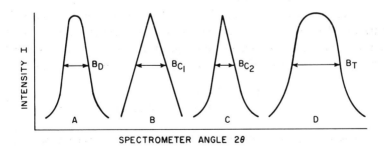

SPECTROMETER ANGLE 2θ

FIGURE 8.12. Intensity distribution and divergence of diffracted and collimated beams. [L. S. Birks, *X-Ray Spectrochemical Analysis*, Interscience Publishers, Inc., p. 22 (1959); courtesy of the author and publisher.] (A) Crystal. (B) One collimator. (C) Two collimators having the same divergence. (D) Resultant distribution of A and B.

assuming the collimator curve to approximate a Gaussian distribution and applying the rule for combining variance [Equation (8.17)],

$$B_T = (B_D{}^2 + B_{C_1}{}^2)^{1/2} \tag{8.59}$$

This equation guides the matching of collimator and crystal. For example, a coarse collimator (10 cm × 0.9 mm, $B_{C_1} = 0.516°$) used with a topaz(303) crystal ($B_D \approx 0.08°$) would pass unnecessarily high background intensity without increasing analyte-line intensity. Similarly, a fine collimator (10 cm × 0.15 mm, $B_{C_1} = 0.086°$) used with a graphite crystal ($B_D \approx 0.6°$) would sacrifice analyte-line intensity without improving resolution.

If a second collimator having the same divergence as C_1 is added, the intensity distribution for the two collimators, shown in Figure 8.12C, is obtained by squaring each ordinate of the curve for a single collimator (Figure 8.12B). The divergence of the squared curve $B_{C_2} \approx 0.6 B_{C_1}$. In general, for two collimators having foil spacings S_1 and S_2, respectively, and distance L' from the specimen end of the source collimator to the detector end of the detector collimator, the divergence is

$$B_{C_2} = (S_1 + S_2)/L' \tag{8.60}$$

In practice, for flat-crystal spectrometers, the crystal rocking curve is small with respect to the collimator intensity distribution; thus, from Equation (8.59), it is evident that the rocking curve has little effect on the total divergence, even with relatively imperfect crystals, and the divergence of the crystal–collimator system is substantially that of the collimator(s) [Equations (8.58) and (8.60)].

Figure 8.13 shows two closely spaced peaks having divergence B and separation $\varDelta 2\theta$. Resolution may be regarded as adequate when $\varDelta 2\theta = 2B$, although two peaks may often be distinguished at much smaller separation, as shown in Figure 8.13. Combining Equations (8.56) and (8.57), one gets

$$\frac{\lambda}{\varDelta\lambda} = \frac{2\tan\theta}{\varDelta 2\theta} = \frac{1}{d\theta\cot\theta} \tag{8.61}$$

where θ is in radians. Substitution of the minimum acceptable value of $\varDelta 2\theta$—that is, $2B$—gives

$$\frac{\lambda}{\varDelta\lambda} = \frac{\tan\theta}{B} \tag{8.62}$$

where B is in radians.

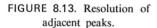

FIGURE 8.13. Resolution of
    adjacent peaks.

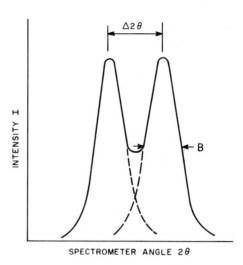

    The left side of Equation (8.62) may be regarded as the required
resolution, the right side as the attainable resolution, which involves θ
and B. The equation shows that resolution increases rapidly with θ (2θ)
because of the tangent function. From the Bragg law, $n\lambda = 2d\sin\theta$ [Equa-
tion (8.56)], it is evident that the smaller the d spacing of the crystal, the
larger is θ, and, thereby, the higher is the resolution. Also, for a specified
crystal, successively higher orders occur at successively higher 2θ angles,
and thus have successively higher resolution. The second-order spectrum
of an element has approximately twice the resolution of the first-order
spectrum.

SUGGESTED READING

BERRY, P. F., "Study of Particulate Heterogeneity Effects in Radioisotope X-Ray Spec-
    trometry," *U. S. At. Energy Comm. Rep.* **ORO-3847-2**, 107 pp. (1971).
BERTIN, E. P., *Principles and Practice of X-Ray Spectrometric Analysis*, 2nd ed., Plenum
    Press, New York (1975); Chaps. 11–13, pp. 459–567.
BIRKS, L. S., *X-Ray Spectrochemical Analysis*, 2nd ed., Wiley–Interscience, New York
    (1969); Chap. 7, pp. 71–83.
BIRKS, L. S., and D. M. BROWN, "Precision in X-Ray Spectrochemical Analysis—Fixed
    Time vs. Fixed Count," *Anal. Chem.* **34**, 240–241 (1962).
BIRKS, L. S., and D. L. HARRIS, "Unusual Matrix Effects in Fluorescent X-Ray Spec-
    trometry," *Anal. Chem.* **34**, 943–945 (1962).
CRISS, J. W., and L. S. BIRKS, *Int. Spectrosc. Colloq.* 15th, Madrid (1969).
DE JONGH, W. K., "Heterogeneity Effects in X-Ray Fluorescence Analysis," N. V. Philips
    Gloeilampenfabrieken, Eindhoven, Neth., *Sci. Anal. Equip. Bull.* 7000.38.0266.11,
    6 pp. (1970).

GAYLOR, D. W., "Precision of Fixed-Time vs. Fixed-Count Measurements," *Anal. Chem.* **34**, 1670–1671 (1962).

JENKINS, R., *Introduction to X-Ray Spectrometry*, Heyden, London (1974); Chap. 6, pp. 107–119.

JENKINS, R., and J. L. DE VRIES, *Practical X-Ray Spectrometry*, 2nd ed., Springer-Verlag New York, New York (1969); Chaps. 5 and 6, pp. 90–125.

LIEBHAFSKY, H. A., H. G. PFEIFFER, E. H. WINSLOW, and P. D. ZEMANY, *X-Rays, Electrons, and Analytical Chemistry*, Wiley–Interscience, New York (1972); Chap. 8, pp. 328–354.

MITCHELL, B. J., "Complex Systems—Interelement Effects"; in *Encyclopedia of Spectroscopy*, G. L. Clark, ed.; Reinhold Publishing Corp., New York, pp. 736–745 (1960).

MITCHELL, B. J., and J. E. KELLAM, "Unusual Matrix Effects in X-Ray Spectroscopy— A Study of the Range and Reversal of Absorption-Enhancement," *Appl. Spectrosc.* **22**, 742–748 (1968).

MÜLLER, R. O., *Spectrochemical Analysis by X-Ray Fluorescence*, Plenum Press, New York (1972); Chaps. 7, 8, and 11, pp. 67–83, 106–129.

POLLAI, G., and H. EBEL, "Tertiary Excitation in X-Ray Fluorescence Analysis," *Spectrochim. Acta* **26B**, 761–766 (1971).

SALMON, M. L., "Factors which Determine Sensitivity;" in *Handbook of X-Rays*, E. F. Kaelble, ed., McGraw-Hill, New York (1967); Chap. 32, 8 pp.

SPIELBERG, N., and M. BRANDENSTEIN, "Instrumental Factors and Figure of Merit in the Detection of Low Concentrations by X-Ray Spectrochemical Analysis," *Appl. Spectrosc.* **17**, 6–9 (1963).

ZEMANY, P. D., "Effects Due to Chemical State of the Samples in X-Ray Emission and Absorption," *Anal. Chem.* **32**, 595–597 (1960).

Chapter 9

Quantitative Analysis

9.1. INTRODUCTION

The first phase of an x-ray spectrometric analysis (Section 3.2) is dual, consisting in the selection of the analytical method (Chapter 9) and specimen preparation and presentation (Chapter 10). The analytical method is chosen mostly to correct absorption–enhancement effects and instrument errors. The specimen preparation-presentation technique is chosen mostly to deal with specimen errors.

This chapter classifies and characterizes the extremely wide variety of methods and techniques of x-ray spectrochemical analysis on the basis of their approach to correction of matrix effects. Determinations of specific elements in specific materials are cited only to illustrate such methods and techniques. Liebhafsky and his coauthors [pp. 535–549 in their book cited in the reading list at the end of the chapter] list 66 elements, plus the lanthanons as a group, and for each cite many literature references for its x-ray spectrometric determination. For some elements, the references are classified by specimen type—metals and alloys, cement materials, minerals and ores, slags, organics, etc. The bibliography has 362 references. Similar listings and bibliographies are given in the biennial reviews in the journal *Analytical Chemistry* and in each quarterly issue of *X-Ray Fluorescence Spectrometry Abstracts* beginning with volume 4 (1973).

Modern x-ray fluorescence spectrometers have very high inherent precision. However, they can give analytical accuracy of the same magnitude only if: (1) calibration standards are available having physical characteristics very similar to those of the samples and accurately known composition, and (2) matrix effects are dealt with effectively.

In Section 8.2, it is shown that were it not for matrix effects, analyte-line intensity $I_{A,M}$ from analyte A in a thick specimen having matrix M would be simply a function of weight fraction $W_{A,M}$ of A in M and the analyte-line intensity $I_{A,A}$ from pure analyte, that is,

$$I_{A,M} = W_{A,M}I_{A,A} \tag{9.1}$$

It has also been pointed out that the principal objective of the analytical "method" or "strategy" is to eliminate, minimize, circumvent, or correct for the effect of the matrix, that is, for the absorption–enhancement effects.

Attempts have been made at absolute x-ray spectrometric analysis by conversion of intensity data to analytical data by wholly mathematical means. The calculations require knowledge of: (1) the intensity–wavelength distribution of the primary x-ray beam; (2) conversion efficiency of each primary wavelength to the analyte line; (3) absorption coefficient of the specimen for each primary wavelength and for the analyte line; (4) conversion efficiency of each primary wavelength to each matrix element line that can excite the analyte line; (5) absorption coefficient of the specimen for each of these lines; (6) conversion efficiencies of each of these lines to the analyte line; and (7) efficiency of the spectrometer (collimators, crystal, path, detector) for transmission, diffraction, and detection of the analyte line. Obviously, the calculations are extremely complex, and absolute methods have had only limited success. Even this has been accomplished only with simple systems having relatively few components, and then only with simplifying assumptions.

Other approaches to the absorption–enhancement problem have been more successful. Most of these methods involve the use of calibration standards. X-ray spectrometric analysis is still largely a calibration method, and intensity data is converted to analytical concentration by use of calibration curves or mathematical relationships derived from measurements on standards.

The many x-ray spectrometric analytical methods may be classified in eight categories with respect to their basic approach to the reduction of absorption–enhancement effects.

1. *Calibration-Standard Methods.* The analyte-line intensity from the samples is compared with that from standards having the same form as the samples and as nearly as possible the same analyte concentration and matrix.

2. *Internal Standardization.* The calibration-standard method is improved by quantitative addition to all specimens of an internal-standard

element having excitation, absorption, and enhancement characteristics similar to those of the analyte in the particular matrix. The calibration function involves the intensity ratio of the analyte and internal-standard lines.

3. *Standardization with Scattered X-Rays.* The intensity of the primary x-rays scattered by the specimen is used to correct the absorption–enhancement effects.

4. *Matrix-Dilution Methods.* The matrix of all specimens is leveled or diluted to a composition such that the effect of the matrix is determined by the diluent.

5. *Thin-Film Methods.* The specimens are made so thin that absorption–enhancement effects substantially disappear.

6. *Standard Addition and Dilution Methods.* The analyte concentration is altered quantitatively in the sample itself. In a certain sense, the sample, subjected to one or more quantitative incremental concentrations or dilutions of the analyte, actually provides its own standard(s) in its own matrix.

7. *Experimental Correction.* Various special experimental techniques have been devised to minimize, circumvent, or compensate for absorption–enhancement effects.

8. *Mathematical Correction.* Absorption–enhancement effects are corrected mathematically by use of experimentally derived parameters.

In this chapter, each of these basic methods is discussed with respect to principles, variations, scope, advantages and disadvantages, and applications. Most of these methods also reduce error from other sources, and Table 9.1 summarizes the effectiveness of each method in compensating the principal sources of error.

In 1971, Jenkins and de Vries classified quantitative x-ray spectrometric methods in four categories and gave an estimate of the fraction of all such analyses constituted by each category: (1) "in-type" methods, in which the sample is analyzed in its as-received form by use of calibration standards in the same form (70%); (2) liquid-solution and solid-solution (fusion) methods (15%); (3) absorption correction methods, including internal standardization, standardization with scattered x-rays, and ratio methods (5%); and (4) mathematical correction methods (10%). Today, the availability of automated computer-controlled instruments has probably nearly doubled the contribution of the last category.

TABLE 9.1. Effectiveness of Analytical Methods in Compensating Errors

Analytical method	Short term	Long term	Operational	Absorption	Enhancement	Inhomogeneity	Position	Surface	Particle size	Density	Bubbles
	Instrumental			Specimen							
Standard addition and dilution	—	C	—	C	C	—	C	—	—	—	—
Thin-film method	—	—	—	C	C	—	—	C	C	—	—
Matrix-dilution											
Dry (low absorber)	—	—	—	C	C	—	—	P	P	C	—
Fusion (low absorber)	—	—	—	C	C	C	—	C	C	C	—
Solution (low absorber)	—	—	—	C	C	C	—	C	C	C	—
High absorber (dry)	—	—	—	X	X	—	—	P	P	X	—
Calibration standardization	S	C	—	C	C	—	—	X	X	P	—
Internal standardization	S	C	P	—	—	—	P	P	P	P	P
Other standardization methods											
Internal control standardization	S	C	P	C	P	—	—	—	—	—	—
Internal intensity-reference standardization	S	C	P	C	P	—	C	—	—	C	C
External standardization	S	C	P	C	P	—	—	—	C	C	—
Scattered x-ray methods											
Background-ratio method	S	C	P	C	—	—	C	C	C	C	P
Scattered target-line ratio method	S	C	P	C	—	—	C	C	C	C	P
Reynolds–Ryland method	S	C	– P	C	—	—	C	C	P	C	—
Briqueting	—	—	—	—	—	—	—	—	P	C	—
Mathematical methods (geometrical, empirical, influence coefficient methods)	—	—	—	C	C	—	—	—	—	—	—

a C: complete or substantial correction; P: partial correction; S: correction only if both the analyte-line and standard intensities are measured simultaneously; X: correction only if standards and samples are very similar with respect to the specified property.

9.2. CALIBRATION-STANDARD METHODS

9.2.1. General Method

With perhaps the exception of the standard addition and dilution methods, *all* x-ray secondary-emission spectrometric analytical methods now in common use are based on comparison of analyte-line intensities measured from the sample(s) and one or more standards. This applies even to the standard addition and dilution methods if one regards the concentrated or diluted samples as standards. However, the other seven basic methods have features that warrant their separate classification distinct from simple calibration standardization. The standards must be similar to the samples with regard to: (1) physical form—solid, briquet, powder, fusion product, solution; or specimen supported on Mylar, filter paper, ion-exchange membrane, etc.; (2) analyte concentration; (3) matrix composition; and (4) physical features, such as surface finish, particle size, and packing density.

In principle, the method consists in application of analyte-line intensity (or some function based on it) measured from the sample(s) to a calibration curve of intensity (or intensity function) *versus* concentration measured from the standards.

A simple calibration curve of peak analyte-line intensity I_P *versus* analyte concentration C is shown in Figure 9.1. If such a curve is linear, its intercept on the intensity axis is the background intensity I_B. Alter-

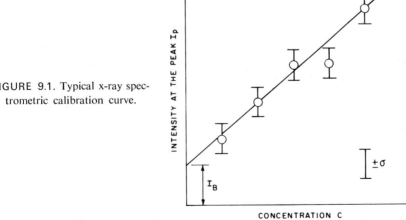

FIGURE 9.1. Typical x-ray spectrometric calibration curve.

natively, background can be measured from each standard and net intensity plotted, in which case the curve should pass through the origin. The figure also indicates the precision of each measured point and shows a point subject to strong systematic error.

Although there is no universally accepted convention, most workers plot x-ray spectrometric calibration curves with the intensity function on the ordinate (vertical axis) and the concentration on the abscissa, as shown in Figure 9.1. If very wide intensity and/or concentration regions are to be bracketed, semilogarithmic or logarithmic scales may be preferable to linear scales.

In fitting a smooth curve to the data points measured from the standards, it is generally agreed that the best fit is the *Cohen least-squares curve*. This is the curve for which the sum of the squares of the deviations of the individual points from the curve is minimal.

The calibration curve need not be linear to be useful, although certainly a linear curve is preferable. In principle, once it has been established with several standards that a curve is linear, two standards define it each time it is used—one if it passes through the origin. More standards are required to plot a nonlinear curve, and the greater the curvature and, especially, the more severely it is skewed, the greater is the number of standards needed. However, if the analytical concentration range of interest is small and occupies only a short segment of a nonlinear curve, and if the standards closely bracket this segment, it may be feasible to assume linearity of the useful interval.

Alternatively, a mathematical calibration factor can be derived from the curve or from the intensity-concentration data of the standards. For example, in Figure 9.1,

$$I_P = mC + I_B \qquad (9.2)$$

where I_P and I_B are peak and background intensities, respectively, C is concentration, and m is the slope of the curve. This equation has the form of the equation for a straight line:

$$y = mx + b \qquad (9.3)$$

The slope of the curve may serve as a *calibration factor*,

$$m = \frac{I_P - I_B}{C} \qquad (9.4)$$

The unit for m is counts per second per unit concentration (%, mg/ml, $\mu g/cm^2$, etc.). Then analytical concentration is given from the net intensity measured from the sample:

$$C = \frac{I_P - I_B}{m} \tag{9.5}$$

Figure 9.2 shows a composite typical x-ray spectrum (2θ scan) showing emitted and scattered radiation that may be used for, or interfere with, an analysis. The significance of each feature of the spectrum is given in the caption. Some features of the figure are discussed in the section on back-

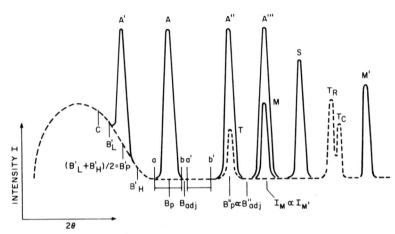

FIGURE 9.2. Composite typical x-ray spectrum (2θ scan) showing emitted and scattered radiation that may be used for, or interfere with, an analysis. Dashed curve: scattered target spectrum. Solid curve: emitted specimen spectrum.

Analytical lines

A	on uniform continuous background
A'	on nonuniform continuous background
A''	on scattered target line T
A'''	on emitted matrix line M

Background

B_{adj}	adjacent to peak
B_P	at peak
B_L', B_H'	adjacent to peak at equal 2θ intervals on the low L and high H sides
C	intensity at a certain 2θ angle in the "hump" of the scattered continuous spectrum

Other lines

M	line of matrix element
M'	line of same matrix element free of spectral-line interference
S	internal standard, internal-control standard, or internal intensity-reference standard line
T	scattered target line
T_C	Compton-scattered target line
T_R	Rayleigh-scattered target line
a—b	limits of integrated-count scan of peak
a'—b'	limits of integrated-count scan of background

ground measurement (Section 5.3.2.2). Here, the figure illustrates many of the various calibration functions that may serve as the basis of an x-ray spectrometric analysis. The intensities or intensity ratios that may be plotted as a function of analyte concentration include the following:

Intensities

1. Analyte-line peak intensity A.
2. Analyte-line net intensity $A - B$.
3. Intensity of a line of an element stoichiometrically combined with the analyte (association analysis, Section 9.8).

Intensity Ratios

4. Analyte line/internal-standard line A/S.
5. Analyte line/internal control-standard line A/S.
6. Analyte line/internal intensity-reference standard line A/S.
7. Analyte line/adjacent background A/B_{adj}.
8. Analyte line/background at the peak A/B_P or A/B_P'.
9. Analyte line/background measured in the scattered continuum "hump" A/C.
10. Analyte line/coherently scattered target line A/T_R.
11. Analyte line/incoherently scattered target line A/T_C.
12. Minor or trace analyte line/line of major matrix element of substantially constant concentration A/M'.
13. Line of analyte A/line of analyte B (binary-ratio method) A/A'.
14. Analyte line from specimen/analyte line from standard.
15. Analyte line from specimen/analyte from pure analyte.
16. Analyte line from specimen/radiation from external standard.

In general, intensity ratios give more reliable analytical results than absolute intensities, and this increased reliability usually more than compensates the increased counting error.

Other intensity functions applicable to films and platings are given in Section 9.10.

9.2.2. Special Calibration Methods

When the samples X have a very narrow range of analyte concentration C bracketing the analyte concentration of a standard S, the following simple proportionality holds reasonably well for the measured net intensities:

$$I_X/I_S = C_X/C_S \tag{9.6}$$

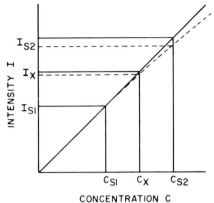

FIGURE 9.3. The two-standard method.

When the analyte concentration in the sample X lies between that in two standards S1 and S2, and the net intensity I *versus* concentration C function is substantially linear in the interval bracketed by the standards,

$$C_X = C_{S1} + \frac{I_X - I_{S1}}{I_{S2} - I_{S1}} (C_{S2} - C_{S1}) \tag{9.7}$$

The solid lines in Figure 9.3 show the intensity–concentration relationships for the two-standard method for the case of a linear calibration function. The dashed lines show how nonlinearity introduces error.

When analyte-line intensity depends on the concentration not only of the analyte A, but also of some matrix element B, a set of calibration curves of A-line intensity *versus* A concentration may be established, each curve for a different B concentration. An analysis for element A requires the previous determination of element B. The A-line intensity is then measured and applied to the particular I_A–C_A curve for the known B concentration. Those B concentrations not represented by a curve are dealt with by interpolation. Figure 9.4 shows a typical set of curves for determination of nickel in stainless steel, each curve for a different chromium concentration. In the analysis of petroleum products, sets of curves may be used, each curve for a different base-stock composition.

The binary-ratio method is applicable to three classes of samples: (1) those in which only two elements A and B are present and both are determinable by x-ray spectrometry—for example, lead–tin alloys; (2) those in which only two such elements vary in an otherwise constant matrix—for example, indium gallium arsenide, (In,Ga)As; and (3) those in which only two elements having atomic number about 19 (potassium) or more are present in a very light matrix—for example, mixtures of chromium and

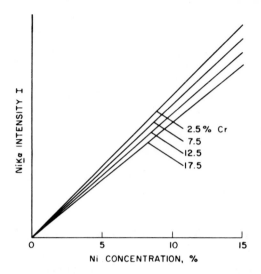

FIGURE 9.4. Set of calibration curves for correction of absorption effect of chromium on determination of nickel in stainless steel.

silver oxides. Classes 2 and 3 may be termed *pseudobinary systems*, which may be defined as multielement systems in which the analyte line intensity of each of two elements depends wholly on its own concentration and that of the other element, or in which two elements vary in concentration in an otherwise substantially constant matrix. For such specimens, the net intensities of the analyte lines of elements A and B are measured from each sample and standard. From the standard data, a calibration curve is established of the function $\log_{10}(I_A/I_B)$ *versus* $\log_{10}(C_A/C_B)$, where I and C are net intensity and concentration, respectively.

The method has many advantages. It has greater sensitivity to change in concentration than the conventional I *versus* C method because I_A increases as I_B decreases, and *vice versa*. Calibration curves are always linear, regardless of how distorted the individual curves (I_A–C_A and I_B–C_B) may be, that is, regardless of how severe the mutual absorption–enhancement effects are. The method is insensitive to reasonable variations in surface texture (porosity, chipping, cracking, grind marks), and the data points usually lie on the linear binary-ratio calibration curve regardless of how widely scattered they are on the individual curves. The method is insensitive to reasonable variations in specimen area and position and to surface contour of small irregular-shaped specimens and fabricated parts. Specimens need not be masked to constant area. In many cases, the standards need not necessarily have the same physical form as the samples. No additional measurements other than the two analyte lines and their backgrounds are required. The principal disadvantage of the method is its inapplicability near the concentration extremes where C_A/C_B and I_A/I_B approach zero or infinity.

The method has been evaluated for tungsten–rhenium, bismuth–tin, and niobium–tin alloys; Cr_2O_3–Ag_2O and Fe_3O_4–CuO powder mixtures; (Ga,In)As; and solutions containing molybdenum and tungsten. Figure 9.5 shows the effectiveness of the method as applied to a relatively unfavorable system—mixtures of Cr_2O_3 and Ag_2O powders. It is evident that the individual *I versus C* curves are very distorted, but the binary-ratio curve is linear.

For adjacent elements in the periodic table, or even for elements differing by ±2 in atomic number, the spectral-line intensity ratio is substantially equal to the concentration ratio. Such systems correspond to curve *A* in Figure 8.2. Of course, the same line ($K\alpha$, $L\beta_1$, etc.) must be measured for both elements, and no specimen, crystal, or detector absorption edge may lie between the measured lines. Rubidium–strontium concentration ratios in minerals have been determined in this way for estimation of geological age, and the method has been applied to copper–zinc (brass), tungsten–rhenium, and tungsten–osmium alloys.

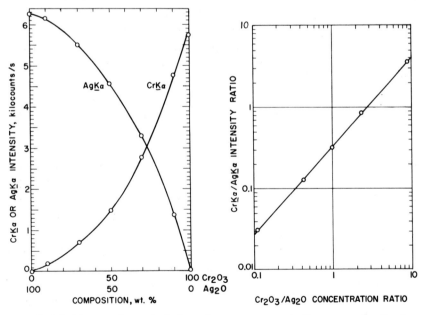

FIGURE 9.5. Binary-ratio method. Intensity *versus* concentration curves for the two individual elements (left) are compared with the binary-ratio curve (right). [E. P. Bertin, *Analytical Chemistry* **36**, 826 (1964); courtesy of the American Chemical Society.]

9.3. INTERNAL-STANDARDIZATION METHODS

9.3.1. Internal-Standard Method

The calibration-standard method is improved by quantitative addition to all specimens (samples and standards) of an internal-standard element IS having excitation, absorption, and enhancement characteristics similar to those of the analyte A in the particular matrix. In such a system,

$$C_A/C_{IS} = I_A/I_{IS} \qquad (9.8)$$

where C and I are concentration and intensity, respectively. In practice, peak and background intensities are measured from the analyte and internal standard from all samples and standards, and a calibration curve of I_A/I_{IS} net-intensity ratio *versus* analyte concentration is established from the standard data. Intensity ratios measured from the samples are applied to this curve to derive analytical concentrations. Internal standardization minimizes absorption–enhancement effects and also reduces the effects of variations in packing density in loose and briqueted powders, surface ripple in glass disks, bubble formation and thermal expansion in liquids, the many problems involved in supported specimens, etc.

The effectiveness of an internal standard is evaluated by the degree of constancy of the I_A/I_{IS} ratio from specimens having the same analyte concentration in matrixes varying over the expected composition range. Table 9.2 compares Cu $K\alpha$ intensity and Cu $K\alpha$/Zn $K\alpha$ intensity ratio for five solutions having copper concentration 2 mg/ml and iron concentration 0, 5, 10, 15, and 20 mg/ml. The Cu $K\alpha$ intensity decreases ~40% while the ratio decreases only ~1%.

The basic requirements of the internal standard have already been given. In practice, the criterion is that the internal standard and analyte lines and absorption edges should have wavelengths as close as possible. Figure 9.6 shows four conditions with respect to the relative wavelengths of the spectral lines and absorption edges of the analyte A, one or two matrix elements M, and the prospective internal standard IS. In every case, the positions of the analyte and internal standard could be interchanged without altering the discussion below. Figure 9.7 shows the wavelengths of the K-absorption edges and $K\alpha$ and $K\beta$ lines of triads of adjacent elements $Z + 1$, Z, and $Z - 1$ in several regions of the periodic table. In this figure, element Z (solid lines) may be regarded as the analyte, and elements $Z + 1$ and $Z - 1$ (dashed lines) as its prospective internal standards.

In Figure 9.6, case A represents the ideal condition. The analyte and

TABLE 9.2. Comparison of Ratio Methods[a]

Iron concentration[b] (mg/ml)	Cu $K\alpha$ net intensity (counts/s)	Ratio $I_{CuK\alpha}/I_{std}$		
		Internal standard, 2 mg Zn/ml	Background ratio, 0.6 Å	Coherently scattered target line, W $L\beta_1$
0	2117	1.045	3.261	3.541
5	1821	1.042	3.258	3.530
10	1628	1.040	3.247	3.522
15	1459	1.035	3.234	3.510
20	1308	1.032	3.221	3.503

[a] From T. J. Cullen (unpublished); presented at *Pittsburgh Conf. Anal. Chem. Appl. Spectrosc.* (1963); reviewed in detail by E. P. Bertin, *Advan. X-Ray Anal.* **11**, 1–22 (1968).
[b] Cu concentration in all solutions 2.0 mg/ml.

internal-standard lines are close and on the same sides of both matrix absorption edges shown, and the analyte and internal-standard absorption edges are close and on the same sides of both matrix lines shown. The analyte and internal-standard lines are absorbed and enhanced to substantially the same extent by each of the two matrix elements. Incidentally, if the two matrix lines and edges are sufficiently remote from the analyte line and edge, internal standardization may not be required. The condition in Figure 9.6A prevails for the $K\alpha$ lines of analyte elements of atomic

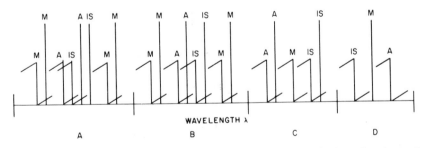

FIGURE 9.6. Specific absorption–enhancement effects and the selection of an internal-standard element. The figure shows spectral lines and absorption edges of analyte A, matrix M, and prospective internal standard IS elements. Figure A is the ideal case, B is acceptable, C and D are unacceptable. In all four cases, interchange of the A and IS lines and/or edges would have no effect.

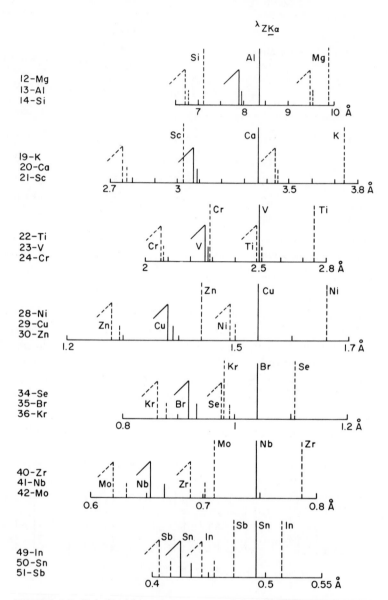

FIGURE 9.7. Relationships among K spectral lines and absorption edges of adjacent elements in various regions of the periodic table. The long and short straight lines represent $K\alpha$ and $K\beta$ lines, respectively, the hooked lines K-absorption edges. The solid lines and edges represent the analyte.

number down to 23 (vanadium) and internal standards $Z \pm 1$, as shown in Figure 9.7.

Case B is not as favorable, but is feasible, and many successful analyte–internal standard pairs are of this type. The same comments apply as for case A regarding the positions of the analyte and internal-standard lines and edges relative to the lines and edges of the matrix elements, and regarding the effect of each matrix element on the analyte and internal standard. However, here the analyte line falls between the analyte and internal-standard absorption edges, and the internal-standard edge falls between the analyte and internal-standard lines. Alternatively, the analyte and internal standard may be interchanged. One might expect case B to be unfavorable. However, difficulty is minimal because the internal-standard concentration is constant in all samples and standards and because the calibration function is based on the intensity ratio. Nevertheless, this condition should be avoided if case A is feasible. An example of case B is the use of Nb $K\beta$ as an internal standard for Mo $K\beta$. Figure 9.7 shows that the niobium K-absorption edge lies between these lines.

Case C is unacceptable: The matrix absorbs the analyte line strongly, the internal-standard line weakly. Interchanging of the analyte and internal standard would only reverse the effects and do no good. Similarly, case D is unacceptable. The matrix line is absorbed weakly by the internal standard, but strongly by the analyte, thereby enhancing the analyte line. Again, interchanging of the analyte and internal standard would do no good.

Note that the intensities of the analyte and internal-standard lines may be severely affected by the matrix. What is important is that they both be affected the same way (absorption or enhancement) and to substantially the same extent.

The selection requirements are usually best filled by an element having atomic number one or two above or below that of the analyte. Then the analyte and internal-standard lines can be of the same series and have about the same intensity for the same concentration; also, the analyte and internal standard have about the same mass-absorption coefficients for the primary and secondary x-rays. As already mentioned, and as is evident in Figure 9.7, this "rule" applies for analytes down to atomic number 23 (vanadium). Among the lighter elements, element $Z - 1$ strongly absorbs the $K\alpha$ line of elements having atomic number 22 (titanium) or lower, so element $Z + 1$ is preferable. The $Z + 1$ elements enhance the $Z K$ lines, but the effect is relatively small in the low-Z region. Also, among the lighter elements, case B (Figure 9.6) is unavoidable. In Figure 9.7, for analytes calcium and aluminum, an absorption edge lies between the analyte and internal-standard lines regardless of whether element $Z + 1$ or $Z - 1$ is chosen as the internal standard.

Rather surprisingly, there are many successful applications of an L internal-standard line for a K analyte line and *vice versa*; for example, Y $K\alpha$ (0.83 Å) serves as an internal-standard line for U $L\alpha_1$ (0.91 Å). However, such selections are to be avoided if possible. An internal standard having L lines at about the same wavelength as an analyte K line would necessarily have much higher atomic number than the analyte and therefore very different absorption coefficients for the primary and secondary x-rays. An analogous appraisal can be made for an internal standard having K lines at about the same wavelength as an analyte L line. Also, for the same analyte and internal-standard concentration, K lines are 5–10 times as intense as the L lines. An internal-standard line should be chosen from a series different from that of the analyte line only when a satisfactory internal-standard element cannot be found having a line in the same series.

Of course, the less severe the absorption-enhancement effects are, the less critical are the requirements for the internal standard. It follows that the selection of the internal standard is also less critical for dilute and thin-layer specimens.

The principal advantages of the internal-standard method are the following. It compensates absorption-enhancement and long-term instrument drift for all types of specimen. It partially compensates variations in density in powder and briquet specimens. It compensates density, evaporation, and bubble formation in liquids. Because the analyte and internal-standard lines are very close, their backgrounds are usually substantially the same and often need not be measured, in which case the peak-intensity ratio is used.

The principal disadvantages are the following. The method is not applicable to many types of sample, such as bulk solids, foils, and small fabricated parts. It is also not applicable to high analyte concentrations—above, say, 25%. Analysis time is greatly increased and the analysis is made much less convenient by the requirement for quantitative and thorough admixture of the internal standard and for measurement of a second line (and perhaps background) from each sample and standard. However, all measurements can be made simultaneously with a multichannel instrument.

The internal standard must be intimately and homogeneously mixed with the specimens. This is most easily accomplished with solutions and fusion products, less easily with loose and briqueted powders and supported specimens. The internal standard should be added at a concentration level equivalent to about the middle of the expected analyte concentration range.

It is this requirement that limits the internal-standard method to analyte concentrations up to $\sim 25\%$ at most. The effectiveness of the internal standard in compensating absorption–enhancement effects increases as analyte concentration decreases, or as dilution increases.

For greatest convenience and precision, it is preferable to measure analyte and internal-standard lines—and their backgrounds, if necessary— simultaneously. This is no problem with multichannel spectrometers, but is impossible with unmodified single-channel instruments.

9.3.2. Other Standardization Methods

An internal control standard is similar to an internal standard in some ways. It is added in constant concentration to all samples and standards, unless a suitable element is already present in the samples in appropriate concentration. The calibration function is the intensity ratio of the analyte and internal control-standard lines *versus* analyte concentration. However, the control standard does not necessarily have the same excitation, absorption, and enhancement properties as the analyte in the specimen matrix, and it may serve for several analytes in the same specimen. The method has been applied to solutions where it compensates variations in volume, temperature, density, specimen plane, acid matrix, and, to a smaller degree, composition of the original matrix.

An external standard, or external intensity-reference standard, is a specimen from which some intensity value is measured to be ratioed with the analyte-line intensity, principally to compensate long-term instrument drift. The standard may be one of the samples or standards retained for this purpose, or it may be any stable specimen not necessarily bearing any relation to the samples to be analyzed—such as a piece of copper alloy or nickel alloy. The measured intensity may be emitted or scattered from the standard. The standard may be measured once after each sample and calibration standard, or, more likely, from time to time during a long series of measurements. A particularly interesting example of the external-standard method is reported by Hirokawa and his colleagues, who used the emission lines from specimen-mask plates of zinc or lead as external-standard lines in the determination of chromium, nickel, and tungsten in steel.

Another way to use an external standard is to set the excitation parameters to reproduce the intensity emitted or scattered from it. For example, in the determination of aluminum, silicon, and iron in clays and bauxites, day-to-day instrument drift was compensated by adjustment of the x-ray

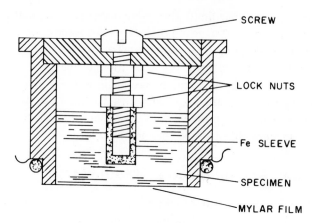

SCREW

LOCK NUTS

Fe SLEEVE

SPECIMEN

MYLAR FILM

FIGURE 9.8. Liquid-specimen cell with built-in iron intensity-
reference standard.

tube current and/or potential to reproduce intensity values established at
the time of the original calibration from selected aluminum and iron plates.

In the *internal intensity reference standard method,* a rod or disk of pure
metallic element is mounted in a liquid-specimen cell, as shown in Figure
9.8. The line intensities are measured from the analyte(s) in the liquid specimen and from the intensity reference standard. The calibration function is
the ratio of the analyte and reference standard lines. The method has been
used to compensate variations in carbon/hydrogen ratio in the determination of additives and trace elements in gasoline and other hydrocarbon
liquids. The method is most effective when the intensity reference standard
has the characteristics required for a true internal standard element.

9.4. STANDARDIZATION WITH SCATTERED X-RAYS

These methods use the continuum or spectral lines emitted by the x-ray
tube or other excitation source and scattered by the specimen to compensate
absorption–enhancement effects, particle-size and surface-texture effects,
packing density, and instrument errors. Three basic methods are in common
use: the background-ratio, scattered target-line ratio, and coherent/in-
coherent scattered target-line intensity ratio methods.

The first two methods are similar to the internal-standard method,
except that the calibration function is I_A/I_{sc} *versus* analyte concentration.
I_A may be peak or net analyte-line intensity. I_{sc} may be the intensity of:

(1) the scattered continuous background of the analyte-line peak measured adjacent to the peak or in any of the ways given in Section 5.3.2.2; (2) a specified wavelength in the scattered continuum "hump"; or (3) a coherently or incoherently scattered target line. A wavelength in the hump or a scattered target line is used when: (1) the background intensity at or adjacent to the peak is too low to be measured with acceptable statistical precision in a reasonable counting time; or (2) such a wavelength gives better standardization than the background at the peak. These methods are sometimes referred to as "internal standardization with scattered x-rays," and, in a sense, this is a valid description because the scattered intensity, like the emitted internal-standard intensity, is matrix-dependent.

The basis of these methods is illustrated in Figure 9.9, which shows the relative intensities of emitted Ni $K\alpha$ and scattered 0.6-Å x-rays measured from minerals having the same nickel concentration but 10–60% iron. Both individual intensities are strongly dependent on iron concentration, but the curves are both hyperbolic and more or less parallel, indicating that the Ni $K\alpha$/0.6 Å intensity ratio should be about constant with change in iron concentration. This prediction is strikingly confirmed in Figure 9.10. The data points on the Ni $K\alpha$ curve are widely scattered. Plotting of the same data as the intensity ratio of Ni $K\alpha$ and 0.6-Å scatter reduces the scatter in the points to insignificance. Figure 9.11 strikingly shows the effectiveness of the method in compensating variations in x-ray tube potential and current, specimen plane, and particle size. Figures 9.9 and 9.10 are examples of an intensity ratio of an analyte line (Ni $K\alpha$, 1.66 Å) and a wavelength (0.6 Å) from the scattered continuum hump. Figure 9.11 is an example of an intensity ratio of an analyte line (Cu $K\alpha$, 1.54 Å) and its adjacent background (∼1.4 Å). Table 9.2 gives some typical data for the scattered continuum method.

FIGURE 9.9. Background-ratio method; substantial constancy of Ni $K\alpha$/0.6-Å scattered-continuum intensity ratio from specimens having the same Ni concentration but 10–60 wt% Fe. [G. Andermann and J. W. Kemp, *Analytical Chemistry* **30**, 1306 (1958); courtesy of the authors and the American Chemical Society.]

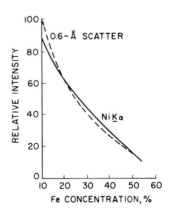

Compared with the internal-standard method, the background-ratio method has the advantages that no addition need be made to the specimens, that it is applicable to specimens having practically any physical form, and that it is applicable to many analytes in the same specimen. The method is especially applicable to minor and trace analytes in low-Z matrixes, where the scattered intensity is high and can be measured precisely in short counting times. The method is largely ineffective against enhancement. In specific cases, it may not be as effective in compensating absorption effects as the internal-standard method, where standardization is effected by an emission line originating *in* the specimen. Also, all scattered wavelengths are not equally effective in the ratio, and principles for selection of the optimal wavelength are difficult to establish, so the choice is usually left to trial and error. No absorption edge should lie between the analyte line and the scattered standard x-rays, including that of the analyte itself. Thus,

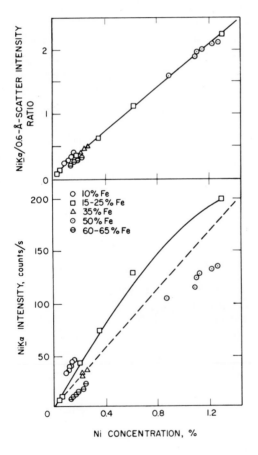

FIGURE 9.10.   Background-ratio method; effectiveness in correcting for absorption–enhancement effects of iron on Ni $K\alpha$ intensity from iron–nickel minerals. [G. Ander-mann and J. W. Kemp, *Analytical Chemistry* **30**, 1306 (1958); courtesy of the authors and the American Chemical Society.]

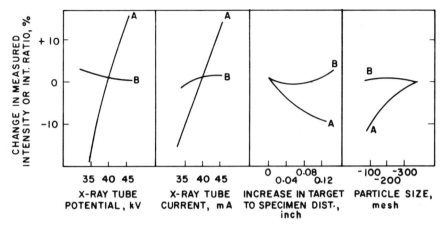

FIGURE 9.11. Background-ratio method; effectiveness in correcting for instrumental variables and the particle-size effect. Curves A show variation of Cu $K\alpha$ intensity. Curves B show variation of the intensity ratio of Cu $K\alpha$ and its adjacent background. [G. Andermann and J. W. Kemp, *Analytical Chemistry* **30**, 1306 (1958); courtesy of the authors and the American Chemical Society.]

in this method, the analyte itself may actually constitute an interfering element for scattered wavelengths shorter than its own line, unless analyte concentration is low. However, there are successful analyses in which this condition does exist.

The third method cited above is the coherent/incoherent scattered target-line ratio method, which includes scattered lines from other excitation sources.

It has been shown that in light matrixes,

$$I_A \propto \frac{C_A}{(\mu/\varrho)_{\lambda_{pri}} + (\mu/\varrho)_{\lambda_A}} \tag{9.9}$$

where I_A and C_A are analyte line intensity and concentration, respectively, and $(\mu/\varrho)_{\lambda_{pri}}$ and $(\mu/\varrho)_{\lambda_A}$ are mass-absorption coefficients of the specimen for the primary and analyte-line x-rays, respectively. To a close approximation, this proportionality reduces to

$$I_A \propto C_A/(\mu/\varrho)_{\lambda_A} \tag{9.10}$$

It has also been shown that in light matrixes, μ/ϱ is a linear function of the ratio of the intensities of the coherently (Rayleigh) I_R and incoherently

(Compton) I_C scattered target lines, that is,

$$(\mu/\varrho)_{\lambda_A} \propto I_R/I_C \propto R \qquad (9.11)$$

Then, from Equation (9.10),

$$I_A \propto C_A/R \qquad (9.12)$$

A very successful absorption correction method for light matrixes is based on this proportionality.

Analyte-line intensity is measured from the sample I_X and from a single standard I_S having known analyte concentration C_S in a matrix similar to that of the sample. Rayleigh- and Compton-scattered target-line intensities are measured from both the sample and standard and used to calculate R_X and R_S [Equation (9.11)]. Analyte concentration in the sample is then given by

$$C_X = C_S \frac{I_X}{I_S} \frac{R_X}{R_S} \qquad (9.13)$$

The method was applied to the determination of elements having atomic number 13–82 (aluminum to lead) in samples having average atomic number 5–10 in the form of briquets, glass disks, and water and organic solutions.

9.5. MATRIX-DILUTION METHODS

The term *matrix-dilution methods* is applied to a group of methods that: (1) correct for absorption–enhancement effects by leveling the absorption coefficients of the samples and standards to a common value determined by an added diluent; and/or (2) correct for inhomogeneity and particle-size effects by subjecting the as-received specimens to dissolution in a solvent or fusion in a flux. In general, analyte-line intensities from diluted specimens are substantially proportional to analyte concentrations, and the calibration curves are substantially linear.

The efficiency of the primary beam of effective wavelength λ_{pri} for exciting analyte line λ_L is given by a relationship derived from Equation (4.12):

$$\frac{I_L}{I_{0,\lambda_{\text{pri}}}} = P_A \frac{C_A(\mu/\varrho)_{A,\lambda_{\text{pri}}}}{\sum C_i[(\mu/\varrho)_{i,\lambda_{\text{pri}}} + A(\mu/\varrho)_{i,\lambda_L}]} \qquad (9.14)$$

where

I_L is analyte-line intensity

$I_{0,\lambda_{pri}}$ is incident primary intensity

C_A is analyte concentration

C_i is concentration of an individual matrix element i

$(\mu/\varrho)_{A,\lambda_{pri}}$ is mass-absorption coefficient of the analyte for the primary x-rays

$(\mu/\varrho)_{i,\lambda_{pri}}$, $(\mu/\varrho)_{i,\lambda_L}$ are mass-absorption coefficients of an individual matrix element for the primary and analyte-line x-rays, respectively

P_A, A are defined by Equations (4.5) and (4.6), respectively, and are constant for any given analysis and need not concern us here

For the present discussion, Equation (9.14) may be expanded as follows:

$$\frac{I_L}{I_{0,\lambda_{pri}}} = P_A \frac{C_A(\mu/\varrho)_{A,\lambda_{pri}}}{C_M[(\mu/\varrho)_{M,\lambda_{pri}} + A(\mu/\varrho)_{M,\lambda_L}] + C_D[(\mu/\varrho)_{D,\lambda_{pri}} + A(\mu/\varrho)_{D,\lambda_L}]}{+ \sum C_i[(\mu/\varrho)_{i,\lambda_{pri}} + A(\mu/\varrho)_{i,\lambda_L}]} \qquad (9.15)$$

where M refers to a specific matrix element, D to the added diluent, and i to the individual elements in the remainder of the matrix; M and D are discussed below.

If absorption–enhancement effects occur, variations in matrix composition among samples and standards may change the denominator in Equation (9.15) and thereby the value of $I_L/I_{0,\lambda_{pri}}$ for the same analyte concentration C_A. Suppose that a certain matrix element M undergoes a change in concentration ΔC_M from specimen to specimen, causing a change in its contribution to the denominator of Equation (9.15), the first term in the denominator. If the change in this contribution is very small, the effect on the denominator—and therefore on the analyte-line intensity—may be negligible. This is the case if: (1) C_M is very small to begin with, so that even a *relatively* large ΔC_M and consequent *relatively* large change in its contribution may have little effect on the denominator; (2) ΔC_M is very small, so that the change in the term is small; or (3) $(\mu/\varrho)_{M,\lambda_{pri}}$ and $(\mu/\varrho)_{M,\lambda_L}$ are not substantially different from the average absorption coefficient of the specimen.

However, the effect of ΔC_M on the analyte-line intensity is substantial when a relatively large ΔC_M occurs in a relatively large C_M of a matrix element having absorption coefficients very different—either higher or lower—from the average absorption coefficient of the specimen.

The denominator of Equation (9.15) may be made relatively insensitive to changes in concentration of the matrix element by judicious selection and addition of a diluent D in either or both of two ways: (1) A high concentration C_D may be added of a diluent having very low absorption coefficients $(\mu/\varrho)_{D,\lambda_{pri}}$ and $(\mu/\varrho)_{D,\lambda_L}$ relative to the average absorption coefficient of the specimens; and/or (2) a low concentration C_D may be added of a "diluent" having high absorption coefficients relative to the average absorption coefficients. In either case, the diluent largely determines the absorption coefficient of the specimen for both primary and analyte-line radiation, and the variations in absorption coefficient caused by variations in concentrations of matrix elements are reduced, often to insignificance.

Dilution also minimizes enhancement, either by actual reduction of the concentration of the enhancing element, or by increasing the absorption of its enhancing spectral lines.

Frequently, when a low-absorption diluent (solvent or flux) is used to reduce inhomogeneity or particle-size effects, it is necessary to add a high-absorption "diluent" to level residual absorption–enhancement contributions of the original matrix. The absorber is added in concentration such that reasonable variations in concentrations of elements in the specimen do not substantially affect the analyte-line intensity from the prepared specimen.

It is evident from the preceding section that matrix-dilution methods may be classified in two ways. On the basis of whether the diluent is simply admixed with the specimen or actually destroys the original specimen form, there are, respectively, *dry-dilution methods* and *solution, fusion,* and *glass-disk methods*. On the basis of the relative values of the absorption coefficients of the diluent and original specimen matrix, there are *low-absorption* and *high-absorption methods*.

Low-absorption dilution is applied in the solution and fusion (including glass-disk) methods of specimen preparation, and in addition of powders having low mean atomic number to loose and briqueted powder specimens. In a sense, specimen preparation methods in which the analytical material is supported on filter paper, Millipore and Nuclepore filters, ion-exchange paper and membrane, etc., constitute low-absorption dilution methods. High-absorption "dilution" is applied in powder, briquet, fusion, and solution preparation methods. These preparation methods are considered in Chapter 10.

9.6. THIN-FILM METHODS

   Absorption–enhancement effects substantially disappear in thin-film-type specimens because neither primary nor analyte-line x-rays are significantly absorbed in the extremely short path lengths in the thin layer. Each atom absorbs and emits substantially independently of the other atoms without significantly altering the incident beam. Thus, for very thin films of constant thickness, analyte-line intensity is directly proportional to analyte concentration, and for films of constant composition, intensity is proportional to thickness. These relationships provide the basis for a method free from absorption–enhancement effects for analysis of films containing more than one element, and a method for measurement of thickness of films of known composition. Of course, many samples are thin films to begin with, but for those that are not, reduction to a thin film is one way to deal with absorption–enhancement effects. In general, most supported specimens (Section 10.8) constitute thin specimens.

   Figure 9.12 shows spectral-line intensity measured from a thin film as a function of thickness. The curve has three regions. For extremely thin films, attenuation of incident primary and emergent secondary spectral-line radiation is extremely small. The curve is linear, and intensity is simply proportional to thickness. For films of intermediate thickness, attenuation of both primary and secondary x-rays increases with depth. The longer-wavelength components of the primary beam are absorbed preferentially, so that as the beam penetrates progressively deeper, it becomes both weaker and of shorter effective wavelength. The emergent secondary spectral x-rays continue to increase in intensity with thickness, but at a decreasing rate. At infinite (critical) thickness, t_∞ or t_C, secondary x-rays are excited at

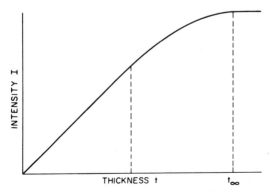

FIGURE 9.12. Infinite (critical) thickness.

depths from which they cannot emerge to the surface. Further increase in thickness results in no further increase in intensity. For example, infinite thickness for chromium, iron, and nickel is 30–40 μm.

Infinite thickness can be calculated from the equation

$$I_t/I_\infty = 1 - \exp[-\overline{(\mu/\varrho)}\varrho t] \qquad (9.16)$$

or

$$\log_e[1 - (I_t/I_\infty)] = -\overline{(\mu/\varrho)}\varrho t \qquad (9.17)$$

where I_t and I_∞ are analyte-line intensities from a film of thickness t and infinite thickness, respectively; ϱ and t are density (g/cm^3) and thickness (cm) of the film, respectively; and $\overline{(\mu/\varrho)}$ is defined by Equation (2.6). At the infinite (critical) thickness, $I_t/I_\infty = 1$. To calculate the thickness at which I_t just becomes I_∞, a value of, say, 0.999 is arbitrarily chosen for I_t/I_∞, and Equation (9.16) or (9.17) solved for t.

In films consisting of more than one element, the absorption coefficient of the film is different for the spectral line of each element. Thus, infinite thickness is different for each spectral line.

A "thin" specimen may be defined as one in which $m(\mu/\varrho)' \lesssim 0.1$, where m is mass per unit area (g/cm^2), and $(\mu/\varrho)'$ is the sum of the mass-absorption coefficients (cm^2/g) of the specimen for both primary and analyte-line x-rays [Equation (2.5)]. The principal advantages of thin specimens are: (1) linearity of the analyte-line intensity–concentration function over several orders of magnitude; (2) elimination of absorption-enhancement effects; (3) reduction of particle-size effects, if any; (4) increased ratio of analyte-line to scattered primary intensities, that is, increased peak-to-background ratio; and (5) low and substantially "flat" background, that is, low and constant background over a wide wavelength region.

It has already been pointed out that some specimens already have the form of thin films. These include true evaporated films, platings, corrosion layers, and sublimation deposits, as well as microgram amounts of specimen evaporated on Mylar or absorbed in filter paper or ion-exchange membrane. Extremely small particles or flakes and extremely thin fibers also act as thin specimens. Birks refers to all specimens having less than infinite thickness as "specimens of limited quantity." Alternatively, in order to reduce absorption–enhancement effects, bulk specimens may be reduced to thin films by vacuum evaporation, rolling and beating, etching, etc., or by grinding to extremely fine powder, suspension in an organic vehicle-binder, and casting to a thin film with a casting knife. Metals can be electroplated on flat copper or other electrodes, which then constitute the samples.

The intensities from extremely thin films are very low, so thin-film methods are applicable principally to major constituents. Moreover, both composition and thickness are variables. Thick films may begin to show absorption–enhancement effects. An internal standard may be added to such films by volumetric addition of a standard solution of an internal-standard element.

For specimens less than infinitely thick, as thickness increases, absorption effects increase more rapidly for emitted analyte-line x-rays than for scattered primary x-rays. Thus, sensitivity decreases as infinite thickness is approached.

9.7. STANDARD ADDITION AND DILUTION METHODS

In principle, in these methods, a weighed portion of sample X is treated with analyte or diluent to alter the analyte concentration C by a known amount $\pm\Delta C$, that is, from C_X to $C_{X\pm\Delta C}$; analyte-line intensity is then measured from the untreated I_X and treated $I_{X\pm\Delta C}$ samples. Then,

$$\frac{I_X}{I_{X\pm\Delta C}} = \frac{C_X}{C_{X\pm\Delta C}} \tag{9.18}$$

If the calibration function—analyte-line intensity I_A versus analyte concentration C_A—is linear in the region of analytical interest, I_A is proportional to C_A, and an incremental or decremental change in C_A gives a proportional change in I_A. If such a linear relationship is questionable, a second change and new measurements may be made to ascertain linearity. If the calibration curve is nonlinear, the slope of the curve is established by repetitive additions. Most standard addition and dilution methods also assume that the calibration curve passes through the origin. The terms *doping* or *spiking* are sometimes used for addition, and *dopant* for additive. The terms *active* and *inert* addition refer to addition of analyte and other material, respectively.

In practice, a weighed portion of the sample X is treated with a weighed portion of a standard material S having known analyte concentration C_S to form a mixture XS containing concentrations $C_{(X)}$ and $C_{(S)}$ of *sample* and *standard*, respectively. Incidentally, $C_{(X)}$ is actually the dilution factor: weight of X/weight of X + S, or weight of X/weight of XS. In the *standard-addition method*, analyte-line intensity is measured from the untreated sample I_X and mixture I_{XS}. Analyte concentration in the original sample is given by

$$C_X = \frac{(I_X/I_{XS})C_{(S)}}{1 - [(I_X/I_{XS})C_{(X)}]} \qquad (9.19)$$

$$= \frac{(I_X/I_{XS})C_S}{1 + \{(w_X/w_S)[1 - (I_X/I_{XS})]\}} \qquad (9.20)$$

where C is weight fraction or percent, and w_X and w_S are the weights of sample and added standard, respectively. For analysis of liquids,

$$C_X = \frac{(I_X/I_{XS})C_S}{(\varrho_S/\varrho_X) + \{(v_X/v_S)[1 - (I_X/I_{XS})]\}} \qquad (9.21)$$

where C is weight/volume fraction or percent (g/ml); ϱ_X and ϱ_S are densities of sample and standard, respectively; and v_X and v_S are volumes of sample and added standard, respectively. In practice, the density ratio is usually ~ 1.

In the standard-dilution method, weighed portions of the sample X and of a standard material S having known analyte concentration C_S of the same order as that of the sample are separately mixed in the same known proportion with an inert diluent containing no analyte(s). Analyte-line intensity is measured from the untreated sample I_X and standard I_S and from the diluted sample $I_{X'}$ and standard $I_{S'}$. Analyte concentration in the untreated sample is given by

$$C_X = C_S \frac{I_S - I_{S'}}{I_X - I_{X'}} \qquad (9.22)$$

Standard addition and dilution methods usually require no standards or calibration curves, so the methods are useful for sample types analyzed too infrequently to warrant preparation of standards and calibration curves, and for samples for which there is no prior knowledge or inadequate knowledge about the matrix. The methods are applicable only to trace and minor analytes (less than 5–10%) where the I_A–C_A calibration function is likely to remain linear for suitable incremental additions or dilutions. In standard-addition methods, an addition must be made for each analyte appropriate for its estimated concentration. In standard-dilution methods, analytes having the same order of concentration may be determined from the same diluted sample. However, usually a portion of the sample must be treated at a different level for each analyte or group of analytes having a different order of concentration. The methods are applicable only to samples to which additions can be made—principally powders, fusion products, and solutions.

The addition of analyte increases the analyte concentration, but, in effect, also reduces the original concentration in proportion to the dilution factor. Thus, to simplify the calculations, the actual weight of substance added should be small compared with the total sample weight. In other words, the analyte concentration in the additive should be high, and the pure analyte itself is preferable whenever its use is feasible. Intimate homogeneous admixture of the additive is required. This condition is easily attained for solutions and melts. For powders, the particle size and density of the additive must be similar to those of the samples.

9.8. MISCELLANEOUS EXPERIMENTAL CORRECTION METHODS

There are many x-ray spectrochemical methods that cannot be classified in any of the categories in the earlier sections of this chapter. Three examples are considered in this section: emission–absorption methods, indirect or association methods, and the application of x-ray spectrometry to supplement other methods of chemical analysis.

In emission–absorption methods, a portion of the sample itself is used as an absorption filter to obtain data for correction of absorption–enhancement effects on the analyte-line emission measured from another portion of the sample in the specimen chamber. A thin, uniform layer of sample is prepared to serve as an absorption filter. This filter may be mounted on the exit end of the source collimator or the entrance end of the detector collimator, or it may be superimposed on the emission sample itself. If a known mass of sample is formed into a briqueted disk or other uniform layer of known thickness and area, the area-density ϱt (g/cm²) in Equation (2.2) is known directly. The mass-absorption coefficient of the sample for the primary beam is evaluated in one of three ways. The goniometer may be set at 0° 2θ and a piece of paraffin placed in the specimen chamber to scatter primary x-rays into the detector, or the x-ray tube turned so that its window faces the detector. In the latter technique, very low x-ray tube current and a small limiting aperture are used to reduce the primary intensity to a measurable value, and considerable rearrangement of the goniometer may be required. Alternatively, one may place in the specimen compartment a secondary target having a strong spectral line at wavelength about the same as the effective wavelength of the primary beam and the goniometer set at 2θ for this wavelength. In any case, the intensity is measured directly and through the filter, giving incident I_0 and transmitted

I intensities, respectively, in Equation (2.2). The mass-absorption coefficient for the analyte line is evaluated by measuring analyte-line intensity, emitted by the sample in the specimen chamber and diffracted by the crystal, directly and through the filter. In any of the foregoing methods, by placing the filter between two identical two-window flow counters, incident and transmitted intensities can be measured simultaneously from the front and back detector, respectively. The combined mass-absorption coefficient for both primary and secondary x-rays is evaluated by measuring analyte-line intensity from the emission sample directly and with the filter placed over the sample so that both primary and secondary x-rays must pass through it, in opposite directions, of course.

An extremely ingenious technique in x-ray spectrometric analysis is the indirect determination—or determination by association—of the following classes of analytes: (1) elements relatively inconvenient to determine by x-ray spectrometry—atomic numbers 9–17 (fluorine through chlorine); (2) elements not determinable at all on standard commercial spectrometers—atomic numbers less than 9; (3) organic compounds and other specific chemical compounds and substances; and (4) analytes that, although readily determinable, give low intensities due to a combination of small total amount and low analytical sensitivity. Such analyses are sometimes conveniently done by stoichiometric association with a readily determined element.

Examples of the first group (above) are the determination of trace chlorine in titanium by precipitation as silver chloride and measurement of the silver; the determination of silicon by formation of silicomolybdate-8-quinolinol complex (Si:Mo = 1:12) and measurement of the molybdenum; and the determination of sodium by precipitation of sodium zinc uranyl acetate $[NaZn(UO_2)_3(OOCCH_3)_9]$ and measurement of U $L\alpha_1$.

A striking example of the second group is the microdetermination of nitrogen in amounts 0.2–5 μg in biological materials by converting it to ammonia, absorbing the ammonia in filter paper impregnated with Nessler's reagent, washing out excess reagent, then determining the mercury in the mercury–iodine–amine complex thus formed. Boron can be precipitated as barium borotartrate and the Ba $L\alpha_1$ line measured; lithium and beryllium can be determined as their phosphates or arsenates and measurement of the P $K\alpha$ or As $K\alpha$ lines.

Three techniques have been evaluated for stoichiometric association of a determinable element with organic compounds: bromination with bromine water, chelation with 5-chloro-7-iodo-8-quinolinol, and salt-formation with barium or lead. These techniques reverse the usual procedure of using an organic reagent, such as 8-quinolinol or dithizone, to specifically separate and precipitate an element prior to its determination by x-ray spectrometry.

Montmorillonite is determined in montmorillonite–kaolinite clay mixtures by prolonged agitation of the samples in strontium chloride solution, filtration, and measurement of the strontium in the recovered clay. Only the montmorillonite undergoes ion exchange with the strontium to a significant extent. The method is calibrated by use of standards having known montmorillonite–kaolinite ratios. An alternative method would be to use strontium chloride solutions of known concentration, then determine the strontium depletion in the recovered filtrate.

The carbon/hydrogen ratio in hydrocarbon liquids can be determined on an unmodified x-ray fluorescence spectrometer. The intensity ratio R of the coherently and incoherently scattered W $L\alpha_1$ line from the x-ray tube is measured from the sample(s) and from a series of pure hydrocarbon liquid compounds having various known carbon/hydrogen ratios. A calibration curve of R *versus* %C is established from the known liquids; analytical concentrations are derived from this curve.

The significance of these indirect methods cannot be overestimated. They extend the applicability of x-ray spectrometry to the entire periodic table, to organic chemistry, and to other specific materials—all with standard commercial instrumentation and accessories and with conventional preparation and presentation techniques. Many of these methods are applicable to micro and trace analysis.

X-ray spectrometry has been used to complement many other analytical methods, including x-ray diffraction, radiochemistry, infrared and ultraviolet spectrophotometry, paper and gas chromatography, electron diffraction, and neutron activation.

9.9. MATHEMATICAL CORRECTION METHODS

Having considered primarily experimental methods of dealing with absorption–enhancement effects, we now consider primarily mathematical methods. Most of these methods are so complex that a comprehensive treatment cannot be given here. Moreover, no good, comprehensive, critical review of the scope, advantages, and limitations of all of the various mathematical methods has appeared, and this chapter does not provide one. However, we can set forth the principles of some of the outstanding methods.

It is explained in Section 8.2 that, disregarding particle-size, heterogeneity, and surface-texture effects, were it not for absorption-enhancement effects by matrix elements, measured analyte-line intensity would be directly proportional to analyte concentration. In fact, the ratio of analyte-line intensities from sample and pure analyte would equal the analyte concentration. In practice, this ideal condition rarely obtains, and one must deal with absorption-enhancement effects by judicious choice of the analytical method (Sections 9.2–9.8) or by mathematical methods. In the latter, the analyst measures one of three quantities: (1) analyte-line intensity from the sample; (2) ratio of analyte-line intensities from the sample and pure analyte; or, preferably, (3) ratio of analyte-line intensities from the sample and a standard similar to the samples and having analyte concentration in the middle of the expected range. Analyte concentration is derived from this measured quantity mathematically by one of six basic approaches: (1) geometric methods; (2) empirical-correction methods; (3) absorption-correction methods; (4) influence-coefficient methods; (5) variable takeoff-angle method; (6) fundamental-parameters method; and (7) multiple-regression method. An example of each of these methods is given in the following sections.

9.9.1. Geometric Methods

Certainly the outstanding examples of the geometric approach to the correction of absorption-enhancement effects are the methods for ternary systems. The chemical composition of a ternary system can be represented by a triangular composition diagram. Each point in the triangle uniquely represents a set of three values for the concentrations of the three components C_A, C_B, C_C. Each such point may also represent the corresponding analyte-line intensities I_A, I_B, I_C that would be measured from this composition.

Taylor *et al.* have developed the graphic method for ternary systems. A description of their work follows. In a ternary system of elements A, B, and C having atomic weights A_A, A_B, and A_C in weight fractions W_A, W_B, and W_C and atom fractions A_A', A_B', and A_C', composition of any specimen is given by equations of the form

$$(W_A)_{ABC} = \frac{A_A' A_A}{A_A' A_A + A_B' A_B + A_C' A_C} \tag{9.23}$$

$$(A_A')_{ABC} = \frac{W_A/A_A}{(W_A/A_A) + (W_B/A_B) + (W_C/A_C)} \tag{9.24}$$

The corresponding relationships in the three binary systems A–B, A–C, and B–C have the forms

$$(W_A)_{AB} = \frac{A_A' A_A}{A_A' A_A + A_B' A_B} = \frac{A_A' A_A}{A_A' A_A + (1 - A_A') A_B} \tag{9.25}$$

$$(A_A')_{AB} = \frac{W_A/A_A}{(W_A/A_A) + (W_B/A_B)} = \frac{W_A/A_A}{(W_A/A_A) + (1 - W_A)/A_B} \tag{9.26}$$

All these equations are readily expressed in terms of percent as well as fraction.

Suppose that the triangular composition diagram for the ternary system A–B–C is plotted in terms of atom concentration. Then Equation (9.25) and the analogous equations for the A–C and B–C binary systems can be used to mark off weight concentrations along the A–B, A–C, and B–C sides of the triangle. A skew grid may be constructed on the triangle by drawing lines joining equal or complementary concentrations on the three sides—for example, 90A–10C on A–C with 90A–10B on A–B and 90B–10C on B–C, and 50A–50C on A–C with 50A–50B on A–B and 50B–50C on B–C. The grid permits reading of weight concentrations directly on the atom composition diagram. Similarly, Equation (9.26) and its analogs can be used to construct a skew grid to permit reading of atom concentrations on a weight composition diagram.

Now, if the spectral-line intensities emitted by the three elements in the ternary system A–B–C are I_A, I_B, and I_C, the "count fraction" F for each intensity is given by an equation of the form

$$(F_A)_{ABC} = I_A/(I_A + I_B + I_C) \tag{9.27}$$

In the corresponding binary systems, the count fraction equations have the form

$$(F_A)_{AB} = I_A/(I_A + I_B) \tag{9.28}$$

Like weight and atom fractions, count fractions can also be expressed in percent.

Taylor *et al.* have based a geometric method for correction of absorption-enhancement effects in ternary systems on the obvious similarity of the equations for weight, atom, and count fractions. Their procedure is as follows:

1. Three sets of standards are prepared, one for each of the three binary systems.

2. Three calibration curves (count percent F *versus* weight percent

W) are established from the intensity data from the standards: $(F_A)_{AB}$ versus W_A in A–B, $(F_B)_{BC}$ versus W_B in B–C, and $(F_C)_{CA}$ versus W_C in C–A.

3. These curves are used to calibrate a triangular coordinate graph having corners $100F_A$, $100F_B$, $100F_C$ and sides graduated in linear 0–100% *F* scales. From the calibration data, scales of 0–100% W_A, W_B, and W_C are plotted along the A–B, B–C, and C–A sides, respectively.

4. A weight-concentration skew grid is then drawn on the triangle over the linear count-percent grid.

The result is a ternary calibration graph. An analysis then consists in: (1) measuring I_A, I_B, and I_C from the sample; (2) calculating F_A, F_B, and F_C from equations having the form of Equation (9.27); (3) finding the point having these three *F* values on the linear count-percent grid; and (4) reading the concentrations off the skewed composition grid.

Figure 9.13 shows the application of the method to the germanium–tellurium–bismuth system. The sides of the triangle are graduated linearly in counts percent, as shown by the equally spaced "ticks." In practice, the triangle is a standard triangular coordinate diagram, but in the figure, the triangular grid is omitted for simplicity. X-ray intensity data measured from germanium–tellurium, germanium–bismuth, and tellurium–bismuth binary standards is used to construct calibration curves of counts percent *versus* atom percent on the sides of the triangle. These curves are used to graduate the triangle in concentration units, as shown by the dashed lines and numbers along each side. Finally, a concentration skew grid is drawn on the triangle.

An example of the application of Taylor's method follows. Consider a sample that gives count fractions for Ge $K\alpha$, Te $K\alpha(2)$, and Bi $L\beta_1$ of 58, 27, and 15%, respectively. This point is located on the triangular diagram of Figure 9.13 by reference to the uniformly spaced ticks along the sides of the triangle, which are linear scales of count percent. For simplicity, the point was selected to fall on an intersection on the atom concentration skew grid, and is circled on the figure. The atom concentrations of the three elements are now read off the skew grid as follows: Ge—20 at% on the Ge–Te scale, Te—70 at% on the Te–Bi scale, and Bi—10 at% on the Bi–Ge scale.

9.9.2. Empirical-Correction Methods

An excellent example of the empirical-correction method is provided by the work of Mitchell, who, using empirical correction factors, successfully analyzed systems having up to seven components. The basic principle of the

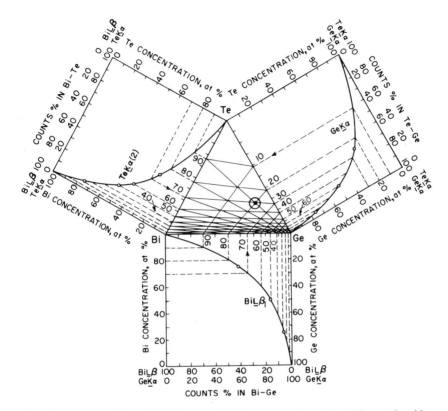

FIGURE 9.13. Graphic method for analysis of ternary systems. The ticks on the sides of the triangle are linear scales of count percent. The skew grid is derived from the calibration curves constructed on the sides of the triangle and gives concentration. The dark spot is an analytical point (see text). [Data from A. Taylor, B. J. Kagle, and E. W. Beiter, *Analytical Chemistry* **33**, 1699 (1961).]

method is as follows. In the A–B–C–D–E–F–G system, line intensity of element A is measured, corrected for absorption-enhancement effects by C, D, E, F, and G, then applied to a calibration curve of A-line intensity *versus* A concentration in a B matrix. In other words, after correction, A is determined as if it were in an A–B binary system. Mitchell applied the method to determination of all seven metallic elements in the system TiO_2–V_2O_5–Fe_2O_3–ZrO_2–Nb_2O_5–Ta_2O_5–WO_3. The procedure follows, but for simplicity, the discussion is in terms of the elements, rather than the oxides.

The calibration procedure is as follows. For each element, one of the other six is selected as its binary complement. The complement may be the element having highest concentration or most severe absorption-enhance-

TABLE 9.3. Plan of Analysis of Seven-Component System[a]

Analyte	Binary complement for basic calibration curve	Interfering elements for which calculation of correction factors is required
Ti	Ta	V or Fe, Zr or Nb, W
V	Ti	Fe, Zr or Nb, Ta or W
Fe	Ta	Ti or V, Zr or Nb, W
Zr	Ta or W	Ti or V, Fe, Nb
Nb	Ta or W	Ti or V, Fe, Zr
Ta	Zr or Nb	Ti or V, Fe, W
W	Ta	Ti or V, Fe, Zr or Nb

[a] From B. J. Mitchell, *Anal. Chem.* **32**, 1652 (1960).

ment effect, or it may simply be an element for which binary standards are available. The complement is not necessarily the same for all seven analytes. Table 9.3 gives Mitchell's choice of binary complements. Standards are prepared for each of the seven binary systems in the composition range of interest, and the analyte-line intensities are measured from all standards. A calibration curve of analyte-line intensity *versus* concentration is established for each of the seven systems. Then, for each analyte binary system, correction factors are derived for the other five elements, except that similarity of certain elements may preclude the need for separate factors, as shown in column 3 of Table 9.3. The derivation of the correction factors is described below.

The analytical procedure is as follows. Line intensity is measured for each of the seven elements from the sample(s). Mitchell measured the intensities for all seven elements simultaneously on an Applied Research Laboratories model XIQ multichannel x-ray Quantometer. The measured line intensity of each of the seven analytes is applied to its binary calibration curve to derive its approximate concentration. This approximate concentration is then corrected by the following procedure, outlined for niobium as an example.

For each interfering element for the niobium system (Table 9.3), the correction factor is derived by applying the approximate analyte and interferant concentrations to the appropriate table or curve in the manner discussed below. The total correction for niobium in tantalum F_{NbTa} is then calculated from

$$F_{NbTa} = \sum F_{NbTa,i} = F_{NbTa,Ti} + F_{NbTa,V} + F_{NbTa,Fe} + F_{NbTa,Zr} \quad (9.29)$$

where F_{NbTa} is the correction factor to be applied to the Nb $K\alpha$ intensity measured from the sample, and $F_{NbTa,i}$ is the correction factor for the effect of element i on Nb $K\alpha$ intensity in a tantalum matrix. Because of similarity of the effect of titanium and vanadium on Nb $K\alpha$ intensity in a tantalum matrix (column 3 of Table 9.3),

$$F_{NbTa,V} = F_{NbTa,Ti} \quad (9.30)$$

Because of similarity of the effect of tantalum and tungsten, no correction is required for tungsten, and these two elements form a common matrix. The niobium-line intensity measured from the sample is then corrected:

$$I_{Nb,cor} = I_{Nb,measd}/F_{NbTa} \quad (9.31)$$

This corrected intensity is applied to the niobium calibration curve to derive corrected niobium concentration. This same procedure is followed for each of the other six elements.

The correction factors are derived as follows. Figure 9.14A shows calibration curves of Nb $K\alpha$ intensity *versus* Nb_2O_5 concentration in TiO_2 (or V_2O_5), Fe_2O_3, ZrO_2, and Ta_2O_5 (or WO_3) matrixes. These curves could be used for determination of Nb_2O_5 in the respective binary systems. For a Nb–Ti–Ta oxide ternary system, Nb $K\alpha$ intensities for all possible

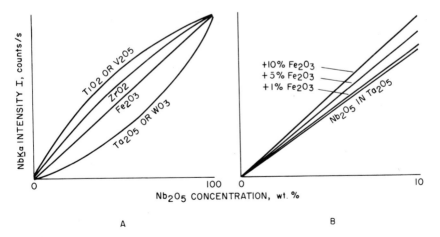

FIGURE 9.14. Absorption–enhancement effects of several elements on Nb $K\alpha$ intensity.

compositions lie on or between the Nb–Ti and Nb–Ta curves. Likewise, in the Nb–Fe–Zr system, all possible Nb $K\alpha$ intensities lie on or between the Nb–Fe and Nb–Zr curves. Of course, the entire "intensity area" may not be of analytical interest. Mitchell derived correction factors for multi-component systems on the basis of such ternary relationships in two ways.

In general, for the multicomponent system A–B–C–D–···, let A be the analyte and B its designated binary complement. Then the correction factor for the effect of, say, C on A in a B matrix is

$$F_{\text{AB,C}} = \frac{I_\text{A} \text{ in a B} + \text{C matrix}}{I_\text{A} \text{ in a B matrix}} \tag{9.32}$$

Note that the factor F is really analogous to relative intensity (the ratio of analyte-line intensities from the specimen and pure analyte) with the ternary and pure binary equivalent to the analytical sample and pure analyte, respectively. For absorption, $F < 1$; for enhancement, $F > 1$. Standards having appropriate A–B–C compositions are prepared, and the $F_{\text{AB,C}}$ value is derived for each. A similar procedure is followed for A–B–D, A–B–E, etc.

The other method for derivation of correction factors is based on the fact that within a limited low-concentration region, curves of analyte-line intensity *versus* concentration are substantially linear. Figure 9.14B shows a set of curves of Nb $K\alpha$ net intensity as a function of Nb_2O_5 concentration in a Ta_2O_5 matrix having 0, 1, 5, and 10% Fe_2O_3 as the interferant (element C). These curves have the form

$$I = mC_\text{A} + b \tag{9.33}$$

where m and b are slope and I-axis intercept, respectively. If I is net intensity, $b = 0$, as shown. Then, for a given C_A, correction factors are calculated for interferant concentrations 1, 2, 3, 4 (in Figure 9.14B, 0, 1, 5, 10% Fe_2O_3, respectively) as follows:

$$\begin{aligned}
F_1 &= m_1/m_1 = I_1/I_1 = 1 \\
F_2 &= m_2/m_1 = I_2/I_1 \\
F_3 &= m_3/m_1 = I_3/I_1 \\
F_4 &= m_4/m_1 = I_4/I_1
\end{aligned} \tag{9.34}$$

Regardless of which method is used, for each analyte binary system, a set of curves is prepared for each interfering element in the multicomponent

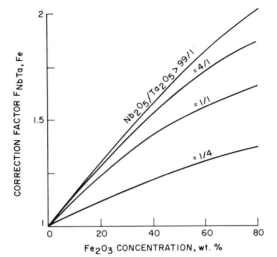

FIGURE 9.15. Correction factors for iron on Nb $K\alpha$ intensity from a Ta_2O_5 matrix. [Data from B. J. Mitchell, *Analytical Chemistry* **30**, 1894 (1958).]

system. These curves relate F with interferant concentration, each curve for a different analyte concentration. The curves have the form shown in Figure 9.15. Alternatively, the data may be prepared in tabular form similar to Table 9.4.

TABLE 9.4. Correction Factors for Effect of Fe_2O_3 on Nb $K\alpha$ Intensity from Nb_2O_5 in a Ta_2O_5 Matrix[a]

Fe_2O_3 concn. (wt%)	Correction factor F, for various Nb_2O_5/Ta_2O_5 weight ratios			
	>40/1	4/1	1/1	1/4
1	1.015	1.010	1.005	1.002
5	1.09	1.07	1.05	1.04
10	1.17	1.14	1.10	1.08
15	1.27	1.21	1.15	1.12
20	1.35	1.29	1.21	1.17
25	1.43	1.355	1.27	1.215
30	1.52	1.43	1.33	1.265
35	1.60	1.50	1.40	1.315
40	1.69	1.575	1.47	1.37
45	1.77	1.65	1.55	1.425
50	1.85	1.73	1.63	1.48

[a] From B. J. Mitchell, *Anal. Chem.* **32**, 1652 (1960).

9.9.3. Absorption-Correction Methods

Absorption–enhancement effects often consist simply of absorption effects, that is, the effects of the absorption coefficients of the matrix elements. In Table 9.5 and Figure 9.16B, the Sn $K\alpha$ intensity data from tin–lead alloys shows negative deviation from linearity (defined in Section 8.2.2.1) for this reason. At the primary wavelengths most efficient in exciting Sn $K\alpha$—those near the tin K-absorption edge (0.425 Å)—the mass-absorption coefficients of tin and lead are 40 and 25 cm²/g, respectively; at Sn $K\alpha$, they are 12 and 52 cm²/g. Then the total mass-absorption coefficients for both primary and Sn $K\alpha$ radiation are, for tin and lead, 52 and 77 cm²/g, respectively.

If it is assumed that only absorption effects are present,

$$I_X = C_{Sn} I_{100Sn} \frac{(\mu/\varrho)'_{100Sn}}{(\mu/\varrho)_X'} \tag{9.35}$$

where C_{Sn} is tin concentration (weight fraction); I_X and I_{100Sn} are Sn $K\alpha$ net intensities from the sample X and pure tin 100Sn, respectively; $(\mu/\varrho)_X'$ and $(\mu/\varrho)'_{100Sn}$ are total mass-absorption coefficients of the sample and pure tin, respectively, for both primary and Sn $K\alpha$ radiation. Equation (9.35) is derived from Equation (4.12) by substitution of a monochromatic effective wavelength for the term $J(\lambda_{pri})$, which represents the total primary spectrum. Equation (9.35) may now be used to correct the Sn $K\alpha$ data in Table 9.5. Values of $(\mu/\varrho)_X'$ for samples 1–6 are calculated by use of the

TABLE 9.5. X-Ray Intensity Data for Tin–Lead Alloys[a]

Sample	Sn concn. (wt%)	Sn $K\alpha$ net intensity (counts/s)	Pb concn. (wt%)	Pb $L\alpha_1$ net intensity (counts/s)
1	100	497	0	4
2	74.98	314	24.97	177
3	59.97	227	39.92	260
4	40.00	144	59.89	373
5	20.18	66	79.84	473
6	0	2	100	568

[a] From R. Jenkins and J. L. de Vries, *Practical X-Ray Spectrometry*, 2nd ed., Springer-Verlag New York, New York, p. 138 (1969).

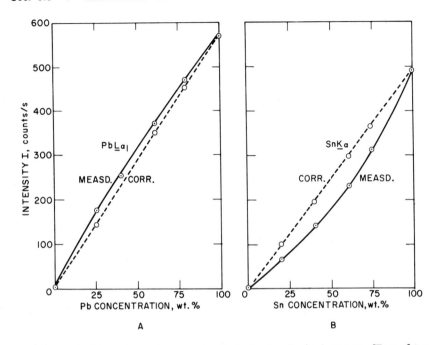

FIGURE 9.16. Absorption–enhancement effects in the tin–lead system. [Data from R. Jenkins and J. L. de Vries, *Practical X-Ray Spectrometry*, 2nd ed., Springer-Verlag New York, pp. 137–143 (1967).]

$(\mu/\varrho)'$ values for tin and lead given above. The Sn $K\alpha$ intensities from samples 2–5 are multiplied by the ratio $(\mu/\varrho)_X'/(\mu/\varrho)_{100Sn}'$, that is, $(\mu/\varrho)_X'/(\mu/\varrho)_1'$, where 1 refers to sample 1, which is pure tin. Then,

$$(I_X)_{cor} = (I_X)_{measd} \frac{(\mu/\varrho)_X'}{(\mu/\varrho)_{100Sn}'} \qquad (9.36)$$

where $(I_X)_{cor}$ and $(I_X)_{measd}$ are, respectively, corrected and measured Sn $K\alpha$ net intensities from the sample. The $(I_X)_{cor}$ values are plotted in Figure 9.16B, giving the dashed linear curve. The Pb $L\alpha_1$ data in Table 9.5 and Figure 9.16A is discussed in Section 9.9.4.

Of course, $(\mu/\varrho)'$ for a sample cannot be calculated before the sample composition is known. This problem may be circumvented in several ways, including the following:

1. If some accuracy may be sacrificed, an approximate analytical concentration may be derived from the uncorrected intensity data, and this value used to calculate $(\mu/\varrho)_X'$ to a first approximation.

2. Analyte-line intensity may be measured from pure analyte directly and with a sample disk of known mass and thickness placed over it. Since both primary and analyte-line radiation must pass through the sample disk, total $(\mu/\varrho)'$ for the matrix may be calculated.

3. If primary absorption can be neglected, the sample disk may be used to attenuate the analyte-line radiation from a sample of pure analyte.

4. This third technique can also be used to evaluate primary absorption if an effective wavelength can be assigned to the primary spectrum. A secondary target having a strong spectral line at this wavelength may be placed in the specimen compartment and its intensity measured through the sample disk.

5. If total (primary and secondary) absorption is not measured, the best approach is iteration. Approximate analyte concentrations are derived by assumption of a linear relationship between concentration and uncorrected analyte-line intensity. If all elements present are determined, the concentrations so derived may be normalized to 100%. These normalized concentrations are then used to calculate mass-absorption coefficients of the sample for all the analyte lines. So far, the procedure is simply that of method 1 above. These absorption coefficients are then used to correct the intensities, and a second set of concentrations is obtained and normalized. This iteration process is repeated until satisfactory convergence is realized.

9.9.4. Influence-Coefficient Methods

The Pb $L\alpha_1$ data in Table 9.5 and Figure 9.16A provides a convenient and simple introduction to the influence-coefficient method for correction of absorption–enhancement effects. The Pb $L\alpha_1$ curve shows positive deviation from linearity (defined in Section 8.2.2.1), but for a reason different from that which explained the deviation in the Sn $K\alpha$ data (Section 9.9.3). The Pb $L\alpha_1$ line is not absorbed to very different degrees by tin and lead (μ/ϱ 134 and 128 cm^2/g, respectively). The Pb $L\alpha_1$ line, which is associated with the Pb LIII absorption edge (0.95 Å), is not strongly enhanced by Sn $K\alpha$ (0.49 Å). The deviation is due to the difference in absorption coefficients of tin and lead for the primary wavelengths most efficient in exciting Pb $L\alpha_1$—those near Pb LIII$_{ab}$; for example, at 0.9 Å, the absorption coefficients of tin and lead are 64 and 145 cm^2/g, respectively; at 0.75 Å, they are 40 and 160. Thus, the excitation efficiency of the primary beam for Pb $L\alpha_1$ and the slope of the calibration curve increase with tin concentration. If it is assumed that the slope increases in proportion to tin concentration, the following equation obtains:

$$C_{\text{Pb}} = \frac{I_{\text{Pb}}}{m_{\text{Pb}}}(1 + a_{\text{PbSn}}C_{\text{Sn}}) \tag{9.37}$$

where C is concentration; I is net intensity; m is the slope of the curve, or Pb $L\alpha_1$ intensity per 1% lead for pure lead; and a_{PbSn} is the absorption–enhancement influence factor of tin for the Pb $L\alpha_1$ line. From this,

$$(I_{\text{Pb}})_{\text{cor}} = (I_{\text{Pb}})_{\text{measd}}(1 + a_{\text{PbSn}}C_{\text{Sn}}) \tag{9.38}$$

The value of a_{PbSn} is determined by substitution of calibration data from standards into Equation (9.37). In the present example, a_{PbSn} was -0.0021. The $(I_{\text{Pb}})_{\text{cor}}$ values are plotted in Figure 9.16A, giving the dashed linear curve. Alternatively, Equation (9.37) may be put in the form

$$C_{\text{Pb}}m_{\text{Pb}}/I_{\text{Pb}} = a_{\text{PbSn}}C_{\text{Sn}} + 1 \tag{9.39}$$

which has the form $y = mx + b$, so that a plot of $(Cm/I)_{\text{Pb}}$ *versus* C_{Sn} is substantially linear and has slope a_{PbSn}. If tin concentration in the standards is not known, Equation (9.37) may be expressed in terms of Sn $K\alpha$ net intensity I_{Sn}:

$$C_{\text{Pb}} = \frac{I_{\text{Pb}}}{m_{\text{Pb}}}(1 + a_{\text{PbSn}}I_{\text{Sn}}) \tag{9.40}$$

An equation analogous to Equation (9.38) is readily obtained from Equation (9.40), and a_{PbSn} is obtained from this equation in the same way as from Equation (9.38).

In general terms,

$$C_i = \frac{I_i}{m_i}[1 + \sum (a_{ij}C_j)] \qquad (j \neq i) \tag{9.41}$$

where C, I, and m have the significance given above; i is the analyte element; j is a matrix element; and a_{ij} is the absorption–enhancement effect of element j on the analyte i. The factor a is negative for calibration curves having negative deviation, $(\mu/\varrho)_j > (\mu/\varrho)_i$, and is positive for positive deviation, $(\mu/\varrho)_j < (\mu/\varrho)_i$. The influence factor corrects for both absorption and enhancement because the latter may be regarded as negative absorption.

For a ratio method,

$$C_{iX} = \left(\frac{I_X}{I_S}\right)_i \{1 + \sum_i [a_{ij}(C_X - C_S)_j]\}C_{iS} \qquad (j \neq i) \tag{9.42}$$

where C and I are concentration and net intensity, respectively, and S and X refer to sample and standard, respectively.

Removal of the summation sign from Equation (9.42) gives the effect of a single matrix element j on analyte i. Rearrangement then gives

$$\left(\frac{C_X}{I_X} \frac{I_S}{C_S} \right)_i = a_{ij}(C_X - C_S)_j + 1 \qquad (9.43)$$

which has the form $y = mx + b$. Thus, a plot of $[(C_X/I_X)(I_S/C_S)]_i$ *versus* $(C_X - C_S)_j$ should be linear, intercept the ordinate at 1, and have slope a_{ij}. This serves as a test of the validity of a_{ij} values.

Beattie and Brissey were among the first to successfully analyze multi-element systems by use of sets of linear simultaneous equations using influence coefficients a relating intensities and mass concentrations of pairs of elements. For a binary system A–B, these coefficients have the form

$$a_{AB} = \frac{W_A}{W_B} (R_A' - 1) \qquad (9.44)$$

$$a_{BA} = \frac{W_B}{W_A} (R_B' - 1) \qquad (9.45)$$

where a_{AB} is the influence coefficient of element B for element A; a_{BA} is the influence coefficient of element A for element B; W is weight fraction; and R' is defined by

$$R_A' = I_{AA}/I_{AS}; \qquad R_B' = I_{BB}/I_{BS} \qquad (9.46)$$

where I_{AA} and I_{AS} are A-line intensity from pure A and from the specimen, and I_{BB} and I_{BS} have similar significance for B-line intensity.

The quantity a_{ij} (some workers use the symbol α_{ij}) is termed the absorption–enhancement *influence coefficient, influence factor, interaction coefficient,* or *interaction factor* of matrix element j on analyte i. The influence coefficients may be calculated from certain known parameters, but are usually—and preferably—evaluated experimentally with data derived from standards and applied to sets of linear equations. Usually the coefficients are evaluated by inserting in the equations *concentrations* of matrix elements in the standards, but sometimes by inserting *intensities* of matrix-element lines measured from the standards. The intensity methods are applicable to wider matrix composition ranges and tend to reduce third-element effects, but require larger numbers of standards. The concentration methods are applicable to relatively more narrow matrix composition ranges.

The individual linear equations in the Beattie–Brissey method have the form

$$a_{i1}W_1 + a_{i2}W_2 + \cdots + [1 - (I_{100i}/I_i)]W_i + \cdots + a_{in}W_n = 0 \quad (9.47)$$

where i is the analyte, and elements $1, 2, \ldots, n$ are matrix elements, including i; a_{in} is the influence coefficient of element n on analyte i; W is concentration (weight fraction); and I_i and I_{100i} are i-line intensities from the specimen and pure i, respectively. An equation of this form is written for each element in the multielement system. If we let

$$I_{100i}/I_i = R_i' \quad (9.48)$$

the following set of simultaneous equations is obtained:

$$(1 - R_1')W_1 + a_{12}W_2 + a_{13}W_3 + \cdots + a_{1n}W_n = 0 \quad (9.49\text{-}1)$$

$$a_{21}W_1 + (1 - R_2')W_2 + a_{23}W_3 + \cdots + a_{2n}W_n = 0 \quad (9.49\text{-}2)$$

$$\cdots\cdots$$

$$a_{n1}W_1 + a_{n2}W_2 + a_{n3}W_3 + \cdots + (1 - R_n')W_n = 0 \quad (9.49\text{-}n)$$

$$W_1 + W_2 + W_3 + \cdots + W_n = 1 \quad (9.50)$$

Each individual member of the set of Equations (9.49) corresponds to one analyte i and contains expressions of the interactions of all the other constituents in the system—that is, of all the matrix elements j—with that analyte. Each analyte is seen to constitute a matrix element for every other analyte. Equation (9.50) is required because the n equations (9.49) are not mutually independent. In practice, if minor constituents are known to be present, but are of no interest or significance, Equation (9.50) may be set equal to the sum of the concentrations of the major and other significant constituents—for example, 0.98. These equations are used both for calibration, that is, evaluation of the a_{ij} coefficients from measurements on suitable standards and for determination, that is, evaluation of the analytical concentrations by use of the a_{ij} values.

Table 9.6 gives the data of Beattie and Brissey for analysis of chromium–iron–nickel–molybdenum alloys to a first approximation. Part A of the table gives the compositions of the 12 binary standards in weight percent W. The influence coefficients were evaluated as follows. Cr $K\alpha$ net intensity was measured from the three chromium binaries and from pure chromium; Fe $K\alpha$ net intensity was measured from the iron binaries and from iron; etc. The values of R_i' in part B of the table were calculated from these intensities. The influence coefficients in part C were obtained by insertion

TABLE 9.6. Data for Analysis of Cr–Fe–Ni–Mo Alloys to a First Approximation by the Influence-Coefficient Method[a]

Element i	Element j			
	Cr	Fe	Ni	Mo
A. Composition of binary standards W (wt %)				
Cr	—	48.24	48.07	74.28
Fe	50.83	—	51.53	65.33
Ni	48.19	46.65	—	53.70
Mo	23.53	34.44	46.27	—
B. Intensity ratios for binary standards, $R_i' = I_{100i}/I_i$				
Cr	1.000	1.760	1.815	1.841
Fe	3.360	1.000	1.613	1.835
Ni	2.860	3.670	1.000	1.954
Mo	3.770	2.787	2.461	1.000
C. Influence coefficients a_{ij} from Equations (9.49)				
Cr	1.000	0.721	0.813	2.660
Fe	2.482	1.000	0.676	1.582
Ni	1.863	2.420	1.000	1.108
Mo	0.877	0.944	1.260	1.000
D. Intensity ratio for an analytical sample, $R_i' = I_{100i}/I_i$				
	R'_{Cr}	R'_{Fe}	R'_{NI}	R'_{Mo}
	5.04	2.346	6.640	15.34

[a] From H. J. Beattie and R. M. Brissey, *Anal. Chem.* **26**, 980 (1954).

of the R_i' values into a set of four simultaneous equations having the form of Equations (9.49). Then the simultaneous equations for analysis of chromium–iron–nickel–molybdenum alloys are

$$
\begin{aligned}
(1 - R'_{Cr})W_{Cr} + \quad 0.721 W_{Fe} \ + \quad 0.813 W_{Ni} + \quad 2.660 W_{Mo} &= 0 \\
2.482 W_{Cr} \ + (1 - R'_{Fe})W_{Fe} + \quad 0.676 W_{Ni} \ + \quad 1.582 W_{Mo} &= 0 \\
1.863 W_{Cr} \ + \quad 2.420 W_{Fe} \ + (1 - R'_{Ni})W_{Ni} + \quad 1.108 W_{Mo} &= 0 \\
0.877 W_{Cr} \ + \quad 0.944 W_{Fe} \ + \quad 1.260 W_{Ni} \ + (1 - R'_{Mo})W_{Mo} &= 0
\end{aligned}
\tag{9.51}
$$

An analysis of an alloy of this quaternary system consists in measurement of the Cr $K\alpha$, Fe $K\alpha$, Ni $K\alpha$, and Mo $K\alpha$ net intensities from the sample and from the four respective elements. The four intensity ratios R_i'

are then calculated; ratios for a typical alloy are given in part D of Table 9.6. These ratios inserted in Equations (9.51) permit calculation of composition of the sample.

The Beattie–Brissey method described above is a typical influence-coefficient method. Many workers have published variations of the method, particularly Burnham, Hower, and Jones; Lucas-Tooth and Price; Lucas-Tooth and Pyne; Marti; Rasberry and Heinrich; Shermann; and Traill and Lachance.

9.9.5. Variable-Takeoff-Angle Method

The availability of suitable standards for quantitative x-ray spectrometric analysis is often a very serious problem, and the analyst always welcomes a method that requires no standards: The methods of standard addition and dilution are considered in Section 9.7. This section considers another method.

The fundamental excitation equations, such as Equation (4.4), show that analyte-line intensity depends not only on analyte concentration, primary intensity distribution, and mass-absorption coefficients of the specimen for the primary and analyte-line x-rays, but also on the incident and takeoff angles. Ebel has developed a method of quantitative x-ray spectrometric analysis based on variation of the takeoff angle. Bulk and film specimens can be analyzed without chemical standards, and film thickness (or area density) can be measured without thickness standards. The principal specimen requirements are homogeneity and a smooth, plane surface.

The geometry of the variable-takeoff-angle method is shown in Figure 9.17, in which the analyte-line intensity emitted from layer dt that actually enters the source collimator is given by

$$\Delta I_L = I_{0,\lambda_\mathrm{pri}} \frac{d\Omega}{4\pi} \, \omega_A g_L \, \frac{r_A - 1}{r_A} \, \frac{n_{A,X}}{n_{A,A}}$$

$$\times \left\{ \exp\left[-\left(\frac{\mu_{X,\lambda_\mathrm{pri}}}{\cos \alpha} + \frac{\mu_{X,\lambda_L}}{\cos \beta} \right) t \right] \right\} d\lambda_\mathrm{pri} \, dt \qquad (9.52)$$

FIGURE 9.17. Geometry of the variable takeoff-angle method.

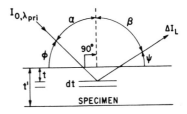

where

$d\Omega/4\pi$ is the useful solid angle of secondary analyte-line x-rays accepted by the x-ray optical system (Figure 4.2)

A   is the analyte element

g_L is the relative intensity of the analyte line in its series [Equation (4.2)]

I_L is the intensity of the analyte line

$I_{0,\lambda_{pri}}$ is the intensity of the incident primary beam

$n_{A,A}$ is the number of analyte atoms per cm^3 in pure analyte

$n_{A,X}$ is the number of analyte atoms per cm^3 in the specimen

r_A is the absorption-edge jump ratio of the analyte

t is the depth of layer dt below the specimen surface

t' is the specimen thickness

α is the angle between the normal to the specimen surface and the central ray of the incident primary beam ($\alpha = 90° - \phi$)

β is the angle between the normal to the specimen surface and the central ray of the secondary beam ($\beta = 90° - \psi$); this is the *takeoff angle* of Ebel's method

λ_L is the wavelength of the analyte line

λ_{pri} is the wavelength of the primary x-rays

μ_{X,λ_L} is the linear absorption coefficient of the specimen for the analyte line

$\mu_{X,\lambda_{pri}}$ is the linear absorption coefficient of the specimen for the primary x-rays

ω_A is the fluorescent yield of the analyte line

Equation (9.52) is substantially the same as Equation (4.4) except that Ebel uses linear absorption coefficients and measures incident and takeoff angles from the normal to the specimen surface (α and β, respectively), rather than from the specimen surface itself (ϕ and ψ, respectively).

The following two substitutions are now made in Equation (9.52):

$$C_A(\varrho_A'/\varrho_A) = n_{A,X}/n_{A,A} \qquad (9.53)$$

$$P_A = \frac{d\Omega}{4\pi}\, \omega_A g_L\, \frac{r_A - 1}{r_A} \qquad (4.5)$$

where C_A is analyte concentration, and ϱ_A' and ϱ_A are the densities of analyte in the specimen and in pure analyte, respectively. By integration and application of the first mean-value theorem of integral calculus to the substituted equation, one gets

$$I_L = C_A \frac{\varrho_A'}{\varrho_A} \frac{1 - \exp\left[-\left(\dfrac{\mu_{X,\bar{\lambda}_{\mathrm{pri}}}}{\cos\alpha} + \dfrac{\mu_{X,\lambda_L}}{\cos\beta}\right)t'\right]}{\dfrac{\mu_{X,\bar{\lambda}_{\mathrm{pri}}}}{\cos\alpha} + \dfrac{\mu_{X,\lambda_L}}{\cos\beta}} \int_{\lambda_{\min}}^{\lambda_{A_{ab}}} I_{0,\lambda_{\mathrm{pri}}} P_A \, d\lambda_{\mathrm{pri}}$$

$$(9.54)$$

where $\bar{\lambda}_{\mathrm{pri}}$ is the mean primary wavelength (see below): $\mu_{X,\bar{\lambda}_{\mathrm{pri}}}$ is the linear absorption coefficient of the specimen for $\bar{\lambda}_{\mathrm{pri}}$; $\lambda_{\min}$ is the short-wavelength limit of the primary beam; and $\lambda_{A_{ab}}$ is the analyte absorption edge. Ebel derives four working equations from Equation (9.54):

1. Analyte-line intensity from infinitely thick ($t' = t_\infty$) multielement specimens is given by

$$I_{L,\infty} = C_A \frac{\varrho_A'}{\varrho_A} \frac{1}{(\mu_{X,\bar{\lambda}_{\mathrm{pri}}}/\cos\alpha) + (\mu_{X,\lambda_L}/\cos\beta)} \int_{\lambda_{\min}}^{\lambda_{A_{ab}}} I_{0,\lambda_{\mathrm{pri}}} P_A \, d\lambda_{\mathrm{pri}} \quad (9.55)$$

2. For pure analyte, $C_A = 1$ and $\varrho_A' = \varrho_A$, so that $C_A(\varrho_A'/\varrho_A) = 1$; then analyte-line intensity from infinitely thick pure analyte is given by

$$I_{L,\infty} = \frac{1}{(\mu_{A,\bar{\lambda}_{\mathrm{pri}}}/\cos\alpha) + (\mu_{A,\lambda_L}/\cos\beta)} \int_{\lambda_{\min}}^{\lambda_{A_{ab}}} I_{0,\lambda_{\mathrm{pri}}} P_A \, d\lambda_{\mathrm{pri}} \qquad (9.56)$$

where $\mu_{A,\bar{\lambda}_{\mathrm{pri}}}$ and μ_{A,λ_L} are the linear absorption coefficients of the analyte for $\bar{\lambda}_{\mathrm{pri}}$ and λ_L, respectively.

3. For thin-film specimens,

$$t' = m/a\varrho_A' \qquad (9.57)$$

where a and m are the area (cm²) and mass (g) of the effective area of film from which analyte-line emission is intercepted by the x-ray optical system, and ϱ_A' is the density (g/cm³) of analyte in the film. Analyte-line intensity from multielement film specimens is given by

$$I_{L,f} = C_A \frac{\varrho_A'}{\varrho_A} \frac{1 - \exp\left[-\left(\dfrac{\mu_{X,\bar{\lambda}_{\mathrm{pri}}}}{\cos\alpha} + \dfrac{\mu_{X,\lambda_L}}{\cos\beta}\right)\dfrac{m}{a\varrho_A'}\right]}{\dfrac{\mu_{X,\bar{\lambda}_{\mathrm{pri}}}}{\cos\alpha} + \dfrac{\mu_{X,\lambda_L}}{\cos\beta}} \int_{\lambda_{\min}}^{\lambda_{A_{ab}}} I_{0,\lambda_{\mathrm{pri}}} P_A \, d\lambda_{\mathrm{pri}}$$

$$(9.58)$$

4. By following the same line of reasoning as for pure bulk analyte specimens (item 2 above), one gets for analyte-line intensity from pure analyte film specimens

$$I_{L,f} = \frac{1 - \exp\left[-\left(\dfrac{\mu_{A,\bar{\lambda}_{pri}}}{\cos \alpha} + \dfrac{\mu_{A,\lambda_L}}{\cos \beta}\right)\dfrac{m}{a\varrho_A}\right]}{\dfrac{\mu_{A,\bar{\lambda}_{pri}}}{\cos \alpha} + \dfrac{\mu_{A,\lambda_L}}{\cos \beta}} \int_{\lambda_{min}}^{\lambda_{A_{ab}}} I_{0,\lambda_{pri}} P_A \, d\lambda_{pri} \qquad (9.59)$$

In all the foregoing equations, mass-absorption coefficients μ/ϱ multiplied by the appropriate densities ϱ may be substituted for the linear coefficients μ.

A discussion of the application of these equations is beyond the scope of this introductory book. The reader is referred to Ebel's papers and pp. 631–638 in the second edition of Bertin's *Principles and Practice of X-Ray Spectrometric Analysis*.

9.9.6. Fundamental-Parameters Method

The fundamental-parameters method permits the calculation of analytical composition from the measured analyte-line intensity and the tabulated values of three fundamental parameters—primary spectral distribution, absorption coefficient, and fluorescent yield. Tables of these parameters are stored in the computer required to make the extremely complex calculations involved in the application of this method. No calibration standards are required. Several workers have developed fundamental-parameters methods, but the treatment in this section is taken largely from a paper by Criss and Birks. This method, like the influence-coefficient method, assumes that the specimen is homogeneous and infinitely thick and has a plane surface. However, instead of assuming that the incident primary spectrum can be described by a single average effective wavelength, the method uses the actual spectral distribution of the primary beam. This distribution can be estimated theoretically, or it may be measured for a given x-ray tube target, potential, and type of power supply (full-wave or constant-potential). Matrix absorption and analyte-line excitation by matrix elements enter the equation explicitly for each specimen. Analyte concentration is calculated by iteration. The advantages of the fundamental-parameters method are that it avoids, at least in principle, the limited composition ranges of regression methods and that no intermediate standards or empirical coefficients are required for any matrix. The principal limitation is the present uncertainty in mass-absorption coefficient and fluorescent-yield data. However, this shortcoming is being overcome as more laboratories engage in measurement of these parameters. Another limitation is that the spectral distribution of the primary x-ray beam, although determinable experimentally (see below), may vary with the age of the x-ray tube. The calculations are very complex and must be done on a computer.

The fundamental-parameters equation is derived from the geometry of Figure 4.2 and the basic excitation equations, such as Equations (4.4) and (4.12). The equation used by Criss and Birks and given here is substantially similar to that derived by other workers, except that it uses a summation over wavelength instead of the more common integration:

$$
\begin{aligned}
[(I_{\lambda_L})_{\mathrm{pri}} + (I_{\lambda_L})_{\mathrm{matrix}}]x & \\
= g_i C_i \sum_{\lambda_{\mathrm{pri}}} & \left(\left[\frac{D_{i,\lambda_{\mathrm{pri}}}(\mu/\varrho)_{i,\lambda_{\mathrm{pri}}}(I_{\lambda_{\mathrm{pri}}}\,\Delta\lambda_{\mathrm{pri}})}{(\mu/\varrho)_{\mathrm{M},\lambda_{\mathrm{pri}}}\csc\phi + (\mu/\varrho)_{\mathrm{M},\lambda_L}\csc\psi} \right] \right. \\
& \times \left\{ 1 + \frac{1}{2(\mu/\varrho)_{i,\lambda_{\mathrm{pri}}}} \sum_{\lambda_J} [D_{j,\lambda_{\mathrm{pri}}}C_j K_j (\mu/\varrho)_{i,\lambda_J}(\mu/\varrho)_{j,\lambda_{\mathrm{pri}}}] \right. \\
& \times \left[\frac{1}{(\mu/\varrho)_{\mathrm{M},\lambda_{\mathrm{pri}}}\csc\phi} \log_e\!\left(1 + \frac{(\mu/\varrho)_{\mathrm{M},\lambda_{\mathrm{pri}}}\csc\phi}{(\mu/\varrho)_{\mathrm{M},\lambda_J}} \right) \right. \\
& \left.\left.\left. + \frac{1}{(\mu/\varrho)_{\mathrm{M},\lambda_L}\csc\psi} \log_e\!\left(1 + \frac{(\mu/\varrho)_{\mathrm{M},\lambda_L}\csc\psi}{(\mu/\varrho)_{\mathrm{M},\lambda_J}} \right) \right] \right\} \right)
\end{aligned}
\tag{9.60}
$$

where

C_i	is concentration of analyte element i
C_j	is concentration of matrix element j
D_i	has the value 1 if λ_{pri} is short enough to excite λ_L; it has the value 0 for longer wavelengths
D_j	has the value 1 if λ_{pri} is short enough to excite λ_J; it has the value 0 for longer wavelengths
g_i	is a function of the absolute intensity of λ_L from element i; g cancels if a relative intensity function is used (see below)
K_j	is $[1 - (1/r)]\omega$ for the particular spectral line of element j that excites λ_L of element i (see below for r and ω)
I_{λ_L}	is intensity of analyte line of element i
$(I_{\lambda_L})_{\mathrm{pri}}$	is intensity of analyte line excited by the primary spectrum
$(I_{\lambda_L})_{\mathrm{matrix}}$	is intensity of analyte line excited by spectral lines of matrix element j in the specimen

$I_{\lambda_{\text{pri}}}$	is intensity of the primary beam in the wavelength interval $\Delta\lambda_{\text{pri}}$
i	is the analyte element
j	is a matrix element
M	is the specimen matrix, including the analyte i
r	is absorption edge jump ratio
$\Delta\lambda_{\text{pri}}$	is a wavelength interval into which the primary spectrum is arbitrarily divided for summation purposes
λ_J	is the wavelength of a spectral line of element j capable of exciting λ_L
λ_L	is the wavelength of the analyte line of element i
λ_{pri}	is a wavelength in the primary beam
$(\mu/\varrho)_i$, $(\mu/\varrho)_j$, $(\mu/\varrho)_M$	are mass-absorption coefficients of analyte i, matrix element j, and matrix M, respectively
$(\mu/\varrho)_{M,\lambda_J}$, $(\mu/\varrho)_{M,\lambda_L}$, $(\mu/\varrho)_{M,\lambda_{\text{pri}}}$	are mass-absorption coefficients of matrix M for λ_J, λ_L, and λ_{pri}, respectively; similar notations for $(\mu/\varrho)_{i,\lambda}$ and $(\mu/\varrho)_{j,\lambda}$ have analogous significance
ϕ	is the angle between the central ray of the primary cone and the specimen
ψ	is the angle between the central ray of the secondary cone and the specimen, the takeoff angle
ω	is fluorescent yield

The primary spectrum is divided into wavelength intervals $\Delta\lambda_{\text{pri}}$ each having intensity $I_{\lambda_{\text{pri}}}$. The product of g_iC_i and the first summation over λ_{pri} represents $(I_{\lambda_L})_{\text{pri}}$, the analyte-line intensity excited only by the primary x-rays. The remainder of the equation represents $(I_{\lambda_L})_{\text{matrix}}$, the analyte-line intensity excited by spectral lines of matrix elements j; this interelement excitation contributes as much as 30% to the total measured analyte-line intensity.

Tables have been published giving spectral distributions for specified x-ray tubes having chromium, copper, molybdenum, rhodium, and tungsten targets. Any x-ray tubes having the same target element, window material and thickness, operating potential and wave form (full-wave or constant-potential), and takeoff angle should have the same spectral distribution. Differences in window material and thickness are easily corrected by use of the absorption equation. Differences in takeoff angle are not easy to correct, but differences within $\pm5°$ usually may be disregarded. Of course, x-ray tube current (mA) has no effect on the spectral distribution.

The fundamental-parameters equation can be based on analyte-line intensity measured from the sample X, as in Equation (9.60). More often, the equation is based on the ratio R of this intensity with that measured from a reference standard S similar to X or from pure analyte $100i$. In the remainder of this section, the symbol R and the term *intensity function* refer to any of these three alternatives, that is,

$$R = I_X \quad \text{or} \quad I_X/I_S \quad \text{or} \quad I_X/I_{100i} \qquad (9.61)$$

where I is $(I_{\lambda_L})_{\text{pri}} + (I_{\lambda_L})_{\text{matrix}}$.

The denominator of the second and third forms of Equation (9.61) is represented by another equation identical with Equation (9.60), except that C_i, C_j, $(\mu/\varrho)_{M,\lambda_{\text{pri}}}$, $(\mu/\varrho)_{M,\lambda_L}$, and $(\mu/\varrho)_{M,\lambda_J}$ refer to the standard or pure analyte. If the standard is pure analyte, $(I_{\lambda_L})_{\text{matrix}}$ in the denominator becomes 0. Analyte-line intensity ratios from the sample X and a multi-component standard S are readily converted to ratios relative to pure analyte as follows:

$$R = \frac{I_X}{I_{100i}} = \left(\frac{I_X}{I_S}\right)_{\text{measd}}\left(\frac{I_S}{I_{100i}}\right)_{\text{calcd}} \qquad (9.62)$$

where I has the same significance as in Equation (9.61).

Equation (9.60) expresses intensity function in terms of specimen composition. However, it is intensity functions that are measured, whereas it is concentrations that are required. Unfortunately, algebraic operation cannot produce explicit equations for mass functions C_1, C_2, etc. Consequently, mass fractions are found by an iteration procedure that makes successively closer estimates of mass fractions until the intensity functions calculated from the estimated concentrations agree satisfactorily with the measured intensity functions. Obviously, the iteration procedure must be performed on a computer. Following is a simple iteration equation:

$$C_i' = \frac{(R_i)_{\text{measd}}}{R_i} C_i \qquad (9.63)$$

where C_i is the present estimate of concentration of element i; C_i' is the next estimate of i concentration in the iteration; $(R_i)_{\text{measd}}$ is the measured intensity function for the spectral line of i; and R_i is the intensity function calculated for C_i.

The iteration procedure is as follows: (1) Initial estimates are made of all analyte concentrations C_i by normalizing their measured intensity functions $(R_i)_{\text{measd}}$ to add to 1, or by some other method. (2) The corresponding intensity functions R_i are calculated. (3) The next estimate C_i' is made by application of the iteration equation to each element i. (4) The C_i' values are scaled to add to 1. (5) The corresponding expected intensity functions R_i' are calculated. (6) A decision is made whether further iteration is indicated. This decision is based on a comparison of values of R_i' and $(R_i)_{\text{measd}}$ or of C_i' and C_i, that is, the latest and the immediately preceding iterations. For example, it may be decided to stop the iteration when the value of C_i' for every element differs from C_i—the value at the preceding iteration—by less than some arbitrary limit, say 0.1%. (7) If further iteration is indicated, the sequence is repeated starting with step 3 and using C_i' and R_i' as the "present" estimate and its calculated intensity function.

9.9.7. Multiple-Regression Method

Multiple regression is a statistical method used to study correlation among observed data and to form equations relating one dependent variable with several independent variables. In many x-ray analytical situations, regression methods can be applied to correlate analyte-line intensity I (the independent variable) with concentration C (the dependent variable). Mitchell and Hopper have developed this technique, and the following discussion is derived largely from their paper.

When plotted data obviously lies on a straight line, the dependent variable C may be predicted by use of the single independent variable I in a linear equation such as the following:

$$C = a_0 + a_1 I \tag{9.64}$$

When the plotted curve is nonlinear, other functional relationships must be assumed and the best fit determined, such as a three-term polynomial (parabolic or quadratic) curve:

$$C = a_0 + a_1 I + a_2 I^2 \tag{9.65}$$

In a multicomponent system, if the dependent variable C_1, the concentration of one of the elements, is a linear function of each of the independent variables $I_1, I_2, I_3, \ldots$, the line intensities of each of the component elements, a multiple linear equation may be estimated by the least-squares method:

$$C_1 = a_0 + a_1 I_1 + a_2 I_2 + a_3 I_3 + \cdots \tag{9.66}$$

However, if the I–C relationship is nonlinear, a higher-order polynomial may be required. In most cases, x-ray data fits an equation of the form

$$C_1 = a_0 + a_1 I_1 + a_2 I_2 + a_3 I_3 + a_4 I_1{}^2 + a_5 I_2{}^2 + a_6 I_3{}^2 + a_7 I_1 I_2 + a_8 I_1 I_3 + a_9 I_2 I_3 \quad (9.67)$$

The constants are evaluated from intensities measured from numbers of multicomponent standards having accurately known compositions. The calculations can be made on a desk calculator for two- and three-term equations and limited sets of I–C data. However, for more complex equations and large sets of data, a regression program using a digital computer is required. An incidental advantage of the multiple-regression method, compared with the influence-coefficient method, is that it is more easily programmed for digital computers and has smaller storage requirements. The result of the multiple regression procedure is an equation in which the constants a have numerical values giving the best fit to the measured intensity data and the known composition of the standards. Thereafter, the equation is used to provide analytical composition data from intensities measured from samples similar to the standards used to evaluate the constants.

9.10. FILMS AND PLATINGS

For plane, homogeneous, infinitely thick specimens, analyte-line intensity is a function of analyte concentration. However, for plane, homogeneous specimens that are less than infinitely thick, analyte-line intensity is a function of both concentration and thickness; if the specimens consist of pure element or have constant composition, intensity is a function of thickness alone.

Five x-ray fluorescence spectrometric methods are in common use for measurement of films, platings, and other layers that are less than infinitely thick. These methods are represented in Figure 9.18 and defined in the following paragraphs, where the notation has the following significance:

p	refers to the plating or film
s	refers to the substrate or base metal
I_0	is the intensity of the incident primary beam
$I_{p,t}$	is the intensity of the plate-element line from a plating of thickness t
$I_{p,\infty}$	is the intensity of the plate-element line from infinitely thick plate metal

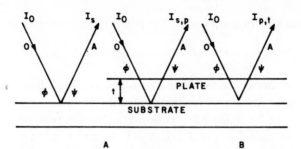

FIGURE 9.18. Methods for x-ray spectrometric determination of film and plating thickness. (A) Attenuation of the line of a substrate element absent in the plating. (B) Emission of the line of a plate element absent in the substrate.

I_s is the intensity of the substrate-element line from unplated substrate

$I_{s,p}$ is the intensity of the substrate-element line after emerging through the overlying plate

$I_{K\alpha}$, $I_{K\beta}$,
    $I_{L\alpha}$, $I_{L\beta}$ are the intensities of the indicated specific lines

t is thickness of the film or plate

$\overline{(\mu/\varrho)}$ is $(\mu/\varrho)_{\lambda_{\mathrm{pri}}} \csc \phi + (\mu/\varrho)_{\lambda_{\mathrm{A}}} \csc \psi$ [Equation (2.6)]

The *substrate-line attenuation method* (Figure 9.18A) is based on the attenuation of a substrate-element line on passing through the overlying layer. The intensity of the substrate line is measured from unplated substrate and the plated sample. Then, from Equation (2.2),

$$\log_e(I_s/I_{s,p}) = \overline{(\mu/\varrho)}_{p,s}\varrho t \qquad (9.68)$$

where ϱ and t are plating density (g/cm³) and thickness (cm), respectively, and $\overline{(\mu/\varrho)}_{p,s}$ refers to plating absorption of primary and substrate-line x-rays. A plot of $I_{s,p}$ or of $I_s/I_{s,p}$ *versus* plating thickness can be established from standards having known layer thickness.

The *plate-line emission method* (Figure 9.18B) is based on the relationship between intensity of a plate-element line and layer thickness on a substrate from which the plate element is absent. The intensity of the plate line is measured from infinitely thick plate element and the plated sample. Then,

$$\log_e[1 - (I_{p,t}/I_{p,\infty})] = -\overline{(\mu/\varrho)}_{p,p}\varrho t \qquad (9.69)$$

where ϱ and t are plating density and thickness, and $\overline{(\mu/\varrho)}_{p,p}$ refers to plating absorption of primary and plate-line x-rays. A plot of I_p or of $I_{p,t}/I_{p,\infty}$ versus thickness can be established from standards.

Many *ratio methods* have been used for determination of plate thickness. Calibration curves or mathematical relationships are established from standards. Any of the following quantities may be plotted as a function of thickness (see the second paragraph of this section for notation): $I_{p,t}/I_B$ (where B refers to the background at or adjacent to the peak or in the continuum hump); $I_{p,t}/I_{p,\infty}$; $(I_{p,\infty} - I_{p,t})/I_{p,\infty}$; $I_{p,\infty}/(I_{p,\infty} - I_{p,t})$; $1 - (I_{p,t}/I_{p,\infty})$; $I_{p,t}/I_{s,p}$; $I_{s,p}/I_s$; $(I_{K\alpha}/I_{K\beta})_p$ (for low-Z platings); $(I_{L\alpha}/I_{L\beta})_p$ (for high-Z platings); $(I_{K\alpha}/I_{K\beta})_s$ (for low-Z substrates); $(I_{L\alpha}/I_{L\beta})_s$ (for high-Z substrates); and $(I_1/I_2)_s$ (where 1 and 2 represent lines of different substrate elements). The intensities may be peak or net. For a single plating on a substrate, the functions $I_{p,t}$, $I_{s,p}/I_s$, or even $I_{p,t}/I_{s,p}$ are of comparable effectiveness. Multiple platings are considered below. The methods involving $(I_\alpha/I_\beta)_s$ and $(I_1/I_2)_s$ are more effective the greater the difference in the absorption coefficient of the plating for the two wavelengths. For example, the ideal case would be the fortuitous occurrence of an absorption edge of the plate metal between the α and β substrate lines; another favorable case would be two widely separated lines λ_1 and λ_2 from different elements in an alloy substrate. In general, the α–β methods usually give calibration curves having relatively shallow slopes. However, two-wavelength methods have the advantage, compared with one-wavelength methods, that they are less prone to error due to pinholes, cracks, scratches, etc. in the plating. They are also applicable to irregular-shaped plated objects.

The *method of variable takeoff angle*, developed by Ebel, permits determination of thickness (area density) of single-element and multielement films without thickness standards. This method is summarized in Section 9.9.5.

If the plating is an alloy of elements A, B, C, ... on substrate s, a line of each plate element and the principal substrate element may be measured and the following ratio used: $(\sum I_i)_p/I_s$ or $(I_A + I_B + I_C + \cdots)/I_s$ versus plate thickness. This function is substantially linear over a wide thickness range, especially if the substrate radiation does not enhance the line of any major plate constituent. The method is particularly useful for platings of alloys of elements of consecutive atomic number, or nearly so.

The relative effectiveness of the substrate-absorption and plate-emission methods depends largely on the specific system, and it is difficult to draw generalizations. For thick layers, the emission method is limited to somewhat less than the thickness through which the plate line can emerge, the absorption method to somewhat less than the thickness through which the substrate line can emerge. (It is assumed that neither method is limited by penetration of the primary beam.) For very thin layers, if the substrate has a short-wavelength line and the layer low absorption—aluminum-clad molybdenum, for example—the emission method is preferable. If the substrate has a long-wavelength line and the layer high absorption—tin-coated steel, for example—the absorption method is preferable. In general, regardless of which of the five methods is used, relatively long-wavelength plate or substrate lines are preferable for thin, low-Z layers, and short-wavelength lines are preferable for thick, high-Z layers.

The thickness of both layers in two-layer platings can be measured. For example, in nickel/copper-plated steel, nickel thickness is determined from a calibration curve of Ni $K\alpha$ intensity *versus* nickel thickness; then copper thickness is determined from Cu $K\alpha$ intensity corrected for Cu $K\alpha$ and perhaps primary absorption in the overlying nickel layer. Alternatively, copper thickness can be determined from a set of curves of Fe $K\alpha$ intensity *versus* copper thickness, each curve for a different nickel thickness. Nondestructive measurement of all of three or more layers is very difficult.

Composition of alloy platings can be determined nondestructively if the plate elements are not present in the substrate and if all samples and standards have the same plate thickness. Mathematical methods have been developed to determine both composition and thickness of alloy platings, but these methods are most successful for binary-alloy platings.

By use of selected-area techniques (Section 10.9.1), analysis of small selected areas and point-by-point measurement of uniformity of composition and/or thickness of films becomes feasible. With the spectrometer set at the spectral line of an element in the film, the specimen may be translated continuously with a motor drive while intensity is recorded as a function of distance in a line across the specimen. Alternatively, the specimen may be moved manually in small increments and intensity measured at each point with the scaler. Successive scans can be made at different places for the same element and/or at the same place for different elements.

Plating thickness can be determined on small fabricated forms and parts (screws, washers, nuts, solder lugs, electronic parts, etc.) by use of selected-area apertures, masks, and wax-impression molds (Figure 10.21).

Plating thickness on wires having diameter 0.01–0.25 mm (0.0004–0.010 inch) has been determined by two methods. One method is applicable to large numbers of specimens consisting of ≳3–5 layers of wire wound on identical spools that fit the spectrometer specimen compartment (Figure 10.2O). With the instrument set at a plate-metal line, each spool is placed in turn in the specimen drawer and the ratemeter allowed to trace plate-metal intensity for 15 s or more. A set of spools wound with standard wires is treated the same way. As many as four specimens can be analyzed each minute. For more accurate analysis, a few meters of wire is wound on a small spool, card, or open frame (Figures 10.2N and O). The intensities of a plate and substrate line are measured and their ratio calculated. A calibration curve of plate/substrate intensity ratio *versus* plate thickness is established from wires of known plate thickness on the same type of support. The ratio method corrects for variations in number, spacing, and slight misorientation of the turns of wire in the specimen. A set of standards is required for each combination of plate metal and substrate, wire diameter, and specimen mount (card, spool diameter, etc.).

Composition and thickness of brass-plated single and stranded steel wires have been determined from measurements on ~40 lengths of wire cemented side by side on a 4×4-cm Lucite card (Figures 10.2M and N). Composition was determined from plots of Cu $K\alpha$/Zn $K\alpha$ net-intensity ratio *versus* wt% copper and Zn $K\alpha$/Cu $K\alpha$ net-intensity ratio *versus* wt% zinc. Plating weight (grams of plating per kilogram of plated wire) was determined from a plot of (Cu $K\alpha$ + Zn $K\alpha$)/Fe $K\alpha$ net-intensity ratio *versus* plating weight.

Plate thickness on irregular-shaped objects is determined most accurately by measurement of the intensity ratio of the $K\alpha$ and $K\beta$ or $L\alpha_1$ and $L\beta_1$ lines of a substrate element, or of two lines from different elements in an alloy substrate. For example, suppose it is required to determine the chromium plate thickness on small irregular-shaped chromium-plated steel parts. The measured Fe $K\alpha$/Fe $K\beta$ net-intensity ratio for the substrate is given by

$$\left(\frac{I_{\mathrm{Fe}K\alpha}}{I_{\mathrm{Fe}K\beta}} \right)_{s,p} = \left(\frac{I_{\mathrm{Fe}K\alpha}}{I_{\mathrm{Fe}K\beta}} \right)_{s} \frac{\exp[-(\mu/\varrho)_{\mathrm{Cr},\mathrm{Fe}K\alpha}\varrho_{\mathrm{Cr}}t]}{\exp[-(\mu/\varrho)_{\mathrm{Cr},\mathrm{Fe}K\beta}\varrho_{\mathrm{Cr}}t]} \qquad (9.70)$$

where $(I)_{s,p}$ is measured substrate-line intensity after emerging through the plating; $(I)_s$ is substrate-line intensity before emerging through the plating; ϱ_{Cr} is the density of chromium; and t is plate thickness. $(I_{\mathrm{Fe}K\alpha}/I_{\mathrm{Fe}K\beta})_{s,p}$ is measured from plated samples; $(I_{\mathrm{Fe}K\alpha}/I_{\mathrm{Fe}K\beta})_s$ is measured from pure, unplated iron, or, preferably, from an unplated specimen; $(\mu/\varrho)_{\mathrm{Cr}}$ for Fe $K\alpha$ and Fe $K\beta$ are known. Thus, all quantities in the equation are known except t.

Uniformity of platings or coatings can be evaluated along the lengths of thin wires and plastic, textile, paper, and thin metal tapes—for example, tapes for audio or video recording and computer memory storage. The wire or tape can be passed through the specimen chamber continuously and the intensity of a plate or substrate line recorded continuously or scaled repetitively. Alternatively, the wire or tape can be indexed and stopped at appropriate intervals and the intensity recorded or scaled. The specimen holder is fitted with feed and takeup spools or reels and wire or tape guides arranged so that the length of wire in the analytical position always occupies precisely the same position and is free of curvature and vibration. The takeup reel is motor-driven. Such work can be done on flat-crystal spectrometers, but the measured intensities are much greater on curved-crystal instruments—even semifocusing ones. Wire and narrow ribbon constitute ideal specimens for curved-crystal spectrometers; wider tapes require a slit.

9.11. TRACE AND MICRO ANALYSIS

Trace analysis is the detection, estimation, or determination of analytes having very low concentrations ($\lesssim 0.1$ wt %)—trace constituents—in relatively large samples. *Microanalysis* is the detection, estimation, or determination of analytes having relatively high concentrations—major and perhaps minor constituents—in very small samples. Detection, estimation, and determination constitute, respectively, qualitative, semiquantitative, and quantitative analysis. In favorable cases, x-ray fluorescence spectrometry, applied on standard, commercial, flat-crystal instruments, can detect concentrations down to ~ 0.0001 wt % in solids and ~ 1 µg/ml in solutions, isolated masses down to ~ 1 ng, and films as "thin" as ~ 0.01 monolayer. Trace and microanalyses approach these levels and consist of the following sample types: (1) very low concentrations, with or without preconcentration; (2) limited total amount; (3) analysis of individual phases in inhomogeneous materials, either after physical separation from the matrix or in place by selected-area techniques; and (4) thin platings, coatings, and films. Preconcentration of trace analytes and physical separation of an individual grain from a heterogeneous specimen usually result in a specimen in the limited-total-amount class.

In x-ray spectrometric trace and micro analysis, all the various sources of error assume increased significance, and errors that may be disregarded in the determination of major and even minor constituents may limit the precision in the determination of traces. Details and refinements of technique unnecessary at higher concentrations must be applied for traces. Thus, x-ray spectrometric trace and micro analyses involve no basically special methods, but rather a refinement of methods used for higher concentrations or larger specimens. This section is intended as a guide to these refinements.

Trace and micro analyses are characterized by low analyte-line intensity, that is, analyte peak intensity approaching background intensity. For 95% (2σ) confidence level, the lower limit of detectability is given by Equation (8.54):

$$C_{DL} = (3/m)(I_B/T_B)^{1/2}$$

where m is the slope of the calibration curve (I_A versus C_A); I_B is background intensity; and T_B is background counting time, which, in this case, is of the same order as analyte-line counting time. Consequently, in trace and micro analysis, background intensity must be measured with the same statistical precision as analyte-peak intensity and the method for measuring background must be chosen carefully (Section 5.3.2.2). The lower limit of determination for quantitative analysis is defined as three times the lower limit of detectability (Section 8.5). It is evident from the equation above that there are three ways to reduce the lower limit of detection and determination: increase analyte-line intensity and thereby m, reduce background intensity, and increase counting time.

The third alternative—increased counting time—is practical only to the point where instrumental instability during the long counting time becomes the factor limiting the precision. It follows that instrumental stability must be high. Methods for increasing the analyte-line intensity are given in Section 8.5. However, of the methods given there, it is necessary to choose those that do not also substantially increase background. Methods for decreasing background are given in Section 5.3.2.3. Here, again, it is necessary to choose methods that do not also substantially decrease analyte-line intensity. Methods for increasing analyte-line intensity and decreasing background fall into two groups: those involving the instrument and its operating conditions and those involving specimen preparation and presentation. If meaningful comparisons of analyte-line intensities are to be made from different samples or from samples and standards, operational and specimen errors must be reduced to a minimum.

The instrumental methods most effective in increasing line intensity for trace analytes include use of: (1) a constant-potential generator for lines having high excitation potential; (2) an x-ray tube target having strong lines close to the short-wavelength side of the analyte absorption edge; (3) x-ray tube operating potential such that the primary continuum hump lies just on the short-wavelength side of the analyte absorption edge; (4) tapered collimation; (5) an analyzer crystal having high "reflectivity" and low emission; and (6) a detector having high quantum efficiency for the analyte line.

The instrumental methods most effective in reducing background intensity for trace analytes include use of: (1) filters to reduce continuum intensity in the region of the analyte line; (2) an analyte line at relatively high 2θ where background is low; (3) an evacuated or helium-flushed specimen chamber to eliminate air-scatter of the primary beam; (4) baffles, collimators, and/or apertures to confine the primary beam and intercept background; (5) spectrometer geometry such that the crystal axis and specimen plane are not parallel; and (6) pulse-height selection.

In the determination of traces homogeneously distributed in a large specimen, invariably the detection limit is reduced by preconcentration by chemical or physical means. The chemical methods include precipitation (selective or collective), selective dissolution, ion-exchange (on granules, membrane, paper, or liquid), liquid–liquid extraction, paper chromatography, paper electrophoresis, electrodeposition, wet or dry ashing, and gas formation. The physical methods include distillation, sublimation (including use of laser beams), prying with needles, microdrilling, and extraction replication.

In the determination of homogeneous traces in large specimens without preconcentration, it is often preferable to measure a small area with a pinhole or slit aperture and curved-crystal optics rather than the large area with flat-crystal optics.

For a trace analyte, emitted spectral-line intensity is directly proportional to concentration, and the calibration curve is linear. The matrix determines only the slope of the curve; the lower the mass-absorption coefficient of the matrix for the analyte line, the steeper is the calibration curve. Trace analytes, unlike those at higher concentrations, can often be determined by use of standards having a matrix entirely different from that of the samples. From Equation (8.4), analyte-line intensities from a sample I_X and standard I_S are related by

$$\frac{I_X}{I_S} = \frac{\overline{(\mu/\varrho)_S}}{\overline{(\mu/\varrho)_X}} \approx \frac{(\mu/\varrho)_S}{(\mu/\varrho)_X} \tag{9.71}$$

where $\overline{(\mu/\varrho)}$ and (μ/ϱ) are the mass-absorption coefficients for, respectively, combined primary and analyte-line x-rays [Equation (2.6)] and analyte-line x-rays alone. From this, trace-analyte concentration in a sample C_X is obtained from a standard having analyte-line concentration C_S in a different matrix from

$$C_X = \frac{I_X \overline{(\mu/\varrho)_X}}{I_S \overline{(\mu/\varrho)_S}} C_S \qquad (9.72)$$

The limited-total-amount category includes isolated particles and fibers microquantities of analyte separated from a matrix or recovered from a solution or gas, and individual phases in heterogeneous specimens. Such specimens present two special problems: (1) They must be supported—and in such a way as to give maximum excitation of the analyte and minimum background, and (2) the x-ray optical system must collect as much of the emitted analyte-line radiation as possible.

In supporting microquantities of analyte, scattered background is minimal when the analyte is suspended as nearly as possible in "free space." The support should preferably be as thin and of as low atomic number as possible to reduce primary beam scatter. Thin filter paper is preferable to thick filter or chromatographic paper, 6- or 3-μm Mylar is preferable to filter paper, and stretched (1-μm) polypropylene or ultrathin Formvar are even better. Millipore filter disks and ion-exchange membrane or impregnated filter paper are also useful. All such x-ray-transparent supports must be stretched over open frames or across the open ends of short cylindrical tubes or liquid-specimen cups. Otherwise, the purpose of the thin substrates is defeated by primary-beam scatter from underlying material. However, enhancement radiators may be used behind such supports to excite additional analyte-line radiation.

Many successful methods have been based on large-area supported specimens used with flat-crystal optics. However, it is almost invariably preferable to confine the preconcentrated analyte(s) to a small area and use selected-area apertures and curved-crystal optics. The small area may be circular (1–4 mm in diameter), as in the "Coprex" method (Section 10.8.1). Or the analyte may be electrodeposited on a pure metal wire or carbonized quartz fiber, held in a thin-walled plastic capillary, absorbed in a cotton fiber, or affixed to a cotton, plastic, or quartz fiber with Vaseline.

SUGGESTED READING

ADLER, I., and J. M. AXELROD, "Internal Standards in Fluorescent X-Ray Spectroscopy," *Spectrochim. Acta* **7**, 91–99 (1955).

ANDERMANN, G., and J. W. KEMP, "Scattered X-Rays as Internal Standards in X-Ray Emission Spectroscopy," *Anal. Chem.* **30**, 1306–1309 (1958).

BEATTIE, H. J., and R. M. BRISSEY, "Calibration Method for X-Ray Fluorescence Spectrometry," *Anal. Chem.* **26**, 980–983 (1954).

BERTIN, E. P., "Intensity-Ratio Technique for X-Ray Spectrometric Analysis of Binary Samples," *Anal. Chem.* **36**, 826–832 (1964).

BERTIN, E. P., *Principles and Practice of X-Ray Spectrometric Analysis*, 2nd ed., Plenum Press, New York (1975); Chaps. 14, 15, and 18, pp. 571–697, 811–828.

BERTIN, E. P., and R. J. LONGOBUCCO, "X-Ray Methods for Determination of Plate Thickness," *Metal Finish.* **60**(8), 42–44 (1962).

BURNHAM, H. D., J. HOWER, and L. C. JONES, "Generalized X-Ray Emission Spectrographic Calibration Applicable to Varying Compositions and Sample Forms," *Anal. Chem.* **29**, 1827–1834 (1957).

CRISS, J. W., and L. S. BIRKS, "Calculation Methods for Fluorescent X-Ray Spectrometry—Empirical Coefficients vs. Fundamental Parameters," *Anal. Chem.* **40**, 1080–1086 (1968).

EBEL, H., "Quantitative X-Ray Fluorescence Analysis with Variable Take-Off Angle," *Advan. X-Ray Anal.* **13**, 68–79 (1970).

JENKINS, R., *Introduction to X-Ray Spectrometry*, Heyden, London (1974); Chap. 7, pp. 120–140.

JENKINS, R., and J. L. DE VRIES, *Practical X-Ray Spectrometry*, 2nd ed., Springer-Verlag New York, Inc., New York (1967); Chap. 7, pp. 126–144.

JONES, R. A., "Determination of Manganese in Gasoline by X-Ray Emission Spectrography," *Anal. Chem.* **31**, 1341–1344 (1959).

LACHANCE, G. R., and R. J. TRAILL, "A Practical Solution to the Matrix Problem in X-Ray Analysis. I. Method," *Can. Spectrosc.* **11**(2), 43–48 (1966).

LARSON, J. A., W. R. PIERSON, and M. A. SHORT, "Corrected Equation for the Standard-Addition Technique in X-Ray Fluorescence Spectrometric Analysis," *Anal. Chem.* **45**, 616 (1973).

LIEBHAFSKY, H. A., H. G. PFEIFFER, E. H. WINSLOW, and P. D. ZEMANY, *X-Rays, Electrons, and Analytical Chemistry*, Wiley–Interscience, New York (1972); Chap. 9, pp. 355–459.

LUCAS-TOOTH, H. J., and B. J. PRICE, "A Mathematical Method of Investigation of Interelement Effects in X-Ray Fluorescent Analysis," *Metallurgia* **64**, 149–152 (1961).

LUCAS-TOOTH, H. J., and E. C. PYNE, "Accurate Determination of Major Constituents by X-Ray Fluorescent Analysis in the Presence of Large Interelement Effects," *Advan. X-Ray Anal.* **7**, 523–541 (1964).

MARTI, W., "Determination of the Interelement Effect in the X-Ray Fluorescence Analysis of Steels," *Spectrochim. Acta* **18**, 1499–1504 (1962).

MITCHELL, B. J., "X-Ray Spectrographic Determination of Zirconium, Tungsten, Vanadium, Iron, Titanium, Tantalum, and Niobium Oxides—Application of the Correction Factor Method," *Anal. Chem.* **32**, 1652–1656 (1960).

MITCHELL, B. J., and F. N. HOPPER, "Digital Computer Calculation and Correction of Matrix Effects in X-Ray Spectroscopy," *Appl. Spectrosc.* **20**, 172–180 (1966).

MÜLLER, R. O., *Spectrochemical Analysis by X-Ray Fluorescence*, Plenum Press, New York (1972); Chaps. 12–17, pp. 133–211.

PAPARIELLO, G. J., H. LETTERMAN, and W. J. MADER, "X-Ray Fluorescent Determination of Organic Substances through Inorganic Association," *Anal. Chem.* **34**, 1251–1253 (1962).

RASBERRY, S. D., and K. F. J. HEINRICH, "Calibration for Interelement Effects in X-Ray Fluorescence Analysis," *Anal. Chem.* **46**, 81–89 (1974).

SHERMAN, J., "Theoretical Derivation of the Composition of Mixable Specimens from Fluorescent X-Ray Intensities," *Advan. X-Ray Anal.* **1**, 231–250 (1958).

TAYLOR, A., B. J. KAGLE, and E. W. Beiter, "Geometrical Representation and X-Ray Fluorescence Analysis of Ternary Alloys," *Anal. Chem.* **33**, 1699–1706 (1961).

TRAILL, R. J., and G. R. LACHANCE, "A New Approach to X-Ray Spectrochemical Analysis," *Geol. Surv. Can. Pap.* **64-57**, 22 pp. (1965).

TRAILL, R. J., and G. R. LACHANCE, "A Practical Solution to the Matrix Problem in X-Ray Analysis. II. Application to a Multicomponent Alloy System," *Can. Spectrosc.* **11**(3), 63–71 (1966).

Chapter 10

Specimen Preparation and Presentation

10.1. INTRODUCTION

A modern, automatic, sequential x-ray spectrometer collects count data at a rate of 30–100 s per element; semiautomatic and manual instruments are somewhat slower because the settings cannot be made as quickly. A modern, automatic, simultaneous spectrometer collects count data for up to 30 elements in 30–200 s; it has been remarked that simultaneous instruments collect data so fast that the readout time may become significant! An energy-dispersive spectrometer with a multichannel analyzer collects data for up to ~20 elements in 30 s to a few minutes. Clearly, specimen preparation is usually the time-limiting step that determines the sample throughput rate, and when the samples can be measured as received—or substantially so—x-ray spectrochemical analysis is very rapid indeed.

X-ray secondary-emission spectrometry is probably the most versatile of all instrumental analytical chemical methods with respect to the variety of specimen forms to which it is readily applicable nondestructively. Many forms of specimen can be analyzed in the x-ray spectrometer without treatment and without any special mounting arrangement. Other specimens require treatment and/or mounting cells or supports.

In general, specimen considerations, requirements, preparation, and presentation are substantially the same for wavelength- and energy-dispersive and nondispersive x-ray spectrometry, provided that the excitation is by x-radiation. However, if the excitation is by electrons, β-radioisotopes, protons, or other ions, the exciting radiation does not penetrate as deeply into the specimen surface. In these cases, surface-texture and particle-size effects become more severe, and the specimen preparation must be directed more toward dealing with these effects. Also, whereas the entire gamut of specimen forms is commonly used in wavelength-dispersive spectrometry, supported specimens of one kind or another are by far the most commonly used form in energy-dispersive spectrometry.

In this chapter, the term specimen *preparation* indicates any treatment—other than simply cutting to appropriate size—to which the as-received sample is subjected prior to its presentation to the spectrometer. The term *presentation* indicates the physical form in which the specimen is placed in its compartment, that is, in which it is "presented" to the spectrometer. Specimen preparation may be physical, chemical, or both, and it may involve treatment of the specimen in its present form or conversion to another form. The treatment must be applicable to both samples and standards. The terms *analyte, specimen, sample,* and *standard,* and *major, minor,* and *trace constituent* are defined in Section 3.1, and the term *nondestructive,* as applied to x-ray spectrometric specimens, is defined in Section 3.5.2.

The primary beam from an x-ray tube operating at 50 kV contains wavelengths as short as 0.25 Å [Equation (1.17)], giving primary-beam penetration and analyte-line excitation in a *relatively* thick layer on the specimen surface. However, only analyte-line x-rays actually emerging from the specimen can be measured. Thus, for x-ray (and γ-ray) excitation, the *effective layer thickness* is determined not by the depth *to* which *primary* x-rays can *penetrate,* but by the depth *from* which *analyte-line* x-rays can emerge. In the practical analysis of metals, minerals, ceramics, etc., the effective layer thickness is typically 10-100 μm. However, for short-wavelength lines in very light matrixes, the effective layer may be much thicker. Effective layer thickness for 0.1–10-Å x-rays in various matrixes is given in Figure 10.1. This figure must be used carefully. For example, the figure shows that the effective layer for Cu $K\alpha$ (1.54 Å) in water is ∼5 mm; however, the effective layer in a water *solution* of heavy solutes would be much less. It is evident that the effective layer is different for different analyte lines, and this layer is thinner the longer the analyte-line wavelength and the higher the absorption coefficient of the specimen for that wavelength.

For electron, proton, and other ion excitation, the effective layer thickness is determined by the depth to which electrons or ions can penetrate with enough energy to excite the specified analyte line. For electron excitation, the layer thickness is typically of the order 0.01–10 μm (Figure 11.2), for ions, as little as a monolayer.

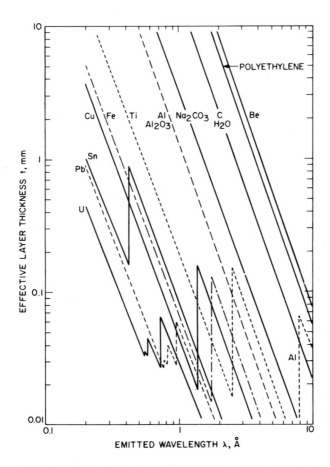

FIGURE 10.1. Effective layer thickness (layer that emits 99.9%
of the measured analyte-line intensity) as a function of wavelength for various matrixes. The curves are calculated for (effective) primary wavelength much shorter than analyte-line
wavelength, both these wavelengths on the same (either) side
of the matrix absorption edge, primary absorption much less
than analyte-line absorption, and incident and takeoff angles
both 45°. [K. Togel, in *Zerstörungsfreie Materialprüfung*, R.
Oldenbourg, Munich, Germany (1961); courtesy of the author
and publisher.]

10.2. SPECIMEN FORMS

The specimen presentation forms may be classified in seven categories as follows:

1. *Solids*, including (a) unfinished ingot, mineral, etc.; (b) finished bar, rod, sheet, foil, ceramic, glass, plastic, etc.; and even (c) rubber, wood, paper, and textile products. Bulk (large) specimens may also be included in this group.

2. *Small fabricated forms and parts*, including rod, wire, shot, screws, nuts, washers, lugs, gears, electron-tube and transistor components, and other parts of regular or irregular shape.

3. *Powders.*

4. *Briquets.*

5. *Fusion products* (solid solutions) resulting from fusion of the sample material with fluxes, such as potassium pyrosulfate, or with glass-forming fluxes, such as lithium or sodium tetraborate.

6. *Liquids and solutions.*

7. *Supported specimens*, including those distributed on or in filter paper, Millipore and Nuclepore filters, Fiberglas, Mylar, Scotch tape, ion-exchange membrane and paper, etc., or on fine quartz or plastic fibers. Thin films and platings may also be included in this group.

Usually, nondestructive x-ray spectrometric analyses of samples in their as-received forms are more rapid and convenient than analyses by any other method. And since the x-ray spectrometer can accommodate almost any specimen form, the analyst is well advised to strive to analyze samples in their original form whenever practical. In fact, one might wonder why it should ever be necessary to alter the specimen form at all—that is, why any preparation other than simple physical treatment of the specimen in its present form should ever be required. Actually, for qualitative x-ray spectrometric analysis, usually the only requirement is that the specimen fit its compartment, and preparation other than, say, provision of a flat surface on a solid may well be unnecessary. However, for quantitative analysis, elaborate preparation and conversion of form may be required to deal with one or more of a large number of conditions, which may be classified in the six groups listed below. When such elaborate preparation is required, much of the speed and convenience of the x-ray spectrometric method is lost.

1. *Specimen errors must be reduced* when: (a) samples are very inhomogeneous; (b) the surface layer is not representative of the bulk sample;

(c) the average particle size or the particle-size distribution varies among the samples; (d) the amount of porosity or pore size varies among the samples; and/or (e) surface texture is unsuitable or varies among the samples.

2. *Additions must be made* to the samples when: (a) absorption-enhancement effects are severe and an internal standard must be added; (b)—or a matrix-masking agent must be added; (c)—or the sample must be diluted; (d) an internal control standard or intensity-reference standard must be added; or (e) a standard-addition or standard-dilution method is to be used. These addition methods are usually not feasible with specimens in the form of solids or fabricated parts.

3. *Chemical separation or other treatment is required* when: (a) the analyte(s) must be preconcentrated or separated; (b) interferants must be separated; or (c) the analyte(s) must be determined by an indirect or association method (Section 9.8).

4. *Standards must be matched.* Sometimes, it is necessary to change the as-received specimen form even though it is entirely suitable for analysis. This is the case when: (a) the sample shape is completely irregular and varies among the samples; (b) standards are not available or readily prepared in the as-received form; or (c) the same substance is routinely analyzed in a number of forms—such as bar, drillings, shot, parts, wire, powder—and conversion to a common form—such as solution or fusion product—permits all analyses to be made with a single set of standards.

5. The x-ray spectrometric method is to complement some other analytical method.

6. The sample may be unsuitable because of chemical instability, hygroscopicity, sensitivity to heating or radiolysis by the x-ray beam, etc.

10.3. STANDARDS

10.3.1. Permanence

Almost all methods of quantitative x-ray spectrochemical analysis (Chapter 9) require calibration standards of accurately known composition. It has already been mentioned that standards for x-ray spectrometry must be similar to the samples to be analyzed, not only in analyte concentration, but in matrix composition, physical form, surface texture, particle size and distribution, packing density, etc. Fortunately, because of the substantially nondestructive nature of x-ray spectrochemical analysis, once one has obtained a set of solid, small-part, briquet, fusion-product, supported, or plated standards, they should last indefinitely if carefully handled and stored. Liquid standards are the exception; they may evaporate or deteriorate on long storage, but even if not, it is customary to discard each portion after exposure to the excitation x-rays. However, there are also many exceptions to the permanence of nonliquid standards, which may deteriorate from many causes, principally the following.

Radiation damage may occur on repeated exposure to the intense primary x-ray beam; this deterioration takes the form of surface pulverization of solids; formation of cracks, voids, and porosity in plastics and glasses; and embrittlement of Mylar, Millipore, filter-paper, ion-exchange membrane, and other supports.

Surface oxidation may occur on repeated exposure to the ozone (O_3) formed by the primary x-ray beam in the specimen chamber.

Chemical change may occur on storage. Deterioration of liquids is mentioned above. All forms of standards may undergo chemical change by reaction with oxygen, moisture, carbon dioxide, and sulfur compounds in the air. These reactions are most serious with thin-film and plated standards and with standards having several phases that may react preferentially.

Mechanical damage may occur on repeated handling, storage, and insertion. Solid standards may become scratched, abraded, or smeared. Briquets may powder, crumble, swell, or crack, but this damage is minimized by use of binders, backings, rings, or metal cups, or by lightly spraying the edges and faces with clear, colorless Krylon or other plastic coating.

Finally, standards may undergo *aging*—deterioration simply on passage of time. Low-melting alloys may undergo diffusion. Glass disks may devitrify, craze, or crack from stress.

Because of these processes, it may be necessary to refinish or reprocess standards occasionally. Solids and glass disks may be resurfaced. Briquets may be resurfaced or repulverized and compacted. Powders may be repacked.

10.3.2. Sources

X-ray spectrometric standards are obtained by purchasing, analysis of selected samples, quantitative concentration or dilution of a material of known composition, or synthesized.

Standards may be purchased from: (1) the National Bureau of Standards and from the corresponding agencies of other countries; (2) optical emission spectrographic supply firms; (3) manufacturers of the type of material to be analyzed; for example, standards for aluminum-base alloys are available from Aluminum Company of America; (4) professional technical societies and institutes concerned with the type of material to be analyzed; and (5) other analytical laboratories engaged in analysis of the same material. A compilation of such sources is available.

Samples analyzed very carefully by independent methods may be used as standards. The "independent method" may be an x-ray spectrometric solution or fusion method using synthetic standards. The homogeneity of the prospective standards should be established; for example, if the samples are metal coupons, both sides should be finished, and x-ray spectrometric measurements of analyte-line intensity should be made from both sides with the coupon in various orientations, or with and without rotation. If the coupons are thus shown to be homogeneous, turnings or drillings are taken from one side for analysis. Alternatively, two or three consecutive adjacent slices may be cut from the sample and all sides finished and tested for uniformity in the same way. Then one slice is retained, and the adjacent one(s) sacrificed for analysis. Standards for nondestructive analysis of small fabricated forms and parts are obtained in a similar way. Several parts from a lot are tested for uniformity of composition by x-ray spectrometric measurement of analyte-line intensities from each. If the parts all give the same intensities, several are sacrificed for analysis, the others retained as standards. Alternatively, standard parts may be fabricated from material of known composition.

If a single powder or liquid of the type to be analyzed and of accurately known composition is available, several standards may be prepared from it by quantitative addition of analyte(s) or of analyte-free matrix material or solvent to form more and less concentrated standards, respectively. Analyte(s) may be added to the powder by: (1) quantitative admixture of powder having accurately known analyte concentration and substantially the same particle size as the base powder; or (2) volumetric addition of a standard analyte solution, followed by homogenizing, drying, and grinding.

Synthesis of standards is most convenient for powders, briquets, fusion products, and liquids. However, small melts of alloys, especially low-melting alloys, and small firings of ceramics and glasses can be made. Powder standards are prepared by admixture of pure individual powder components or from solution by precipitation or evaporation, then firing. Briquet standards are made from standard powders. Fusion products and solutions are most easily synthesized because all inhomogeneities and particle-size effects disappear completely. The fusion-product standards may be in the form of glass disks, or they may be ground and briqueted.

Solution standards are synthesized by dissolution of a solid or powder standard, or pure individual solutes. Metallic elements are added to organic liquids as organometallic compounds, such as naphthenates or salts of organic acids. Silicon, phosphorus, and sulfur, and fluorine, chlorine, bromine, and iodine are added as their organic compounds.

Plastic standards are made by adding analyte(s) in the form of organic compounds to solutions of the plastic or to the monomer, which are then evaporated or polymerized, respectively.

Plated standards are prepared by electroplating a relatively large area of substrate having its back, edges, and a 1-cm border on its front face masked with insulating tape. These precautions avoid useless plating of the back and the nonuniformity that always occurs at the edges. Calculations based on the electrolysis laws and the exposed area permit plating to approximately the required thickness. Several coupons are cut from the plated area. The uniformity of the plating is verified by x-ray spectrometric measurement of the intensity of a plate-metal line. Then one or more of the coupons are sacrificed for chemical analysis, the others retained as standards.

Thin-film standards are prepared in several ways. An analyte solution may be sprayed from an atomizer on a blank substrate and area concentration calculated from gain in weight and composition of the sprayed chemical species. The analyte may be vacuum-evaporated on a blank substrate, and area concentration derived from interference colors or fringes, optical transmission, or gain in weight. Alternatively, several substrate blanks may be exposed at geometrically equivalent positions, the uniformity of the deposits verified by x-ray spectrometric measurement, and one or more of the blanks sacrificed for chemical analysis. A weighed portion of analyte may be evaporated completely and area concentration calculated from the geometry, that is, from the solid angle intercepted by the substrate. A measured volume of a slurry having a known weight/volume analyte concentration can be applied to a large, optically flat glass plate and spread into a thin film by means of a casting knife ("doctor blade"). The area of the film is measured with a planimeter or by tracing the outline of the film on graph paper and counting squares. The area concentration is then calculated. Blanks can be removed from the film by scoring, flooding with water, and "coaxing" squares of the film to float off. Films having various area concentrations are made by varying either the analyte concentration in the slurry or the film thickness, or by stacking various numbers of layers of film. A film may be placed on a substrate and moistened with solvent to affix it. The substrate may be plane, cylindrical, or of other form.

Standards for analysis of supported samples are prepared by volumetric addition of standard analyte solutions to blanks of the support material. The technique is also applicable to standards for analysis of paper products, textiles, etc. Particulate material filtered from gases or liquids is usually retained mostly on the filter surface, and longer-wavelength radiation is adsorbed less than for disks prepared as described above, where the analytical material is distributed throughout the thickness of the support. Standards for such samples may be prepared by filtering dilute suspensions of insoluble analyte compounds onto the filters. Standards for aerosol samples are also prepared by bubbling clean air through solutions of analyte salts for various times and passing the spray-laden air through weighed filter disks, which are then dried and reweighed.

The remaining sections of this chapter consider specimen preparation for solids (including fabricated forms and parts), powders and briquets, fusion products (solid solutions), liquids, supported specimens, and certain special specimen types. For each specimen type, there is a consideration of sources of samples, alternative methods of presentation, favorable and unfavorable features, treatment, precautions and considerations, and, perhaps, descriptions of certain specific techniques. However, emphasis is on generalities, rather than specific applications and materials.

10.4. SOLIDS

10.4.1. Scope, Advantages, Limitations

The term *solid* here applies to bulk solid, as distinguished from powders, granules, drillings, chips, shot, and small pellets. The solid may be of metal, mineral, rock, ore, slag, ceramic, glass, plastic, etc., or it may be of any material that can be formed into a "solid" specimen, such as wood, rubber, etc. The sample may be taken from unfinished ingot, stock, rock, etc., or from finished plate, sheet, bar, rod, tube, wire, or fabricated parts.

Solid specimens may be presented to the spectrometer as such, or they may be reduced to powders, briquets, fusion products, or solutions. This section is limited to presentation as solid.

If the material can be analyzed in solid form, the analyses are usually very rapid and convenient. Small fabricated parts may require no preparation. Finished materials may require only cutting to appropriate size. Unfinished bulk materials may require only the cutting of a coupon and finishing of one of its surfaces.

The principal disadvantage of the solid form is that any analytical method involving an addition is likely not to be feasible—standard addition and dilution, low- and high-absorption dilution, internal standardization, or internal intensity-reference standardization. The last two methods may be feasible if a suitable element is already present in all specimens in appropriate constant concentration. Chemical concentration or separation is also not feasible. Surface texture and composition are sometimes difficult to reproduce. Standards may not be available and are usually difficult to synthesize.

10.4.2. Presentation

For qualitative analysis, it is usually sufficient that the sample fit in the specimen chamber or in a large-specimen accessory. However, for quantitative and even semiquantitative analysis, it is essential that for all samples and standards in a series, the effective specimen surfaces be identical in all respects—area, dimensions, inclination, surface texture and orientation, etc. Most solid x-ray spectrometric samples are objects that already have a flat surface, or they are rectangular or circular coupons cut from stock and finished on one side, if necessary. Some specimens of this form are shown in *A–G* in Figure 10.2. *H* represents polished metallographic sections of small parts or devices "potted" in plastic. These are best analyzed by use of selected-area techniques (Section 10.9.1).

Wax-impression molds (*I* in Figure 10.2) are made by cutting or casting rectangular blocks of sealing wax, warming them on a hot plate to the softening point, then pressing a part of the type to be analyzed into the soft wax. After the wax has hardened, the part is removed, leaving its impression. The mold is then mounted in the spectrometer specimen chamber with the impression centered in the mask window. The mold permits a series of sample and standard parts to be placed reproducibly. The method has been used for determination of plating or "tinning" thickness on small plated parts.

J represents small cylindrical or filamentary parts inserted in wax-filled cups, and *K* represents small disk-like parts cemented on small spindles; the cups and spindles are placed in a specimen drawer fitted for specimen rotation. *L* shows small, stiff rods, which may be placed in V-grooves in the specimen mask. Alternatively, several small rods may be cemented side by side on a stiff card, as shown in *M*. *N* and *O* are fine-wire specimens wound on cards or small spools (Section 9.10).

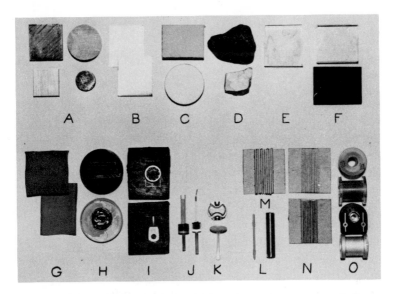

FIGURE 10.2. Typical solid-specimen forms. [E. P. Bertin and R. J. Longo-bucco, *Norelco Reporter* **9**, 31 (1962); courtesy of Philips Electronic Instruments, Inc.] (*A*) Metal coupons and stock. (*B*) Metal sheet and foil. (*C*) Ceramic stock. (*D*) Rock, mineral, ore, etc. (*E*) Glass stock. (*F*) Plastic stock. (*G*) Rubber stock. (*H*) Metallographic sections. (*I*) Small parts held in wax-impression molds. (*J*) Small parts held in wax-filled plastic cups (rotable). (*K*) Small parts held on plastic pedestals (rotable). (*L*) Small rod. (*M*) Small rod or stiff wire cemented on card. (*N*) Wire wound on cards. (*O*) Wire wound on small spools (rotable).

10.4.3. Preparation

The finished stock or wheel-cut surfaces may be satisfactory, but the latter usually require finishing to reduce surface roughness and/or remove work-damaged or nonrepresentative surface layers. Preliminary finishing may be done on a belt grinder. Final finishing may be done in many ways, including the following: (1) polishing with progressively finer wet or dry abrasive papers on a strip grinder; (2) polishing with progressively finer papers or slurries on a manual or vibratory polishing wheel; (3) machining on a lathe or grinder; (4) electropolishing; (5) chemical etching; or (6) spark planing. In general, electropolishing and etching are avoided because of possible preferential removal of certain constituents. A final finish of ~100 μm may be satisfactory for measurement of short-wavelength lines, 10–50 μm for lines of long wavelength down to Mg $K\alpha$ and F $K\alpha$. It is especially important that all samples and standards in a related series be given the same finish.

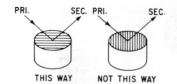

FIGURE 10.3. Orientation of parallel grind marks.

Surface-texture, particle size, and heterogeneity effects are reduced by rotation of the specimens during measurement. If rotation is not provided, specimens must be loaded with parallel grind marks parallel to the plane defined by the directions of the primary and secondary beams, as shown in Figure 10.3.

The surface of a solid may not be representative of the bulk composition, and in such cases, it must be decided whether analysis is required of the surface, the bulk material, or both. Moreover, surface texture and composition may be difficult to reproduce from specimen to specimen. The remainder of this section is devoted to this problem.

Removal of inclusions may occur during the surface treatment, causing low analytical results for the removed constituent(s). Pickup of surface contamination or abrasive may also occur, and all traces of abrasive, lubricant, and cutting fluid must be removed. Use of aluminum oxide, silicon carbide, and cerium oxide abrasives may result in contamination by these elements, and if the abrasive is supported on a lead- or tin-coated polishing wheel, these elements may be picked up. Electropolishing may result in selective removal of certain constituents and/or in contamination by constituents in the electrolyte. Chemical etches are usually avoided for the same reasons.

Oxidation and other corrosion must be removed. This oxidation and corrosion may be preferential, so that their removal may result in a non-representative surface. Pores and other cavities usually decrease analyte-line intensity in proportion to the surface area they represent. However, cavities may increase analyte-line intensity if the walls of the cavities happen to be enriched in the analyte. Cast and rolled surfaces are very poor for x-ray spectrometric analysis. They are very inhomogeneous; they contain oxide layers, impurities, and inclusions; and their texture may be very coarse-grained and porous. When metal specimens are heated, preferential diffusion of one or more constituents to the surface and/or preferential oxidation may occur. If the metal is heated strongly, sublimation of volatile constituents may occur.

A very common surface problem arises when the specimen is a multi-phase system having one or more soft phases. Any surface preparation method involving movement of a cutting tool or abrasive powder across the surface may result in smearing of the soft constituent(s). This causes high analytical results for the smeared constituents and low results for the covered constituents. The latter effect is more severe the higher the absorption coefficient of the smeared material and the longer the analyte-line wavelength of the covered material. Surface smearing effects are also likely to be more severe with a chromium-target x-ray tube than with molybdenum, rhodium, silver, tungsten, platinum, or gold tubes.

Etching, electropolishing, and spark planing do not involve cutting or abrasion and so do not cause smearing. However, etching and electropolishing may act preferentially on certain constituents. Smearing is less likely with machining than with polishing. When it does occur, it may be removed by final light polishes with 6-μm, then 0.25-μm diamond dust with kerosene lubricant on a cloth polishing wheel. Alternatively, the smear may be removed by a light chemical etch specific for the smeared constituent or by electropolishing or spark planing.

Reduction of surface smearing of soft alloys, such as tin–lead–antimony alloys, has also been effected by cooling the blanks to liquid-nitrogen temperature during surface finishing on a lathe, and by pressing the finished disks in vacuum against a mirror-smooth surface.

A good test for such superficial smearing is to measure, for a covered element, the $L\alpha_1/K\alpha$ intensity ratio for elements up to atomic number $\sim$60 and the $M\alpha/L\alpha_1$ ratio for the heavier elements. This is done for the possibly smeared specimen and for another specimen of the same alloy known not to be smeared, or even from the pure element. The intensity of the longer-wavelength line ($L\alpha_1$ or $M\alpha$)—and thereby the ratio—is decreased by any overlying smeared constituent. It follows that the technique also serves as a test for removal of the smear.

Another difficulty with soft constituents is the possibility that fine particles of abrasive may become imbedded and packed in soft surface grains, thereby "plugging" them.

10.5. POWDERS AND BRIQUETS

10.5.1. Scope, Advantages, Limitations

Powder samples are derived from: (1) powders; (2) bulk solids—minerals, rocks, ores, slags, etc., pulverized to reduce heterogeneity, to permit addition of an internal standard, or for some other reason; (3) metal filings, chips, drillings, and turnings; (4) metals converted to oxide;

(5) precipitates and residues from solutions; (6) ground fusion products; (7) ion-exchange resins; and (8) ashed, oven-dried, or freeze-dried (lyophilized) organic and biological material.

Powders can be presented to the instrument as loose powder packed in cells, as briquets, or in thin layers; these three forms are discussed together in this section because they have in common the feature that the powder particles retain their individual identities. Fusion products are usually prepared from powders, but because the particles disappear on fusion, these melts constitute a basically different type and are considered separately in Section 10.6.

The principal disadvantage of the solid specimen form—the inapplicability of addition methods—does not apply to powders. Standard addition and dilution, low- and high-absorption dilution, internal standardization, and internal intensity-reference standardization are all readily applied to powders. The applicability of these methods permits great flexibility in dealing with absorption-enhancement effects. Standards are prepared relatively easily. Wherever applicable, powder methods are usually rapid and convenient.

The principal disadvantages of powder specimens are the following: Trace impurities may be introduced by the grinding and briqueting operations, especially when the specimen powder is abrasive. The surface texture of loose-packed powders may be difficult to reproduce, but the difficulty is surmounted by use of back-loading cells and substantially eliminated by briqueting. Some powders are hygroscopic or react with oxygen or carbon dioxide in the air; such powders are best handled in Mylar-covered cells. Other powders have low cohesion and flow when the specimen plane is inclined; such powders are placed in Mylar-covered cells, mixed with binder and packed in open cells, or briqueted with binder. Powders may outgas in vacuum spectrometers, sometimes ruining the specimen surface and spattering into the chamber; this difficulty is dealt with by use of helium path, Mylar-covered cells, cells covered with microporous film, previous outgassing, or vacuum briqueting.

However, the most serious difficulties with powders arise from particle-size effects. The spectral-line intensity of a specified element in a powder depends not only on the concentration of that element, but also on the particle size. The intensity is constant for extremely small particles up to a certain size, then decreases with increasing particle size. The size above which intensity begins to decrease varies with wavelength and is larger the shorter the wavelength. If the powder is not homogeneous, particles of different size, shape, density, or composition tend to segregate.

10.5.2. Preparation

For upright and inclined specimen planes, loose powders are usually placed in shallow, metal or plastic, circular, dish-like disposable cells. The cell may be slightly overfilled, then leveled by drawing the edge of a spatula across the rim, or it may be made heaping full, then packed and surfaced with the flat side of a wide spatula. In the latter case, if the cells are loaded with the same weight or volume of powder, the specimens will have substantially uniform packing density. For inverted geometry, the filled cells just described may be covered with 6- or 3-μm (0.00025- or 0.00012-inch) Mylar film secured with an O-ring. More conveniently, a liquid-specimen cup may be used. A measured amount of powder is placed in the cell, which is then covered with Mylar film. Microporous film, which permits equalization of pressure inside and outside the cell, may be used for vacuum spectrometers.

The powder may be formed into a self-supporting pellet or *briquet* in a hydraulic press at pressure up to 7000 kg/cm^2 (100,000 lb/inch^2). For a series of samples and standards, uniform packing density is achieved by briqueting the same weight of powder at the same pressure for each. If the briquets are to be measured in a vacuum spectrometer, they should be briqueted under vacuum. The powder may be briqueted as is, with a binder (such as boric acid, starch, or cellulose), with a cellulose or other backing, or in a shallow aluminum cup. The edges and surfaces of briquets may be sprayed with clear, colorless Krylon or 10% (wt/vol) collodion in acetone to prevent flaking and crumbling.

The effects of particle size and inhomogeneity and absorption-enhancement effects disappear if the powder is ground to extremely fine particle size and distributed in a thin layer. Under these conditions, the powder assumes, to a certain extent, the nature of a thin film. Another advantage of the thin-layer presentation is realized when only a small piece of material is available. The measured analyte-line intensity is greater if the small piece is reduced to a fine powder and spread out, as shown in Figure 10.4. Powders may be spread into thin layers dry or as slurries.

The dry technique can be applied as follows. The powder can be dusted on the adhesive surface of a piece of Scotch tape. In another technique, a 6-μm (0.00025-inch) Mylar film is stretched across an empty plastic liquid-specimen cup and secured with an O-ring, in the manner of a drum. A measured amount of powder is placed on the Mylar and distributed as uniformly as possible. A second Mylar film is now stretched over the cup and secured, enclosing the powder layer between the two films. Evacuation

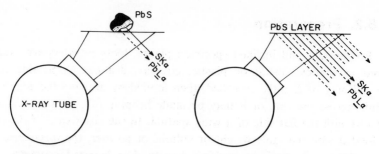

FIGURE 10.4. Effect of specimen presentation on S $K\alpha$ (5.37 Å) and Pb $L\alpha_1$ (1.18 Å) line intensity measured from lead sulfide. For the particulate specimen, only the unshaded region contributes to the measured S $K\alpha$ intensity, the unshaded and dotted regions to the Pb $L\alpha_1$ intensity. If the specimen is dispersed as a film, the entire mass contributes to the measured intensity of both lines. [R. Jenkins and J. L. de Vries, *Practical X-Ray Spectrometry*, Springer-Verlag New York, New York (1967); courtesy of the authors and publisher.]

holes must be punched in the cup for use in vacuum path. A cell for this application is described in Section 10.7.2 and shown in Figure 10.9.

A better way to collect fine powders in thin layers is the "puff technique." The apparatus is shown in Figure 10.5. The specimen, pulverized

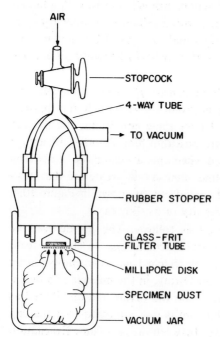

FIGURE 10.5. Apparatus for collection of thin powder layers on filter disks by the "puff" technique.

to −325 mesh (44 μm) or finer, is placed in the jar, and a 0.8-μm Millipore filter 2.5 cm in diameter is affixed to the glass frit. A constant vacuum is maintained on the filter tube while short blasts of air are admitted to the vessel by rapid manipulation of the stopcock. The air disperses the powder as a fine dust, which is collected on and imbedded in the Millipore disk. This technique may be unsatisfactory in some cases where more than one phase is present and fine particles of one phase selectively adhere to the wall so that a nonrepresentative specimen is collected. This may be prevented by fusing the specimen with lithium tetraborate ($Li_2B_4O_7$) to form a homogeneous material, which is then pulverized and prepared as above.

In the slurry technique, a small amount of fine powder is slurried with a solution of 2–5% (wt/vol) nitrocellulose in amyl acetate and spread over a scrupulously clean microscope slide. The film may be scored around the edge and floated off on water, then scooped up and supported some other way. A more versatile slurry technique is that of Finnegan. A weighed portion of powder is mixed with a weighed portion of ethylcellulose or nitrocellulose binder and added to a measured volume of amyl or butyl acetate. The slurry is next mixed and ground by rolling on a roller mill. A measured volume of the slurry is placed on an optically flat glass plate and spread in a uniform thin film with a casting knife ("doctor blade"). Blanks are removed from the film by scoring, flooding with water, and coaxing the pieces of film to float off.

10.5.3. Precautions

Variations in packing density may cause variations in analyte-line intensity unless an internal standard is used, or unless the same weight of each specimen in the analytical series is packed or briqueted under identical conditions. Mixing, grinding, and briqueting may introduce trace impurities, especially when the specimen powder is abrasive. Powders and briquets may outgas or even mildly explode in vacuum spectrometers unless previously outgassed in a vacuum chamber or vacuum oven, or unless briqueted in a vacuum press. Rotation of powder cells and briquets is recommended.

Powders having nonuniform particle size may segregate by size, heterogeneous powders may segregate by density. Particle-size effects are minimized by: (1) grinding to a size so small that the effects substantially disappear; (2) grinding all samples and standards to the same size or distribution by use of a standardized grinding procedure or a grinder that grinds to uniform size; (3) dry dilution; (4) briqueting at high pressure; (5) use of an x-ray tube target having short-wavelength lines to reduce the primary component of the specimen absorption; and (6) mathematical methods.

A good test for a nonrepresentative analytical layer caused by particle-size effects is to measure the $L\alpha_1/K\alpha$ intensity ratio for analytes up to atomic number $\sim$60 (neodymium) and the $M\alpha/L\alpha_1$ ratio for heavier elements. The intensity of the longer-wavelength line—and thereby the ratio—is strongly increased by increasing analyte concentration on the irradiated surface. Some workers grind powders to constant $L\alpha_1/K\alpha$ or $M\alpha/L\alpha_1$ ratio.

Binders have the advantages that they give coherent briquets even for powders having low cohesion, relatively homogeneous briquets even with powders of nonuniform size and density, higher packing density, smoother surfaces, and reduced absorption–enhancement effects. However, binders reduce the mean atomic number of the specimen and thereby increase scattered background, and reduce low-Z analyte-line intensity by dilution.

For a given briqueting pressure, the smaller the particle size, the higher is the analyte-line intensity. For a given particle size, the higher the briqueting pressure, the higher is the analyte-line intensity. These relationships are shown in Figure 10.6. The use of a binder or diluent accentuates the decreased intensity with increased particle size, and reduces the effect of pressure in overcoming this decrease.

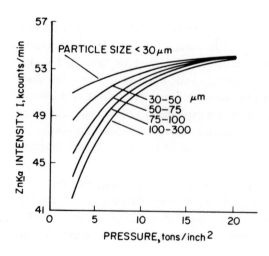

FIGURE 10.6. Effect of briqueting pressure on Zn $K\alpha$ intensity measured from briquets of zinc powder having various particle sizes. (1 ton/inch^2 = 140 kg/cm^2). [K. Togel in *Zerstörungsfreie Materialprüfung*, R. Oldenbourg, Munich, Germany (1961); courtesy of the author and publisher.]

10.6. FUSION PRODUCTS

The material to be analyzed may be fused with a flux for one of three reasons: (1) to render an insoluble or difficultly soluble material readily soluble prior to analysis as a solution; (2) to homogenize a material with respect to composition and density and to eliminate its particulate structure so that it may be ground to a homogeneous powder that may be used as such or as a briquet; or (3) to provide a homogeneous solid or glass pellet suitable for direct use as a specimen. The first application is the concern of Section 10.7; the other two are the subject of this section.

Fusion methods have many advantages regardless of whether the "melt" is used directly or powdered or briqueted. All inhomogeneity of composition, density, and particle size is eliminated. Standard addition and dilution methods and internal standardization are readily feasible. Absorption–enhancement effects are reduced or even eliminated by the leveling effect of the flux. A high-absorption diluent can be added, or a specific, interfering, variable minor matrix element can be added to fix its concentration. Standards are easily prepared—synthetically, if necessary. If the melt is ground and briqueted, it need be ground only to -200 mesh or so because of the homogeneity. If the solidified melt is used as the specimen, additional advantages are realized. The surface texture is smooth and uniform, although some polishing may be required, and standards are very conveniently stored. Properly prepared disks have been known to undergo no noticeable change after 20 h of irradiation in the primary beam.

The most serious disadvantages are that analyte-line intensities from low-Z elements are reduced by dilution and absorption, and that trace and minor constituents are highly diluted. The fusion requires considerable time and effort and, if glass disks are to be prepared, skill. Glass-disk standards may break in storage due to stress or devitrification. The pieces can be remelted and recast if a crucible is used from which the glass pours cleanly without adherence.

In the *potassium pyrosulfate fusion*, the charge typically consists of 200 mg of sample and 10 g of potassium pyrosulfate. If graphite is present, 200 mg of potassium persulfate ($K_2S_2O_8$) is added. For silicate minerals, a few grams of potassium fluoride (KF) or sodium fluoride (NaF) is added to volatilize silicon as silicon tetrafluoride. The charge is heated in a Pyrex or Vycor crucible at $<700°C$ for ~3 min. The melt is allowed to cool, ground to -200 mesh, and briqueted at >840 kg/cm^2 (12,000 lb/inch^2). If the melt is allowed to cool very slowly—usually overnight—a very homogeneous bead results that can be surfaced with fine abrasive paper and used as a specimen.

In the *Claisse sodium tetraborate fusion*, the charge typically consists of 100 mg of sample and 10 g of sodium tetraborate. If a high-absorption diluent is required, 100 mg of lanthanum oxide (La_2O_3) may be added. Claisse used platinum crucibles, but platinum–gold, platinum–rhodium, and vitreous carbon are preferable, and expendable graphite crucibles can also be used. The charge is fused over a Meker burner or in a muffle furnace at 800–1000°C for 5–15 min. By means of platinum-tipped tongs, the crucible is tipped and agitated occasionally to ensure homogeneity and to aid the expulsion of gas bubbles. The melt is poured on a polished aluminum plate heated at 450°C on a hot plate. If the melt is very fluid, an aluminum ring of ∼3-cm ($1\frac{1}{4}$-inch) inside diameter and at room temperature is placed on the aluminum plate and the melt poured into it. The pellet is kept on the hot plate for ∼2 min, then pushed with a Transite rod onto a Transite plate previously heated to 450°C. The pellet is allowed to cool with the Transite to room temperature. The ring is removed. The bottom surface of the glass pellet may be satisfactory as is; if not, it may be polished.

The principal fluxes for preparation of specimens in the form of glass disks are sodium tetraborate ($Na_2B_4O_7$) and lithium tetraborate ($Li_2B_4O_7$). The lithium salt, having the lower mass-absorption coefficient, is preferable to the sodium salt for light-element specimens. In place of sodium tetraborate, a mixture of sodium carbonate (Na_2CO_3) and boron oxide (B_2O_3) in mole ratio 1 : 2 can be used; lithium tetraborate can be replaced with the corresponding mixture of lithium carbonate and boron oxide. Other materials may be added to the charge prior to fusion for various purposes: Sodium carbonate may be added to sodium tetraborate charges to reduce fusion temperature. Sodium carbonate and sodium fluoride (NaF) may be added to render acidic and basic samples, respectively, more soluble. Also, during fusion, the carbonate decomposes with evolution of carbon dioxide (CO_2), which may aid in agitating and mixing the melt, but which may also form bubbles in the glass. Lithium carbonate and lithium fluoride may be added to lithium tetraborate charges for the same purposes. High-absorption diluents include lanthanum oxide (La_2O_3), barium oxide (BaO), barium sulfate ($BaSO_4$), and potassium pyrosulfate ($K_2S_2O_7$). Oxidants, such as sodium nitrate ($NaNO_3$), potassium chlorate ($KClO_3$), barium peroxide (BaO_2), and cerium oxide (CeO_2) may be added to sulfides and other samples to render them less refractory. The amount to be added may be calculated from the amount of oxygen required for the oxidation. A small amount of manganese dioxide (MnO_2) may be added to catalyze the oxidation. For silicate minerals and ceramics, a few grams of potassium fluoride (KF) or sodium fluoride (NaF) may serve to volatilize the silicon as silicon tetrafluoride (SiF_4).

The successful preparation of glass-disk specimens is an art, and each worker has his/her own "tricks"; even a summary of these techniques is beyond the scope of this introductory book. The inexperienced worker should not attempt to prepare glass disks without a detailed description from the literature of the application of the method to the particular type of sample to be analyzed.

10.7. LIQUIDS

10.7.1. Scope, Advantages, Limitations

Samples received in liquid form include the following: (1) liquids and solutions, aqueous and nonaqueous, including hydrocarbons and plating solutions; (2) dissolved solids, powders, and parts; (3) dissolved fusion products; (4) leach and wash liquids; (5) eluates from ion-exchange columns; (6) filtrates; (7) liquid–liquid and liquid–solid extractions; (8) biological fluids; (9) slurries; (10) thin greases, waxes, etc.; and (11) liquid ion-exchange resins.

The principal advantages of specimen presentation in liquid form are the following. Solution techniques are applicable to almost any type of sample, and are particularly advantageous when the same material is received in various forms, such as bar, foil, drillings, powder, wire, and parts; a single liquid-specimen procedure and set of standards can apply to all sample forms. For samples already in liquid, sol, or slurry form, often little or no preparation is required; what preparation is required is usually simple and convenient. For other sample forms, preparation is also usually simple and convenient. Samples and standards are perfectly homogeneous. Standards are usually prepared easily by dissolution of the individual constituents in proper concentrations, dissolution of standard materials, or quantitative variation of the composition of a single dissolved standard or of a standard solution. Blanks for measurement of background and for compensation of contamination are usually prepared easily. Standard addition and dilution techniques are applicable, which preclude the need for standards; such analyses are particularly useful for one-of-a-kind analyses. The analytical results are representative of the bulk sample, not just the surface; the effects of nonrepresentative surface composition, surface texture, and particle size are eliminated. The generally low effective atomic number of liquid matrixes results in relatively high scattered background; this higher background is usually disadvantageous, but may be beneficial in facilitating the use of analytical methods based on scattered x-rays. The absorption–enhancement effects of the total solute are reduced or eliminated by the dilution and leveling effect of the solvent. The matrix (solvent) usually consists of elements that have low atomic number and that contribute little absorption and no enhancement of the analyte lines. When residual absorption–enhancement effects from the original sample persist, a wide choice of techniques is available to deal with them, some of which are applicable only to solutions, others are more convenient with solutions. The low absorption by the matrix leads to high penetration by the primary beam and low absorption of the emitted secondary radiation, and consequently, high absolute sensitivity; unfortunately, this advantage may be canceled by the dilution of the analyte and by the increased background and consequent decreased peak-to-background ratio. Concentration or dilution to appropriate analyte concentration is usually feasible, and chemical separations, if required, are usually made conveniently. Solution and slurry techniques are particularly applicable to dynamic systems and to on-line process control. Dissolution of powder and solid samples opens up many alternative possibilities for subsequent preparation, presentation, and analysis. Even when samples are to be analyzed in another form, a solution method can be used to provide standards; samples of the required form are selected as prospective standards and analyzed by a solution method using synthetic solution standards; other samples in the same lot are then retained as standards.

The principal disadvantages of specimen presentation in liquid form are the following: Except for samples already in the form of liquids, solutions, sols, and slurries, liquid-specimen methods are destructive. For solid and powder samples, a solution technique is usually less rapid and convenient than a technique that uses the sample in its original form. Some elements are very difficult to put into and/or keep in solution, and fusion may be required. The portions of liquid standards actually used for the analysis are usually discarded. When large numbers of samples are to be analyzed and the analysis requires many hours, standards must be measured at intervals throughout the analysis. In such cases, sometimes a cell of standard can be measured only a few times because of deterioration on repeated exposure to the primary x-rays. Thus, relatively large volumes of standards must be prepared and stored, or else standards must be prepared frequently. Moreover, liquid standards may not be stable on long storage. Conversely, solid, powder, briquet, and glass-disk standards are usually permanent and require little storage space. Minor and trace constituents and micro samples may be difficult to analyze because of the high dilution. The intensities from analytes of low atomic number are seriously reduced by dilution and by absorption in the liquid matrix and cell windows. Because of the low effective atomic number of solvents and cell windows, the scattered primary background is relatively high; this increases the minimum detectable concentration, and the coherently and incoherently scattered target lines increase the possibility of spectral interference. Evaporation of solvent may occur during irradiation, especially with volatile solvents and in vacuum spectrometers, with consequent change in concentration and, possibly, in specimen plane. The specimen compartment, mask, and, worse, the beryllium window of the x-ray tube may be subjected to corrosive vapors and leakage; with inverted specimen geometry, perforation or rupture of the cell window may allow specimen liquid in quantity to flood these components. Liquids may heat under prolonged irradiation, causing expansion and, in closed cells, distension of the cell window. Liquids may outgas, spatter, or rupture their cell windows when used in vacuum spectrometers. Because of the outgassing, heating, and expansion, the specimen plane may vary during irradiation. Some liquid cells are difficult to fill properly or are leaky and messy. Bubbles often form in the liquid during irradiation, adhere to the cell window, and reduce the measured analyte-line intensity; the effect is reduced, but by no means eliminated by inverted specimen plane. Some specimens undergo x-ray-induced radiolysis with consequent precipitation. In inverted geometry, the precipitate settles on the window, which is the specimen plane. In upright geometry, the precipitate settles to the bottom of the cell, away from the measured surface. Thus, assuming that the precipitate is analyte-rich, the measured analyte-line intensity is increased in inverted geometry, decreased in upright geometry. Colloidal precipitation that does not flocculate is much less serious than precipitation that settles. However, any precipitation must be regarded as undesirable.

10.7.2. Liquid-Specimen Cells

Typical liquid-specimen cells for inverted geometry (primary x-ray beam directed upward on the bottom specimen surface) are shown in Figure 10.7. The cell bodies are of stainless steel (which may be gold-plated or coated with plastic), polyethylene, or Teflon. Typical inexpensive expendable liquid cells are shown in Figure 10.8. The cell bodies are of

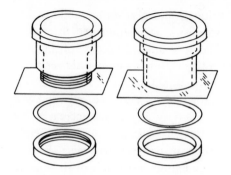

FIGURE 10.7. Liquid-specimen cells for some commercial x-ray spectrometers having inverted geometry. The retaining collars may screw (left) or push (right) on. The cover film is usually 6-μm (0.00025-inch) Mylar.

plastic. For both types of cell, the covers are usually 6-μm (0.00025-inch) Mylar, although 3-μm (0.00012-inch) Mylar is preferable in the 6–20-Å region. Microporous film permits equalization of pressure inside and outside the cell and is useful for vacuum path.

A widely used commercial liquid-specimen cell is the Somar Spectro-Cup (Somar Laboratories, Inc., New York), which is a versatile cell for all liquids and slurries in air or helium path and for nonvolatile liquids and slurries in vacuum. The cell is also useful for powders in any path, either loosely packed or spread in thin layers between Mylar films (Section 10.5.2). Provision is made for equalization of internal and external pressures. The cells are filled rapidly and conveniently, inexpensive enough to be disposable, yet suitable for reuse and for storage of samples and standards for future use. However, the cell is applicable only to instruments having inverted geometry. The cell, shown in Figure 10.9, consists of a cylindrical body *A*, retaining ring *B*, and snap-on cap *C* with handle *D*, all of polyethylene.

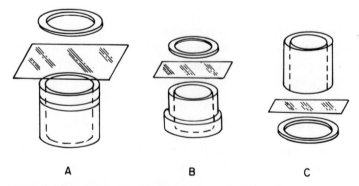

FIGURE 10.8. Disposable liquid-specimen cells made from (A) Spex cups, (B) Caplugs, and (C) cylinders cut from plastic tubing. The rings slip over the outside of the cylindrical bodies.

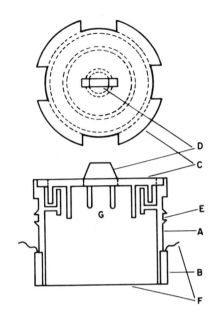

FIGURE 10.9. Somar Laboratories liquid-specimen cell. All components are of polyethylene except the Mylar film. (*A*) Cylindrical cell body. (*B*) Retaining ring for Mylar film *F*. (*C*) Snap-on cap. (*D*) Handle for *C*. (*E*) Groove to receive retaining ring (not shown) for second Mylar film for thin-layer powder method. (*F*) Mylar film. (*G*) Female cylindrical socket for male cylindrical internal intensity reference standard (Section 9.3.2, Figure 9.8).

The labyrinthine structure between the cell body and cap provides a liquid anticreep undercut and prevents escape of specimen material into the specimen chamber. Equalization of internal and external pressures also occurs through the labyrinth, which opens very slightly into the four slots in the cap. The ring groove *E* receives a flat polyethylene ring (supplied, but not shown in the figure), as explained below. For use with liquids, slurries, and loose powders, the cup is inverted, and a Mylar film *F* (6- or 3-μm, 0.00025- or 0.00012-inch) is stretched over the end and secured with the ring *B*. The cup is then set upright as shown, filled, and covered with the cap *C*, which has a socket *G* for insertion of an internal intensity reference standard (Section 9.3.2, Figure 9.8). For use with thin-layer sandwiched powder specimens, the cup is inverted, and a Mylar film is stretched over the end and secured, this time with the polyethylene ring in groove *E*. The powder is spread uniformly over the film. Then a second film *F* is stretched over the powder and secured with ring *B*.

A completely clean, leak-free cell especially useful for instruments having upright, inclined specimen planes is shown in Figure 10.10. The cell body *A* is 1 cm ($\frac{3}{8}$ inch) deep and may be of Lucite, polyethylene, or Teflon. The depression *B* is 0.6 cm ($\frac{1}{4}$ inch) deep and holds 4–5 ml. The Mylar cover *C* may be 6–125 μm (0.00025–0.005 inch) thick, depending on the wavelength to be measured, and is cemented to the cell with Pliobond rubber cement. When the cell is in place in the specimen drawer, the tri-

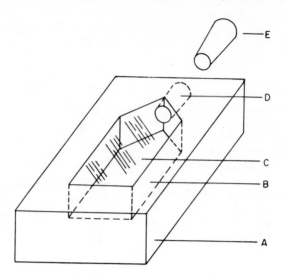

FIGURE 10.10. Leak-free liquid-specimen cell for upright or inverted, horizontal or inclined specimen planes (Bertin). (*A*) Cell body. (*B*) Depression. (*C*) Mylar cover. (*D*) Tapered filling port. (*E*) Tapered plug.

angular space at the neck of the cell is upward and covered by the specimen mask. This space serves to conceal the air volume at the tapered filling port *D* and provides an expansion chamber for the liquid. The plug *E* is required only for volatile or corrosive liquids. Alternatively, when the cell is in use, a short length of Tygon tubing may be attached to the filling port to conduct corrosive vapors out of the specimen compartment. The cells cannot be made of any readily available component and must be made in a machine shop.

Five to ten cells may be covered at a time with Mylar 6–125 μm thick as follows. A strip of Mylar, somewhat wider than the cell length and long enough to accommodate 5–10 cells side by side, is stretched extremely tightly on a smooth plane surface and secured at both ends. Each cell is coated on its top face with a thin layer of Pliobond and laid face down on the Mylar. A heavy weight—2 kg or so—is placed on each cell and left undisturbed for an hour or so. The Mylar is then cut between the cells, and any wrinkles in the window are stretched out. The cells are then replaced face down on the flat surface, weighted again, and left undisturbed overnight. Thicker windows are applied the same way, but less care is required to avoid wrinkling. Windows 250 μm thick never do sag or wrinkle, and windows 6 μm thick have been used up to 100 times at 50 kV, 50 mA before they finally sagged or perforated. When this happens, the window is peeled off, the cell resurfaced on a 400-grit wet paper, and a new window applied.

Filling pipets are made from 5-ml pipets by drawing down the tips and blowing spherical chambers in the upper part of the enlarged chambers. After each use, the cell is emptied, rinsed with one or two portions of the next specimen, then filled with the next specimen.

The principal disadvantage of the cells is that some specimens tend to form adherent deposits on the window. However, often, complexing agents can be added to the specimen to prevent this, or a suitable rinse can be used to remove the deposits between fillings.

10.8. SUPPORTED SPECIMENS

10.8.1. General Techniques

The term *supported specimen* as used here indicates a specimen in which a relatively small amount of analytical material—less than enough to constitute an infinitely thick layer—is supported on some type of substrate. Usually, the analytical material is derived from a solution or some other material by preconcentration, so supported specimens are the most commonly used forms in trace and micro analysis, and also in energy-dispersive analysis.

Some materials inherently constitute supported specimens in that small amounts of analytes are already distributed on or in a substrate, preferably a radiolucent one. In this class are coatings, impregnants, fillers, sizings, pigments, dyes, inks, fire retardants, trace impurities, etc., on or in papers, fabrics, thin plastic or rubber sheet, etc. Interesting examples are the distinguishing of genuine and counterfeit U.S. currency by differences in the x-ray spectra of their ink and paper, and determination of silver, chlorine, bromine, and iodine in photographic emulsions on paper and film bases. Platings, paints, corrosion layers, thin films, etc., may also be regarded as supported specimens (Section 9.10). Other specimens fall into the supported category only as a result of the preparation technique.

Preparation and presentation criteria for supported specimens are as follows: (1) Critical thicknesses for the analyte lines are not exceeded. (2) Absorption–enhancement effects for the analyte lines remain negligible or small. (3) Intensities do not exceed the capabilities of the detection-readout system; this is especially important in energy-dispersive spectrometry. (4) Intensities are sufficiently high so that acceptable statistical precision is obtained in reasonable counting time. (5) Finally, specimen geometry may not vary significantly among samples and standards.

The ideal specimen support should be: (1) thin, so that the analytical material lies in a plane; (2) low in area density (mg/cm²) to minimize x-ray scatter and absorption; (3) free of all elements having atomic number ≥ 9 (fluorine), even as traces; (4) reasonably strong, wet as well as dry; (5) weighable accurately (for certain applications); in effect, this means non-hygroscopic; (6) stable on repeated and prolonged exposure to high-intensity x-rays (for excitation by high-power x-ray tubes); (7) wettable by liquids (if the analytical material is to be applied in liquid form); (8) porous enough to give high flow rates for liquids and/or gases (if samples are to be collected by filtration of these media); this flow rate should not diminish significantly as sample accumulates; and (9) retentive to particulate material from filtration and evaporation.

The most commonly used support media are the following: (1) filter paper, including 2.5-cm disks, bioassay disks 12.5 and 6 mm ($\frac{1}{2}$ and $\frac{1}{4}$ inch) in diameter and $\sim$0.5 mm (0.020 inch) thick, and black filter paper (for better visibility of flecks of white precipitate); (2) confined spot-test papers consisting of small filter-paper squares having centered impregnated wax rings $\sim$1 cm in inside diameter to confine the spread of solutions spotted on the paper; (3) chromatographic paper (some of these papers absorb a greater volume of liquid per unit weight than filter paper, and do not warp when dry); (4) Fiberglas and other glass-fiber filters; (5) Millipore and Nuclepore filters, with or without nylon reinforcement; (6) thin (1–10-μm) Mylar, polyethylene, and polypropylene films; (7) ultrathin ($\leq$0.1-mm) collodion and Formvar films; (8) ion-exchange-resin granules, membranes, and impregnated filter paper; (9) Scotch tape; (10) abrasive (preferably boron carbide) paper and cloth for removing surface layers; and (10) quartz and plastic capillary tubes and quartz, plastic, and textile fibers.

The x-ray scattering power of a substrate, and therefore the background and detection limit, decrease linearly with area density (mass thickness). The most common substrates follow, in order of decreasing area density in milligrams per centimeter squared: filter paper ($\sim$10), Millipore filter ($\sim$5), Nuclepore filter ($\sim$1), 6-μm (0.00025-inch) Mylar ($\sim$1), 3-μm (0.00012-inch) Mylar ($\sim$0.5), 1-μm polypropylene ($\sim$0.2), and 1000-Å collodion or Formvar ($\sim$0.02). With regard to absence of trace impurities, Whatman No. 41 filter paper is best, Millipore filter very good.

The nature of supported specimens leads to several problems peculiar to this specimen form. Such specimens are not as uniform or reproducible as other specimen forms. Consequently, it is usually necessary to provide internal standardization and/or, when feasible, to rotate the specimen in its own plane during measurement. When the specimen is supported on an x-ray-transparent support, such as filter paper or Mylar film, the side containing the material, or the side to which it was pipeted or on which it was collected, should face the primary x-ray beam. Supported specimens tend to be loose and fluffy, and loss of material can occur. Specimens supported on thin Mylar or filter disks may be sprayed lightly with clear Krylon or 10% wt/vol collodion in acetone, or covered with or sandwiched between 6- or 3-μm Mylar films. Such sandwiched specimens may be stretched across the end of a plastic cylinder or the rim of a plastic cup and secured with an O-ring.

Specimens having substrates of thin filter paper, Mylar, etc. must be supported in a way that presents them to the x-ray spectrometer as plane surfaces. A simple way to do this is to back them with Lucite or other rigid plates. However, most thin substrates are highly transparent to the primary x-ray beam, and, if such backings are used, much primary radiation will be scattered back up through the specimen, thereby increasing the background. This is especially true with 6- or 3-μm Mylar, which, by itself, gives very low scattered background. Consequently, such substrates are preferably mounted by stretching them on open, rigid frames, or over the ends of plastic, glass, or metal cylindrical tubes (∼3 cm diameter, 0.6–1 cm high), or over the open ends of (empty) liquid-specimen cups.

Supported specimens are prepared from liquids and solutions by impregnation, evaporation, absorption, and filtration.

Small volumes of liquid are delivered by means of micropipets on filter paper or other filter disks, then evaporated to dryness. Confined-spot papers (see above) limit the spread of liquid to a region ∼1 cm in diameter. Solutions may be subjected to paper chromatography to separate the analytes and the paper chromatogram used as the sample.

Small volumes of liquid can also be placed on Mylar, Formvar, or other plastic films or in small ($\lesssim$2.5 cm in diameter) quartz, plastic, or metal dishes ("planchets"). The plastic and metal planchets are disposable. Uniformity of distribution is improved by use of Desicote or other wetting agent. The amount of analytical material is established by pipeting or by weight gain after evaporation.

Solutes can be removed from liquids by placing ion-exchange resin granules in the liquid and agitating for several hours, then recovering the granules and treating them as a loose or briqueted powder sample. Alternatively, a piece of ion-exchange membrane can be placed in the solution and agitated until the solutes are taken up. The membrane is recovered and used as the sample. Finally, the solution can be passed repeatedly through ion-exchange resin impregnated paper until the solutes have been removed. The filter becomes the sample.

The solutes in a solution can be precipitated selectively or collectively, with or without a gathering agent, then filtered. Other solid matter already suspended in the liquid may be removed the same way.

Gas-borne particulate matter can also be collected by filtration or impingement, that is, by drawing the gas through or directing it on filter-paper, Millipore, or other filter disks. Gas and vapor contaminants in air or other gases can be collected by filtration through filter paper impregnated with reagents specific for each contaminant. For example, in one air-pollution study, air was drawn through a cylindrical cartridge containing a series of four stacked filters, each supported on a separate ring, as follows: a Nuclepore filter followed by three dry Whatman No. 41 filter-paper disks impregnated with orthotolidine, silver nitrate, and sodium hydroxide, respectively; these filters retain particles, chlorine, sulfides, and sulfur dioxide, respectively.

The value of supported specimen techniques in environmental pollution studies is evident.

Luke has highly refined the filtration technique in his "coprex" (*copre-cipitation-x-ray*) method for trace analysis: ~0.5-g solid samples are dissolved in several milliliters of water, acid, or other solvent; organic or other matrix material and interfering elements are separated as required; 50–200 μg of a suitable element is added as a coprecipitant; and the trace analytes and coprecipitant are precipitated with a suitable reagent. The reagent may be specific for one analyte or applicable to a group of analytes. The precipitate is collected on a filter-paper, Millipore, or other filter disk in the special filter apparatus shown in Figure 10.11. Trace analytes in liquids and solutions are treated the same way. Solid material suspended in liquids may be filtered without treatment, or perhaps with a gathering agent. There is no reason why gases cannot be drawn through a similar arrangement to collect suspended solid material. The filter disk constitutes the specimen. Calibration curves are established from standard disks prepared the same way from known amounts of analyte. Reagent-blank disks for background measurement are prepared by filtration of solutions processed the same way but without analyte.

The opening in the bottom of the reservoir in the filter apparatus (Figure 10.11) determines the diameter of the residue spot on the filter and may be 2 cm to 1 mm. The larger areas may be measured in conven-

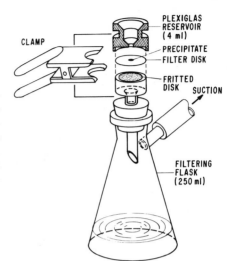

FIGURE 10.11. Coprex filtration apparatus. The diameter of the precipitate spot may be 1 mm to 2 cm and is determined by the diameter of the opening in the bottom of the reservoir. [From J. W. Mitchell, C. L. Luke, and W. R. Northover, *Anal. Chem.* **45**, 1503 (1973).]

tional flat-crystal spectrometers, the smallest in focusing, curved-crystal spectrometers. Any of the sizes is suitable for energy-dispersive instruments.

The great success of the method is attributable to three features: (1) elimination of matrix and possible interfering elements by preliminary separation; (2) substantially quantitative recovery of trace analytes by use of a coprecipitant; and (3) substantial elimination of absorption–enhancement effects by dispersion of the precipitate in a uniform, thin layer on a low-Z substrate.

Sensitivity with supported specimens is of the order 10 ng to 1 µg for large-area specimens in flat-crystal spectrometers, in the picogram region for "microdot" specimens in curved-crystal spectrometers. Calibration curves are usually linear up to 100 ng/cm² or more.

10.8.2. Ion-Exchange Techniques

Ion exchange may be defined as a reversible exchange of ions between a resin and a liquid without substantial change in the structure of the resin. The *ion-exchange resin* consists of a three-dimensional hydrocarbon polymer to which are attached a very large number of ionizable—and therefore exchangeable—chemical groups. Physically, the resin may have the form of granules, liquid, membrane, or sheet, or impregnated filter paper. Chemically, the resin may have cation-exchange (acidic) or anion-exchange (basic) form, and the ions to be absorbed must be in the same form as the resin.

Prior to use, cation-exchange resins are usually saturated with H^+ or Na^+—that is, all the ionizable sites are occupied by H^+ or Na^+. Similarly, anion-exchange resins are usually saturated with OH^- or Cl^-. The ion-exchange reaction may be represented by simple equations of the following forms, where R represents the hydrocarbon polymer matrix:

Cation exchange:

$$(RH) + Ag^+ \leftrightarrows (RAg) + H^+$$
$$(RNa) + Fe^{3+} \leftrightarrows (RFe) + 3Na^+$$

Anion exchange:

$$(ROH) + Br^- \leftrightarrows (RBr) + OH^-$$
$$(RCl) + SO_4^{2-} \leftrightarrows (RSO_4) + 2Cl^-$$

The *affinity* or *retentivity* of the resin varies widely for different ions and is affected by many factors, such as the chemical nature of the resin and solution, pH, concentration, and temperature. However, in general, the affinity of a resin for ions increases as the ionic radius of the hydrated ion increases, that is, as its valence and atomic number increase. However, this "rule" is by no means rigorously applicable, as is evident from the following *typical* decreasing retentivity series for cation and anion exchange:

$$La^{3+} > Nd^{3+} > Eu^{3+} > Y^{3+} > Fe^{3+} > Al^{3+} > Ba^{2+} \sim Ag^+ > Pb^{2+}$$
$$> Hg^{2+} > Sr^{2+} > Ca^{2+} > Cu^+ > Ni^{2+} > Cd^{2+} > Cu^{2+} \sim Zn^{2+}$$
$$> Fe^{2+} \sim Mg^{2+} > Cs^+ > Rb^+ > K^+ > Na^+ > H^+ > Li^+$$

$$I^- > HSO_4^- > Br^- > Cl^- > F^- \sim OH^-$$

Analytes in solution can be taken up in ion-exchange resin by any of three processes:

1. In the *batch technique*, the solution is allowed to stand or is shaken with a portion of ion-exchange granules or liquid, or with a square or circle of membrane or resin-impregnated paper. The technique of allowing a piece of membrane or paper to stand in contact with a solution has been termed "danglation". The granules are separated and washed by filtration, then placed in a powder cell wet or dry, or briqueted. The liquid resin is separated and washed by decantation and used in a liquid cell. The membrane or paper is scooped out, rinsed, dried, and mounted as is.

2. In the *filtration technique*, the solution is filtered several times through an ion-exchange filter paper on a Buchner or other filter funnel in a suction flask until quantitative recovery of the analytes is realized. The filter constitutes the specimen.

3. In the *column technique*, the solution is passed through a column of ion-exchange granules packed in a glass apparatus resembling a short buret. The analytes are retained on the column and removed collectively, or individually in order of increasing retentivity by elution with a solution of complexing agent. The eluate constitutes a liquid specimen.

In the batch and filter techniques, it is wise to establish the agitation time or number of filtrations required to effect complete equilibrium or recovery prior to undertaking the actual analysis. This is done by plotting analyte-line intensity *versus* time or number of filtrations. The point is noted beyond which no further increase in intensity occurs. Equilibrium is established more rapidly with liquid resins than with granules, but sensitivity is lower with liquids. Equilibrium with granules is expedited by presoaking. Care must be taken not to exceed the capacity of the resin, which is usually stated in milliequivalents per gram. In the filtration technique, completeness of recovery can be tested by passing the filtrate through a second, fresh ion-exchange filter disk, which is then measured on the x-ray spectrometer. For use with very dilute analyte solutions, it may be necessary to purify the ion-exchange resin by repeated washing or elution with pure eluant.

Ion-exchange filters may have several forms:

1. Filter paper chemically treated to impart ion-exchange properties.

2. Filter paper impregnated with cation- or anion-exchange resin having low selectivity, that is, useful for a wide range of cations or anions. The weight ratio of paper to resin is usually about 1 : 1 and the capacity $\sim$1.5–2 meq/g. For a 3.5-cm disk, this comes to $\sim$0.2 meq—$\sim$4 mg of Fe^{3+}, for example.

3. Filter paper impregnated with cation- or anion-exchange resin having high specificity for a certain element or group of elements. Such resins are termed *selective* or *chelating* resins. For example, Dowex A-1 resin has relative retentivities for Hg^{2+}, Cu^{2+}, Ni^{2+}, Zn^{2+}, and Mg^{2+} of about 1060, 126, 4.4, 1, and 0.01, respectively. The nonspecific ion-exchange papers can be made selective by appropriate regulation of pH and by use of chemical masking and complexing reagents. Filter paper impregnated with chemical reagents specific for one or more elements, such as lead acetate paper for sulfides, performs the same function in a different way.

Campbell and Green discuss the use of ion-exchange membranes and filter papers in detail, compare filtration and immersion and mechanical and ultrasonic agitation, describe filtering apparatus and specimen holders, and give mass-absorption coefficients.

Internal standards can be added to ion-exchange resins. The internal standard can be added to the sample solution prior to treatment with the resin, but this technique is not recommended because analyte and internal standard may not coabsorb in constant proportion. A better way is to add a solution of internal standard volumetrically to the analyte-loaded resin in the specimen cell.

Ion-exchange techniques are applicable to trace and micro analysis. Microgram quantities of, say, nickel in a liter of water can be collected on 1 cm² of ion-exchange membrane stirred in the water for ~48 h; the membrane is the sample. Eleven trace elements were determined in tungsten by dissolution of the samples in tartrate medium, passage of the solutions through an ion-exchange column to retain the trace analytes, elution of the analytes, and filtration of the eluates through disks of ion-exchange paper. Seven passes through the disks gave >99% retention of all trace analytes. Analytes in quantities as small as 0.01 μg were determined by isolating them chemically, taking them up in 1 ml of acid or water, then collecting them on individual disks of ion-exchange membrane 3 mm ($\frac{1}{8}$ inch) in diameter. The disks were measured on a curved-crystal spectrometer.

10.9. SPECIAL SPECIMEN FORMS

10.9.1. Selected-Area Analysis

If the conventional specimen-mask technique is applied to very small specimens or to small areas on large specimens, analyte-line peak-to-background ratio may decrease prohibitively. As analyte-line intensity decreases with specimen area, primary scatter from the increasing mask area increases, and primary scatter from air in the specimen compartment assumes greater significance. Such cases are best dealt with by confining the excitation to the area of interest or by excluding all secondary x-rays from the spectrometer except those originating in the area of interest.

If a small aperture is placed in the primary beam (I in Figure 3.2), only that incremental specimen area lying in the projection of the aperture receives primary radiation. Alternatively, if the aperture is placed in the secondary beam (J in Figure 3.2), the entire specimen is irradiated, but only secondary radiation originating in the projection of the aperture enters the optical path. Such apertures are the basis for selected-area analysis.

An x-ray spectrometer arranged for selected-area analysis is termed an "x-ray milliprobe" (compare with electron "microprobe") because excitation of the "probed" (selected) area is by x-rays rather than electrons and the probed area has dimensions of the order of millimeters or mils rather than micrometers.

Specimen areas somewhat greater than $\sim$1 mm^2 and substantially smaller than $\sim$1 cm^2 represent a case intermediate between conventional and selected-area techniques and may be measured in a *collimated specimen accessory*. The primary beam is conducted to the specimen and the secondary beam to the optical path in a continuous L-shaped tunnel in a block on the specimen mask. This arrangement reduces primary scatter from the mask and air. The Soller collimators may usually be retained.

Selected-area techniques are applicable to two classes of specimens: very small specimens, which may be regarded as "preselected" areas, and small selected areas on extensive specimens.

The first group includes the following: (1) particles, flakes, very small chips, etc.; (2) fibers, filaments, fine wires, etc.; (3) submilligram weights of powders on fibers or in fine plastic capillary tubes (these are the types of specimen used in x-ray powder diffraction analysis, and selected-area techniques may serve as a preliminary analysis prior to powder diffraction); (4) metallic trace analytes electroplated on fine wire or on carbonized quartz fibers; (5) flecks of purified trace analyte after microchemical separation from a matrix; (6) very small disks of filter paper or ion-exchange membrane supporting microgram amounts of analyte; (7) small spots of analyte residue formed by evaporating droplets of solution on thin Mylar film; and (8) segregations, grains, and inclusions extracted from materials by micro drilling, extraction replication, or preferential dissolution or etching away of the matrix.

The second group (see above) includes samples submitted for the following types of work: (1) mapping of composition distribution of analyte(s) and evaluation of homogeneity of composition; (2) analysis of individual grains of various phases and analysis of inclusions, segregations, etc., in alloys, minerals, ceramics, etc.; (3) analysis of particles filtered from liquids or gases and still in place on the filters; (4) evaluation of composition, thickness, and uniformity of thin films and platings, including wear of platings; (5) measurement of extent of diffusion in diffusion couples;

(6) identification of layers and measurement of extent of diffusion in sections of metal-to-metal and metal-to-ceramic interfaces; (7) measurement of extent of diffusion from liquid to solid phase; (8) investigation of sections of plated, coated, or corroded surfaces; (9) analysis of small components and parts in place—in transistors, for example; and (10) determination and distribution of natural and foreign elements in biological materials.

The selected-area accessory consists of an aperture in the primary or secondary beam and a specimen stage that permits manual and motor-driven translation of the specimen in two mutually perpendicular directions. This accessory may be used with an energy-dispersive or wavelength-dispersive spectrometer. If the latter, the Soller collimators are removed and, ideally, a full-focusing curved-crystal accessory should be provided, that is, one in which either the crystal radius or the specimen-to-crystal and crystal-to-detector distances vary continuously with 2θ to maintain optimal focusing. However, much work can be done by simply replacing the flat crystal with a curved crystal. Earlier wavelength-dispersive spectrometers (Figure 3.2) are easily fitted with selected-area accessories, but modern models are not.

Typical pinhole and slit apertures are shown in Figure 10.12. Pinhole diameters and slit widths are of the order 0.01–1 mm.

Slits are preferable to pinholes for some specimens, such as sections of long, linear inclusions; diffusion couples; metal-to-metal and metal-to-ceramic interfaces; plated, coated, or corroded surfaces; and other layered structures where composition varies in the direction across, but not along, the layers. Figure 10.13 shows a section of a typical ceramic-to-metal system. The ceramic has been metallized, plated, and brazed to the metal. The

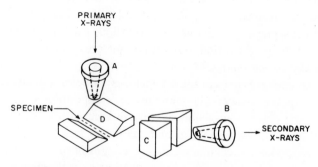

FIGURE 10.12. Forms and positions of selected-area apertures. (*A*) Primary pinhole. (*B*) Secondary pinhole. (*C*) Secondary slit. (*D*) Primary–secondary slit or edge.

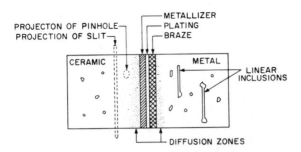

FIGURE 10.13. Applications of pinhole and slit apertures.

metallizer may diffuse into the ceramic, the braze into the metal, and the metallizer, plating, and braze into each other. Inclusions or segregations occur in the ceramic and metal. Selected-area techniques can be applied to identify the metallizer, plating, braze, and inclusions, and to evaluate the extent of diffusion. The figure shows the projections of pinhole and slit apertures on the specimen.

The roughly circular inclusions are identified by use of the pinhole. However, if a pinhole is used to identify the linear inclusions or interfacial layers, or to evaluate the extent of diffusion, only that incremental specimen layer under the pinhole contributes to the measured analyte-line intensity. A slit is advantageous for these applications because it passes secondary radiation from the entire length of the incremental layer. Increased intensity, and therefore sensitivity and statistical precision, are realized without loss of resolution by replacing a pinhole of given diameter with a slit of the same width. Increased resolution is realized without loss of intensity by replacing the pinhole with a slit of smaller width. Moreover, a more representative sampling of a layer is obtained with a slit than with a pinhole.

Figures 10.14 and 10.15 show specimen drawers fitted with secondary-beam pinhole and slit apertures, respectively. The micrometers operate the specimen drives for the X and Y directions.

The small specimen or area can be placed in the projection of the pinhole or slit and a qualitative analysis made by recording a 2θ scan on the ratemeter–recorder, or a quantitative analysis made by scaling the analyte-line intensities. The 2θ scan charts are similar to those shown in Figure 7.2. The procedure is repeated at other points of interest.

The concentration distribution profile of a selected analyte along a line across the specimen can be mapped by setting the spectrometer at the analyte line, moving the specimen manually in small increments, and scaling

FIGURE 10.14. Specimen drawer fitted with secondary-beam pinhole aperture.

FIGURE 10.15. Specimen drawer fitted with secondary-beam slit aperture.

the intensity at each point. For layered specimens, a slit is preferable; otherwise, a pinhole must be used. The data is plotted manually. The same distribution profile can be mapped by translating the specimen with a motor drive while recording analyte-line intensity on the recorder. The two techniques are compared below. Successive scans can be made for different analytes, and, with pinholes, scans can be made along different lines for the same analyte.

The motor-drive should be slow enough to allow the recorder to follow small irregularities in concentration.

The effectiveness of a selected-area analysis in excluding surrounding matrix is shown in Figure 10.16 for a nickel wire 0.04 mm (0.0015 inch) in diameter imbedded in Wood's metal (Bi–Pb–Cd–Sn). Ratemeter–recorder 2θ scans are shown in the Ni $K\alpha$ spectral region for conventional flat-crystal operation and for selected-area operation with a curved-crystal and three pinhole apertures having different diameters. Note the progressive decrease in intensity of matrix lines and background without loss of Ni $K\alpha$ intensity.

Typical performance of a secondary slit aperture (Figure 10.15) with

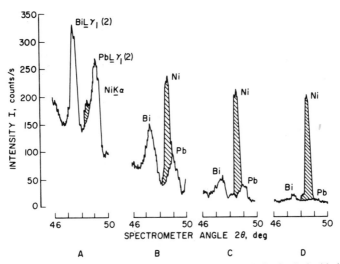

FIGURE 10.16. Partial spectra from a 0.04-mm nickel wire imbedded in Wood's metal (Bi–Pb–Cd–Sn). [T. C. Loomis and K. H. Storks, *Bell Laboratories Record* **45**, 2 (1967); courtesy of the authors and the Bell Telephone Laboratories.] (*A*) Conventional flat-crystal x-ray spectrometer. (*B*). Full-focusing curved-crystal x-ray probe with 1.6-mm pinhole. (*C*) Same as *B* with 0.75-mm pinhole. (*D*) Same as *B* with 0.35-mm pinhole.

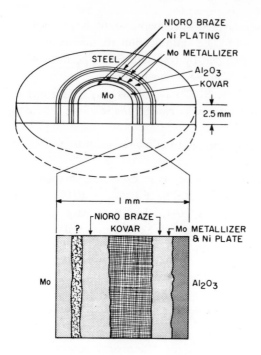

FIGURE 10.17. Diagram and photomicrograph of circular metal–ceramic bonded system, which was sectioned as shown and subjected to selected-area analysis over the 1-mm width indicated (see Figure 10.18). On the diagram, thicknesses of metallizer, plating, and braze layers are exaggerated.

a semifocusing curved crystal is shown in Figures 10.17 and 10.18. The slit probe was applied to an investigation of a disk-shaped system involving both metal-to-metal and metal-to-ceramic bonds. Figure 10.17 shows the various concentric layers in the disk as established by metallography and x-ray-probe spectrometry, and a photomicrograph of the region of specific interest in the x-ray study. Particular curiosity was aroused by the layer indicated by the question mark on the photomicrograph; only a braze was expected between the Kovar and the molybdenum. The disk was sectioned on a chord just intersecting the molybdenum center area, rather than on the diameter as shown in Figure 10.17; this section results in somewhat wider layers in the zone of interest and facilitates the x-ray study. The disk was ~2.5 mm thick, covering substantially less than the full useful height of the slit. The specimen was scanned with a 0.006-mm (0.00025-inch) slit for Fe $K\alpha$, Co $K\alpha$, Ni $K\alpha$, Mo $K\alpha$, and Au $L\alpha_1$ both manually and with the motor drive. The results of both series of scans are shown in Figure 10.18, where the following features are evident: (1) The ceramic was metallized with molybdenum, which diffused a considerable distance into the ceramic; (2) the molybdenum was nickel-plated to improve bonding; (3) the brazing metal was a gold–nickel alloy, probably Nioro (82Au–18Ni); (4) separation of gold and nickel in the brazing alloy has occurred; and (5) the well-defined phase indicated by the question mark on the photomicrograph in Figure 10.17 is seen to contain cobalt and iron, which appear to have been dissolved from the Kovar by the Nioro during brazing and deposited near the molybdenum.

Figure 10.18 permits comparison of the manual and motor-drive methods. It is evident that the information available from the two methods is substantially the same. However, whereas the manual study required nearly half a work day, each of the motor-drive scans required ∼18 min, the entire study ∼1.5 h.

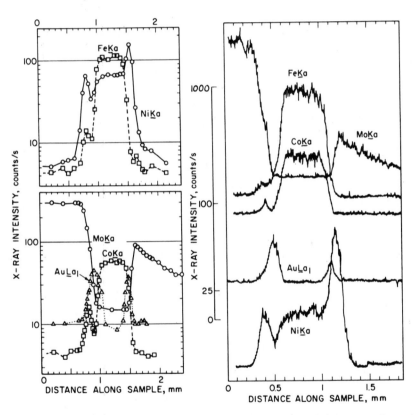

FIGURE 10.18. Intensity profiles across the indicated portion of the system shown in Figure 10.17 made with manual incremental sample translation and the scaler–timer (left) and with motor-driven sample translation and the ratemeter–recorder (right). The selected-area accessory was that shown in Figure 10.15 set at 6-μm (0.00025-inch) slit width. To avoid confusion, the curves at right were traced from the ratemeter charts and displaced vertically with respect to one another. The low-intensity portion of each curve corresponds to 0 on the intensity scale. The scale is linear up to ∼25 counts/s, logarithmic above.

10.9.2. Radioactive Specimens

The problems peculiar to radioactive specimens are personnel safety and prevention of interference with the analyte-line x-rays by radiation of direct and indirect radioactive origin.

Personnel safety involves prevention of ingestion of radioactive material, irradiation of personnel, and contamination of the laboratory environment, including the spectrometer. Ingestion and contamination are avoided by good housekeeping, careful and orderly work practices, provision of glove boxes and/or restricted specimen preparation areas, use of respirators, and encapsulation of radioactive specimens. Eating, drinking, and smoking should be forbidden in laboratories dealing with radioactive specimens.

The precautions required to preclude interference with x-ray measurements depend on the type (α, β, γ, or x), energy (keV), and activity (Ci) of the radioactivity. In general, γ-rays are by far the most penetrating, α-rays the least. Usually α- and β-rays are adequately confined by encapsulation, but they may generate within the specimen x-rays that can penetrate the encapsulation. In general, γ-rays are not confined by encapsulation and may reach the detector by penetration of the specimen chamber or by passing through the source collimator and scattering from the crystal; γ-rays also generate x-rays in the specimen. Some radioactive isotopes emit x-rays directly as a result of their decay process (Section 1.8.8). X-rays originating directly or indirectly from the radioactivity contribute to the x-ray background and reduce the analyte peak-to-background ratio.

The principal techniques used in x-ray spectrometric analysis of radioactive specimens are the following.

1. *Encapsulation.* Preferably, all radioactive specimens are *doubly* encapsulated. Solid specimens can be sealed into plastic envelopes, coated with an adherent plastic layer, or molded in plastic, then sealed in outer plastic envelopes. Liquids can be presented in double cells having Mylar, polystyrene, mica, or glass windows. An inner cell is filled and sealed, then totally enclosed in a larger outer cell or even a thin Mylar or polyethylene envelope. Glass disks are already encapsulated, in effect, by the glass and may be sealed in plastic envelopes. Supported specimens on filter paper or other substrates, or in metal, glass, or plastic planchets, can be sealed in plastic envelopes. Of course, care must be taken that all samples and standards are encapsulated the same way so that the attenuation of the emitted analyte-line and radioactive radiation and scatter of the primary x-rays by the plastic are the same for all specimens.

2. *Shielding* between the specimen chamber and detector may be provided to minimize direct γ-irradiation of the detector.

3. *Pulse-Height Selection.* If a scintillation counter is used and if the γ-rays or α-, β-, or γ-excited x-rays have high energy, some discrimination may occur in the scintillation crystal itself. The crystal has high quantum efficiency for the relatively low-energy analyte-line x-rays, but low quantum efficiency for the more penetrating high-energy γ- or x-rays. The pulse distributions arising from the γ- or x-rays of radioactive origin are likely to be discriminated by the pulse-height selector in the usual way.

4. *"Phoswich" Detection and Pulse-Shape Selection.* X- and γ-rays can be discriminated by a scintillation counter having a composite phosphor consisting of a thin, front phosphor that absorbs x-rays but transmits high-energy γ-rays and a thick, back phosphor that absorbs the γ-rays. The two groups of detector-output pulses are separated on the basis of their shapes by a pulse-shape discriminator.

5. *Radiation Blank.* The intensity at the 2θ angle of the analyte line can be measured from the radioactive specimen in the x-ray spectrometer operating normally, then operating without power applied to the x-ray tube. The latter measurement is the contribution to background by γ-rays and x-rays of radioactive origin, and is subtracted from the intensity measured with x-ray tube power.

10.9.3. Dynamic Systems

If special specimen presentation systems are provided, x-ray spectrometry can be applied to dynamic process control and to other dynamic processes such as the following: dissolution of a solid in a liquid, precipitation of a solid from a liquid, admixture of partially miscible liquids, diffusion of solid in solid or liquid in liquid, diffusion of an element to the surface, diffusion of an element from the surface into the bulk, sublimation and desorption, preferential corrosion of a surface, rate of electrodeposition, and continuous paper and gas chromatography.

10.9.4. Gases and Vapors

Gases and vapors containing x-ray spectrometrically determinable elements may be taken up in solution or in a solid absorber. It may be feasible to determine gases not containing such an element by indirect or association analysis (Section 9.8). Even the group 0 ("inert") gases (helium, neon, argon, krypton, xenon, and radon) may form stoichiometric association and clathrate compounds with certain substances. For helium and neon, an associated substance containing a determinable element is required. It might be possible to adsorb gases in activated charcoal, which can then be placed in the spectrometer. It would be necessary to provide for cooling the charcoal with liquid nitrogen. Some examples of determinations of gases and vapors follow.

Occluded argon in concentration 0.05–7 wt% has been determined in argon-sputtered silicon dioxide (SiO_2) films 100 Å to 5 μm thick. The relationship between Ar $K\alpha$ intensity and argon concentration is established in either of two ways: Ar $K\alpha$ intensity can be measured from weighed specimens, then argon determined by weight loss on heating the specimens in helium at 600°C for several hours. Alternatively, Ar $K\alpha$ intensity from samples can be applied to curves of Cl $K\alpha$ and K $K\alpha$ intensity measured from potassium chloride (KCl) films of known mass and thickness. Chlorine, argon, and potassium have consecutive atomic numbers, 17–19.

Mixtures of iodomethane (CH_3I, boiling point $\sim$42°C) and bromoethane (CH_3CH_2Br, boiling point $\sim$38°C) have been analyzed by measuring I $K\alpha$ and Br $K\alpha$ intensities from their vapors. The liquid samples were placed in a rubber balloon attached to a side tube of a liquid-specimen cell having a glass body and a 50-μm (0.002-inch) Mylar window. The measurements were made on the emission from the vapor in the cell. The I $K\alpha$/Br $K\alpha$ intensity ratios were applied to a calibration curve established from standard mixtures.

The $K\alpha$ lines of sulfur and halogens have been measured directly from gas-chromatograph effluents passing through thin-wall plastic tubes in the x-ray spectrometer specimen chamber.

SUGGESTED READING

BERTIN, E. P., *Principles and Practice of X-Ray Spectrometric Analysis*, 2nd ed., Plenum Press, New York (1975); Chaps. 16, 17, and 19, pp. 701–807, 829–858.

BERTIN, E. P., "Recent Advances in Quantitative X-Ray Spectrometric Analysis by Solution Techniques," *Advan. X-Ray Anal.* **11**, 1–22 (1968).

BERTIN, E. P., and R. J. LONGOBUCCO, "Sample Preparation Methods for X-Ray Fluorescence Emission Spectrometry," *Norelco Rep.* **9**, 31–43 (1962).

BERTIN, E. P., and R. J. LONGOBUCCO, "X-Ray Spectrometric Analysis—Nickel, Copper, Silver, and Gold in Plating Baths," *Metal. Finish.* **60**(3), 54–58 (1962).

CAMPBELL, W. J., "Fluorescent X-Ray Spectrographic Analysis of Trace Elements, Including Thin Films," *Amer. Soc. Test. Mater. Spec. Tech. Publ.* **349**, 48–69 (1964).

CHESSIN, H., and E. H. McLAREN, "X-Ray Spectrometric Determination of Atmospheric Aerosols," *Advan. X-Ray Anal.* **16**, 165–176 (1973).

CLAISSE, F., "Accurate X-Ray Fluorescence Analysis without Internal Standard," *Que. Prov. Dep. Mines. Prelim. Rep.* **327**, 24 pp. (1956); *Norelco Rep.* **4**, 3–7, 17, 19, 95–96 (1957).

CROKE, J. F., and W. R. KILEY, "Specimen-Preparation Techniques;" in *Handbook of X-Rays*, E. F. Kaelble, ed., McGraw-Hill, New York, Chap. 33, 22 pp. (1967).

CULLEN, T. J., "Potassium Pyrosulfate Fusion Technique—Determination of Copper in Mattes and Slags by X-Ray Spectroscopy," *Anal. Chem.* **32**, 516–517 (1960).

DAVIS, C. M., K. E. BURKE, and M. M. YANAK, "X-Ray Spectrographic Analysis of Traces in Metals by Preconcentration Techniques," *Advan. X-Ray Anal.* **11**, 56–62 (1968).

FINNEGAN, J. J., "Thin-Film X-Ray Spectroscopy," *Advan. X-Ray Anal.* **5**, 500–511 (1962).

GIAUQUE, R. D., F. S. GOULDING, J. M. JAKLEVIC, and R. H. PEHL, "Trace-Element Determination with Semiconductor-Detector X-Ray Spectrometers," *Anal. Chem.* **45**, 671–681 (1973) (the "puff" technique).

GREEN, T. E., S. L. LAW, and W. J. CAMPBELL, "Use of Selective Ion-Exchange Paper in X-Ray Spectrography and Neutron Activation—Application to the Determination of Gold," *Anal. Chem.* **42**, 1749–1753 (1970).

HIRT, R. C., W. R. DOUGHMAN, and J. B. GISCLARD, "Application of X-Ray Emission Spectrography to Air-Borne Dusts in Industrial Hygiene Studies," *Anal. Chem.* **28**, 1649–1651 (1956).

JENKINS, R., and J. L. DE VRIES, *Practical X-Ray Spectrometry*, 2nd ed., Springer-Verlag New York, New York (1969); Chap. 8, pp. 145–164.

LUKE, C. L., "Determination of Trace Elements in Inorganic Materials by X-Ray Fluorescence Spectroscopy," *Anal. Chim. Acta* **41**, 237–250 (1968).

MATOCHA, C. K., "New Briqueting Technique," *Appl. Spectrosc.* **20**, 252–253 (1966).

MICHAELIS, R. E., "Report on Available Standard Samples, Reference Samples, and High-Purity Materials for Spectrochemical Analysis," *Amer. Soc. Test. Mater. Data Ser.* **DS-2**, 156 pp. (1964).

MICHAELIS, R. E., and B. A. KILDAY, "Surface Preparation of Solid Metallic Samples for X-Ray Spectrochemical Analysis," *Advan. X-Ray Anal.* **5**, 405–411 (1962).

MIZUIKE, A., "Separation and Preconcentration Techniques"; in *Modern Analytical Techniques for Metals and Alloys*, Part 1, R. F. Bunshah, ed., Techniques of Metals Research Series, vol. 3, part 1; Wiley–Interscience, New York, pp. 25–67 (1970).

MUELLER, J. I., V. G. SCOTTI, and J. J. LITTLE, "Fluorescent X-Ray Analysis of Highly Radioactive Samples," *Advan. X-Ray Anal.* **2**, 157–166 (1959).

NATIONAL BUREAU OF STANDARDS (U.S.), "Catalog of Standard Reference Materials," *Nat. Bur. Stand. (U.S.) Spec. Publ.* **260**, 77 pp. (1970).

NATIONAL BUREAU OF STANDARDS (U.S.), "Standard Reference Materials—Sources of Information," *Nat. Bur. Stand. (U.S.) Misc. Publ.* **260-4**, 18 pp. (1965).

SALMON, M. L., "Simple Multielement-Calibration System for Analysis of Minor and Major Elements in Minerals by Fluorescent X-Ray Spectrography," *Advan. X-Ray Anal.* **5**, 389–404 (1962).

TOGEL, K., "Preparation Technique for X-Ray Spectrometry"; in *Zerstörungsfreie Materialprüfung* ("Nondestructive Material Testing"), E. A. W. Muller, ed.; Oldenbourg, Munich, Germany, sec. U152 (1961).

Chapter 11

Electron-Probe Microanalysis

11.1. INTRODUCTION

Electron-probe microanalysis (EPMA), or electron microprobe analysis (EMA), is essentially (1) a nondestructive instrumental method of qualitative and quantitative analysis for chemical elements (2)—based on measurement of the wavelengths and intensities of their characteristic x-ray spectral lines (3)—excited by an electron beam having diameter of the order 0.1–1 μm. The method permits in-place determination of composition and spacial variation of composition on a microscopic scale.

The electron-probe microanalyzer is essentially an electron-probe x-ray primary-emission *spectrometer*. A fine-focus electron beam excites the x-ray spectra of the elements in a specimen region ∼1 μm in diameter and ∼1 μm³ in volume. The x-rays are read out in one or more curved-crystal wavelength-dispersive x-ray spectrometers and/or in an energy-dispersive spectrometer. However, the interaction of the electron beam and specimen results in other phenomena—principally backscattered, secondary, and absorbed electrons and cathodoluminescence—that may provide other types of information. Thus, the more general term, electron-probe *microanalyzer*, is more appropriate.

For *qualitative analysis*, in favorable cases, the method is applicable to all chemical elements down to atomic number 4 (beryllium); concentrations down to ∼0.01 wt%; isolated masses down to ∼1 fg (10^{-15} g); depths down to several micrometers; films as "thin" as a fraction of a monolayer; and specimens in the form of bulk solid, film, powder (including individual particles), small fabricated forms and devices, and metallographic sections, all without serious limitations on surface finish.

For *quantitative analysis*, the minimum determinable concentration is likely to be greater (less sensitive) than for qualitative analysis, and there are severe limitations on specimen form and finish. In favorable cases, precision may be of the order ± 0.5–2% of the amount present for major and minor constituents, but much greater (less precise) as concentration decreases. Calibration standards are preferably specimens similar to the samples but of known composition; analytical concentrations are derived by applying intensity data from samples to a calibration curve established from intensity data from the standards. Alternatively, the standards may be the pure constituent elements, or compounds or alloys of these elements different from the samples but of known composition; analytical concentrations are derived from intensity data from samples and standards by means of complex mathematical equations. In either case, samples and standards must have identical fine surface finishes and must be homogeneous on a micrometer scale.

11.2. INSTRUMENTATION

Two types of essentially similar instrument for electron-probe microanalysis must be distinguished: the electron-probe microanalyzer and the scanning electron microscope (SEM).

Electron-probe microanalyzers are designed primarily for x-ray spectrometric analysis and so operate at relatively high electron-beam currents (usually 0.01–10 μA) at the expense of high spacial resolution ($\gtrsim$0.5 μm) and magnification ($\lesssim$5000×). Cathode-ray tube displays of backscattered and secondary electrons are of secondary importance and are used principally to examine the specimen surface as an aid to locating features to be analyzed. The high electron-beam current excites analyte-line intensities adequate for wavelength-dispersive (crystal) spectrometers. Excepting the next paragraph, this chapter is devoted to instruments of this class, but the principles apply to scanning electron microscopes as well.

Conversely, scanning electron microscopes are designed primarily to give electron images of the specimen surface at the highest possible magnification ($\lesssim$100,000×) and resolution ($\gtrsim$150 Å) and so operate at low electron-beam currents (0.01–1 nA). X-ray spectrometric analysis, when provided for at all, is of secondary importance. The low electron-beam current gives analyte-line intensities usually too low for wavelength dispersion but adequate for energy dispersion.

An electron-probe microanalyzer consists of six systems, as shown schematically in Figure 11.1: electron optics, specimen stage, light microscope, x-ray spectrometer(s), electron analysis components, and a vacuum system. These components are housed in or near an evacuated electron-optic column similar to that of an electron microscope, one or more attached x-ray spectrometers, and an electronics console.

The *electron-optic system* consists of the following components: (1) an electron gun comprising a V-shaped tungsten filament, grid cup (Wehnelt cylinder), and anode; (2) two electromagnetic reducing electron lenses; (3) two electron-beam limiting apertures; (4) a fluorescent viewing screen; (5) an electron-beam scanning system; (6) an astigmatism compensator, and (7) an electron-beam current regulator.

The tungsten filament, heated to incandescence by electric current, emits electrons, which are formed into a thin beam and accelerated to high velocity by the electron gun operated at up to 50 kV dc. The two reducing lenses demagnify and focus the beam to 0.1–1-μm diameter at the specimen surface, where excitation occurs. The column may be regarded as a highly sophisticated x-ray tube with the specimen as target. The lenses consist of soft iron cylinders having cross-sectional form similar to that shown in Figure 11.1. The lower portions of the cylinders are formed into *pole pieces* having apertures to limit the electron-beam diameter. The cylinders house electromagnetic coils through which pass the lens currents.

The electron beam can be made to scan the specimen surface by means of two horizontal, mutually perpendicular deflection coils connected to sawtooth sweep (time-base) generators, as shown. The X-sweep coils deflect the beam forth and back in the X direction while the Y-sweep coils deflect it in the Y direction. By appropriate use of the deflection system, the intersection of the electron beam and specimen surface can be made to assume any of three forms—point, line, or raster (rectangular area).

Provision is made to place a fluorescent screen in and out of the specimen plane to aid in aligning, focusing, and positioning the electron beam. The astigmatism compensator permits adjustment of the beam spot on the specimen plane to circular shape. The electron-beam current regulator feeds back a portion of the current intercepted by one of the beam apertures so as to vary the first-lens current and regulate fluctuations in the beam current.

The *specimen stage* holds one or more specimen mounts having diameter up to ~2.5 cm (1 inch), each of which may contain a large number of individual specimens. The stage provides for precise coarse and fine linear specimen translation in two mutually perpendicular directions (X, Y) in its own plane and normal to its own plane (Z). It also provides for rotation of the specimen in its own plane to any required orientation.

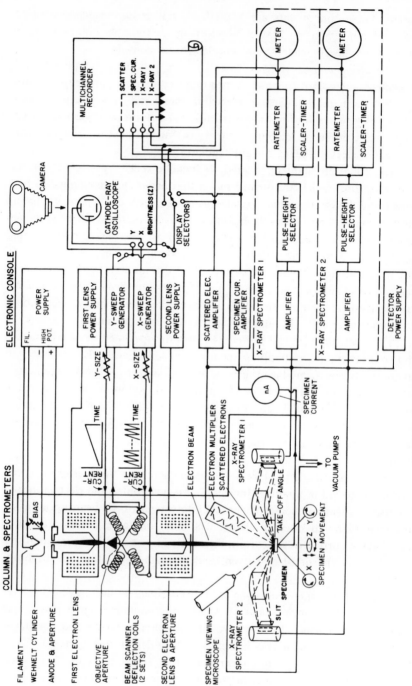

FIGURE 11.1. Electron-probe microanalyzer.

The *light microscope* permits visual observation of the specimen at magnification up to ~400× to permit placement of the specimen feature to be analyzed.

The *x-ray analysis system* consists of one to four full-focusing, curved-crystal, wavelength-dispersive x-ray spectrometers, each having its own slits, crystal, detector, preamplifier, and manual and motor-driven wavelength drive. Gas-filled proportional counters are used almost exclusively for this application. Incidentally, the x-ray spectrometer planes are usually arrayed vertically (axes horizontally) instead of horizontally as shown for convenience in Figure 11.1. Each spectrometer has a complete readout system consisting of an amplifier, pulse-height selector, scaler–timer, and recorder channel. An energy-dispersive x-ray spectrometer may also be provided having a lithium-drifted silicon detector (in its liquid-nitrogen vacuum cryostat), multichannel analyzer, and cathode-ray tube display.

The *electron-analysis system* consists of detectors for the backscattered and secondary electrons, and amplifiers and readout and display units for these and the specimen current.

The *vacuum system* consists of a rotary mechanical rough and fore pump, an oil-diffusion pump, an automatic valving arrangement, and protective devices against failure of power, diffusion-pump cooling water, and fore vacuum. The working vacuum is of the order 10^{-5} torr.

The electronics console houses the following components: (1) filament and 1–50-kV dc power supplies for the electron gun; these supplies must be extremely well regulated ($\pm 0.01\%$ or less), not so much for constancy of x-ray emission as for constancy of electron spot diameter and position; (2) two electron lens current supplies; (3) X- and Y-sweep generators; (4) backscattered electron current amplifier; (5) secondary electron current amplifier; (6) specimen current amplifier; (7) a complete electronic readout system for each x-ray spectrometer, including amplifier, pulse-height selector, ratemeter, and scaler; (8) power supplies for x-ray and electron detectors; (9) multichannel recorder, having a channel and pen for each x-ray ratemeter and the specimen current microammeter; and (10) cathode-ray display oscilloscope. If an energy-dispersive x-ray spectrometer is provided, the console also houses the spectrometer control unit, multichannel analyzer, and cathode-ray display unit.

The functions of positioning the specimen stage, setting the x-ray spectrometers, scaling the x-ray intensities, and printing out all pertinent analytical data, can all be automated. Instruments are available commercially that operate for many hours substantially unattended. Moreover, analytical concentrations can be calculated from the intensity data measured from samples and standards; even the complex intensity corrections (Section 11.6) can be made. The instrument may include a "dedicated" computer for this purpose, or the data may be transmitted to a central time-shared computer.

11.3. INTERACTION OF THE ELECTRON BEAM AND SPECIMEN

Many phenomena occur at the point of intersection of the electron beam and specimen. Five of these are commonly used in electron-probe microanalysis.

1. *Backscattered Electrons.* Some of the incident electrons are scattered back in the general direction from which they arrived. The fraction so scattered varies with the mean atomic number of the bombarded specimen spot: about half are scattered by the heaviest elements, very few by the lightest. Backscatter is also strongly influenced by surface topography. Backscattered electrons have relatively high energy, ranging up to that of the incident electrons, and their paths have high angles with respect to the specimen surface.

2. *Secondary Electrons.* The incident electrons eject some true secondary electrons from the specimen surface; these include electrons ejected from atomic orbitals by the beam electrons, and Auger electrons. The number of secondary electrons has no simple relation to atomic number, but they are even more strongly influenced by topography than scattered electrons. They have low energy ($\lesssim 50$ eV), and their paths have low angles with respect to the specimen surface.

3. *Absorbed Electrons.* Some of the incident electrons are collected by the specimen and constitute what is termed variously the *absorbed-electron, target,* or *specimen current.* The fraction so collected varies inversely with mean atomic number: nearly all are collected by beryllium, about half by uranium. The backscattered and absorbed electrons are essentially complementary, and the absorbed current approximates the beam current *minus* the backscattered and secondary electron currents. The beam current is measured most conveniently in a Faraday cup. A *Faraday cup* is a cylindrical hole ~ 1 mm in diameter and ~ 2 mm deep in a metal block, covered with a thin metal foil having a circular aperture of diameter slightly larger than that of the electron beam. The aperture and hole axis are collinear and aligned with the electron beam when the

cup is in use. The principle is that not only all the incident beam electrons, but any backscattered and secondary electrons are all trapped in the cup. Thus, the cup permits accurate measurement of the beam current.

4. *Cathodoluminescence.* The electron beam may excite visible and ultraviolet luminescence, particularly in mineral and ceramic specimens.

5. *Continuous and Characteristic X-Ray Spectra.*

Figure 11.2A shows electron diffusion zones in cross sections of the specimen surface as functions of electron-beam acceleration potential and specimen atomic number. These are the regions throughout which the incident electrons diffuse, lose energy decrementally, and finally come to rest. The figure shows several features. (1) The electron diffusion zone deepens as atomic number decreases and acceleration potential increases. (2) The zone widens, that is, electrons scatter laterally in the specimen farther and in greater numbers, as atomic number and potential increase. (3) The area from which backscattered electrons (solid arrows directed

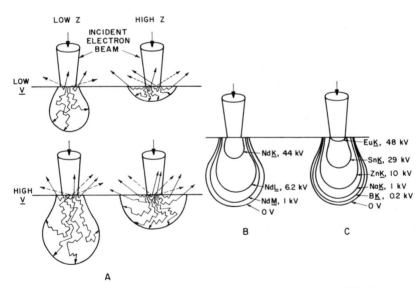

FIGURE 11.2. Sections through the specimen surface with electron-diffusion zones, showing variation of electron penetration, electron scattering, and depth of origin of x-ray spectral lines as functions of atomic number and acceleration potential. (A) Electron penetration and scatter; solid and dashed arrows directed upward represent backscattered and secondary electrons, respectively. (B) Depth of origin of electron-excited x-ray lines of different series of the same element. (C) Depth of origin of electron-excited x-ray lines of the same series of different elements. In B and C, the numbers are excitation potentials.

upward from the specimen surface) arise is wider than the area actually bombarded, and the higher the atomic number and/or lower the potential, the wider it is. (4) Finally, the area from which secondary electrons (dashed arrows directed upward) arise is the same as the bombarded area.

Figure 11.2B shows that *electron-excited K, L,* and *M* x-ray spectral lines of the same element arise from successively deeper zones. As the electrons diffuse into the specimen surface, their initial energy is lost decrementally and successively falls below the *K, L,* and *M* excitation potentials. Figure 11.2C shows that the *K* (or *L* or *M*) lines of elements of successively lower atomic number arise from successively deeper zones for the same reason.

The total zone from which characteristic x-rays arise has two to three times the volume of the electron-excitation zones shown in Figures 11.2B and C; this is because electron-excited primary continuum penetrates beyond the electron diffusion zones and excites more characteristic x-radiation.

11.4. MODES OF READOUT AND DISPLAY

The electron-probe microanalyzer can detect analytical data in the form of characteristic x-rays, backscattered, secondary, and absorbed electrons, and cathodoluminescence. The data can be collected from the specimen surface at a point, along a line, or over a raster (rectangular area). The data can be read out or displayed on meters, strip-chart or *X-Y* recorders, digital scalers, printers, or cathode-ray oscilloscopes. The cathode-ray images can be photographed. Cathodoluminescence can be observed visually in the light microscope. Table 11.1 summarizes these features.

11.4.1. Cathode-Ray Tube Displays

Before discussing the various modes of readout and display, we digress to consider the principles of cathode-ray tube display of analyte concentration distribution along a line or over a raster on the specimen surface. Figures 11.3 and 11.4 represent the electron-optic column of the electron-probe microanalyzer, an x-ray spectrometer, the cathode-ray display tube, and the horizontal and vertical deflection generators. These generators are common to the electron beams in the column and the cathode-ray tube. Thus, the electron beam in the display tube scans a line or raster (see below) in synchronism with the electron beam on the specimen surface. These

TABLE 11.1. Classification of Modes of Operation of the Electron-Probe Microanalyzer

Phenomena (principal)
Characteristic x-ray spectra
Electrons
 Backscattered
 Secondary
Absorbed (specimen current)
Cathodoluminescence

Readout Modes
Visual, through the light microscope (cathodoluminescence)
Meter (x-ray intensity and specimen current)
Recorder (x-ray intensity and specimen current)
Digital scaler (x-ray intensity and specimen current)
Printer (x-ray intensity and specimen current)
Cathode-ray oscilloscope (x-ray intensity, backscattered and secondary electrons, and specimen current)
Photographic camera (images on cathode-ray oscilloscope and in light microscope)

Spatial distribution:

Extent	Scanning mode	Movement	Distance	Principal applicable modes of readout and display	Figures
Point	—	—	—	Ratemeter, scaler	—
Line	Stepwise	Manual	1 cm	Scaler	
		Motor-driven	1 cm	"	} 11.6C
		Electronic	1 mm	"	—
	Continuous	Manual	1 cm	Ratemeter	
		Motor-driven	1 cm	"	} 11.6B
		Electronic	1 mm	Oscilloscope	11.7, 11.8B
Area (raster)	Stepwise	Manual	$(1\ cm)^2$	Scaler	
		Motor-driven	$(1\ cm)^2$	"	} 11.9
		Electronic	$(1\ mm)^2$	"	—
	Continuous	Motor-driven	$(1\ cm)^2$	Ratemeter	11.10
		Electronic	$(1\ mm)^2$	Oscilloscope	11.11

figures are essentially similar to Figure 11.1 with respect to these components except that the column is shown to have electrostatic deflection in Figures 11.3 and 11.4, electromagnetic deflection in Figure 11.1.

11.4.1.1. The Cathode-Ray Tube

The structure of the cathode-ray tube (CRT) is shown in highly simplified form at the right in Figures 11.3 and 11.4. The tube consists of the following components in an evacuated funnel-shaped glass envelope: (1) an *electron gun*, represented by the cylinder in the neck of the funnel, and consisting of an electron source (*cathode*) followed by a series of coaxial electrodes that define the electron beam, accelerate it, and focus it at a point on the circular end of the funnel; (2) an electron-beam intensity control electrode or *grid*, represented in the figures by the apertured disk just outside the electron gun; (3) two mutually perpendicular pairs of

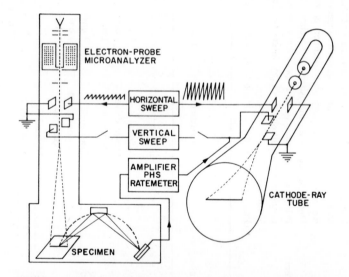

FIGURE 11.3. Arrangement for line-scanning and cathode-ray tube display of intensity and concentration distribution of an element in a line across the specimen in an electron-probe microanalyzer. The electron beams in the column and cathode-ray tube synchronously line-scan the specimen and display screen, respectively. The instantaneous x-ray spectral-line intensity excited by the electron beam on the specimen is detected, amplified, and applied to the vertical deflection plates of the cathode-ray tube, varying the instantaneous vertical deflection of the electron beam on the display screen (*deflection or Y-axis modulation*).

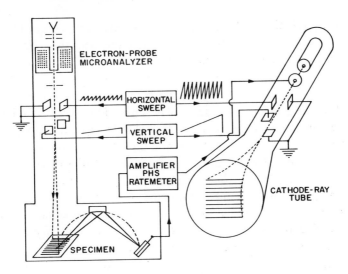

FIGURE 11.4. Arrangement for raster-scanning and cathode-ray tube display of intensity and concentration distribution of an element over a rectangular area on the specimen in an electron-probe micro-analyzer. The electron beams in the column and cathode-ray tube synchronously raster-scan the specimen and display screen, respectively. The instantaneous x-ray spectral-line intensity excited by the electron beam on the specimen is detected, amplified, and applied to the grid of the cathode-ray tube, varying the instantaneous *brightness* of the luminous spot on the display screen (*intensity*, *density*, or *Z-axis modulation*).

parallel *deflection plates* to deflect the electron beam horizontally and vertically; these are known as *horizontal* and *vertical*, or *X* and *Y*, deflection plates, respectively (the cathode-ray tube described here is the electrostatic-deflection type; the electromagnetic-deflection type has no deflection plates, and deflection is effected by external coils of wire in a "yoke" around the neck of the tube); and (4) a *display screen*, consisting of a cathodoluminescent phosphor coating on the inside surface of the flat end face of the funnel.

In operation, the cathode is heated to incandescence by an electric current and emits electrons. These electrons pass through an electron gun consisting of a series of coaxial apertures in disk electrodes separated by coaxial cylindrical electrodes, each operated at successively higher positive dc potential up to several kilovolts. The gun has the effect of forming the electrons into a fine beam, accelerating it to high velocity, and focusing it at a point on the fluorescent screen, where it excites a luminous spot.

The grid controls the intensity of the electron beam and thereby the brightness of the luminous spot. The more positive the grid, the more electrons it passes and the brighter the spot is. The more negative the grid, the fewer electrons it passes and the dimmer the spot is. If no electric potential is applied to the deflection plates, the electron beam passes along the tube axis and strikes the center of the display screen. However, if the left plate of the pair nearer the gun (Figures 11.3 and 11.4) is made positive, the electron beam is deflected to the left, and the higher the positive potential, the greater is the deflection. Similarly, positive potential to the right plate —or negative potential to the left plate—deflects the beam to the right. The second pair of plates can be used in the same way to deflect the electron beam up and down. It is evident that by application of suitable potentials to the X and Y plates, the electron beam, and thereby the luminous spot, can be placed at any specified point on the screen.

11.4.1.2. Line and Raster Scanning

In Figure 11.3, assume that a potential of $+10$ V must be applied to the left horizontal (X) deflection plate to deflect the beam to the extreme left of the display screen. Assume now that a linear *sweep* potential having a "sawtooth" or "ramp" wave form of the type shown in the figure is applied to the left horizontal plate, and that this wave form varies with time from -10 to $+10$ V, then rapidly returns to -10 V, etc.: The beam will repeatedly scan a horizontal line from right to left on the display screen. If the connections to the plates are reversed or if the polarity of the sawtooth ($+10$ to -10 V) is reversed, the scan direction will be left to right. Since the wave form returns rapidly to -10 V, the beam *flyback* is also rapid. The length of the line is determined by the sweep *amplitude* (volts). The scan frequency, that is, the number of times the line is scanned each second, is determined by the sweep *frequency*, that is, the number of "sawteeth" per second. In the same way, application of sweep potentials to the vertical (Y) deflection plates permits a vertical line to be scanned on the display screen.

In Figure 11.4, now assume that a sweep having amplitude $+10$ to -10 V and frequency 10 Hz (10/s) is applied to the horizontal-deflection plates *and* a sweep having the same amplitude but frequency 1 Hz (1/s) is applied to the vertical plates. If each of these wave forms were applied *alone* to its respective pair of plates, the display would consist of a horizontal line scan at 10/s or a vertical line scan at 1/s. However, with the two sweeps applied simultaneously, the fluorescent spot moves across the display screen from left to right 10 times each second, but not at the same place on the horizontal diagonal, as before. Instead, the first line is across the top of the screen because it occurs when the top vertical plate is most positive. Each successive horizontal line is lower as the vertical deflection becomes less positive. Finally, the 10th line is across the bottom of the screen because it occurs when the top plate is most negative. The horizontal flyback quickly returns the spot to the beginning of each new line. The vertical flyback quickly returns the spot to the top left, and the sequence is repeated. Because the flybacks are so rapid, they are very dim compared with the lines themselves and may be invisible. This parallel-line scan in known as a *raster* or *frame* (Figure 11.4). The raster width is determined by the horizontal sweep amplitude, the height by the vertical sweep amplitude. The number of lines in the raster is determined by the ratio of the horizontal and vertical sweep frequencies. The raster or frame frequency is determined by the vertical sweep frequency.

11.4.1.3. Deflection Modulation Displays

In Figure 11.3, assume that a linear sawtooth sweep of frequency 60 Hz is applied to the horizontal-deflection plates and that the 60-Hz ac line, appropriately attenuated to just give full vertical deflection, is applied to the vertical plates. These sweeps are shown in Figures 11.5A and B,

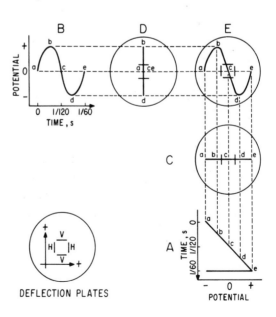

FIGURE 11.5. Deflection modulation of a cathode-ray tube line scan. (A) Horizontal (X-axis) sweep waveform (60 Hz). (B) Vertical (Y-axis) waveform to be displayed (60-Hz ac sine wave). (C) Display with horizontal sweep (A) only. (D) Display with vertical waveform (B) only. (E) Display with both horizontal sweep and vertical waveform (A and B) in synchronism as shown.

respectively; of course, Figure B is a *sine* wave. If each of these wave forms were applied *alone* to its respective pair of plates, the display would consist of a horizontal or vertical line, as shown in Figures 11.5C and D, respectively. However, if both wave forms are applied simultaneously and in synchronism, as the horizontal sweep displaces the beam horizontally and linearly, the sine wave displaces it vertically and sinusoidally. Thus, the sine wave form is traced on the display screen 60 times each second, as shown in Figure 11.5E. This process is known as *deflection* or *Y-axis modulation*.

The sine-wave pattern can be expanded or contracted horizontally or vertically by variation of the horizontal or vertical amplitude, respectively. If the horizontal sweep frequency were decreased by a factor of $\frac{1}{2}$ (30 Hz), $\frac{1}{3}$ (20 Hz), $\frac{1}{4}$ (15 Hz), etc., two, three, four, etc., sine waves would appear on the display; the same effect would result from *increasing* the ac frequency by a factor of 2, 3, 4, etc. Alternatively, if the horizontal sweep frequency were doubled (120 Hz) or the ac frequency halved (30 Hz), the positive ac alternations would be traced during the odd-numbered line scans, the negative alternations during the even-numbered scans, giving a more or less oval-shaped pattern with its long axis vertical.

We can now understand how the cathode-ray tube line scan can be used to display the concentration distribution of an element along a line on the specimen surface. In Figure 11.3, assume that the electron beam in the electron-probe microanalyzer scans a line on the specimen surface in synchronism with the beam on the cathode-ray display screen. This is effected by using a common horizontal sweep generator for both the column and cathode-ray tube. Of course, the *amplitude* of the specimen sweep must be smaller because the specimen is smaller ($\sim$1 mm) than the cathode-ray tube display screen ($\sim$100 mm). This also provides the basis for magnification: $100/1 = 100\times$; larger magnifications are attained by scanning smaller distances on the specimen while leaving the cathode-ray scan the same length.

In Figure 11.3, now assume that the x-ray spectrometer is set for, say, Fe $K\alpha$ and that the amplified detector output is applied to the vertical deflection plates and provides the vertical deflection for the cathode-ray display. Then, when the electron beam line-scans the specimen, as iron concentration varies along its path, the instantaneous Fe $K\alpha$ intensity varies, so does the detector output, and so does the vertical displacement of the cathode-ray spot. Thus, the cathode-ray beam traces the instantaneous Fe $K\alpha$ intensity from the corresponding point on the specimen. Displays of this type are shown in Figures 11.7 and 11.8B.

A variation of the deflection modulation technique is shown in Figure 11.8C. A raster is scanned on the cathode-ray display screen and specimen surface, and the x-ray detector output is then superimposed on the linear vertical sweep, giving the deflection-modulated raster image shown in Figure 11.8C. Each line in this raster scan shows Au $M\alpha_1$ intensity distribution along a different line across the specimen described in the figure caption.

11.4.1.4. Intensity Modulation Displays

In Figure 11.4, assume that the electron beam in the electron-probe microanalyzer scans a raster on the specimen surface in synchronism with the beam on the cathode-ray tube display screen. This is effected by using common horizontal and vertical sweep generators for both the column and cathode-ray tube. As is the case for deflection modulation (above), the amplitude of the specimen sweep must be smaller and provides the basis for magnification. Now assume that the x-ray spectrometer is set for, say, Fe $K\alpha$ and that the amplified detector output is applied to the grid of the cathode-ray tube and therefore determines the instantaneous electron-beam intensity and luminous-spot brightness. Then, when the electron beam scans the specimen, as iron concentration varies along its path, the instantaneous Fe $K\alpha$ intensity varies, so does the detector output, and so does the brightness of the cathode-ray spot. Thus, the cathode-ray beam traces the instantaneous Fe $K\alpha$ intensity from the corresponding point on the specimen. This process is known as *intensity, density,* or *Z-axis modulation.* Displays of this type are shown in Figures 11.7 and 11.11A. Figures 11.8A and 11.11A show similar displays in which specimen current and backscattered electrons, respectively, rather than x-rays, produce the modulation. Figure 11.11B shows a color display derived from intensity-modulated x-ray raster images.

11.4.2. Measurement at a Point

With neither X nor Y deflection, the electron beam intersects the fixed specimen surface at a fixed point. The specimen is placed so that the beam falls at a point of interest. A qualitative analysis is made by recording the full emitted x-ray spectrum using the x-ray spectrometer(s). If three wavelength-dispersive spectrometers are available, they may be fitted with crystals and detectors for the short-, intermediate-, and long-wavelength regions, respectively. These three regions can be recorded simultaneously on a multipen recorder. Alternatively, an energy-dispersive spectrometer having a cathode-ray tube display can be used. A quantitative analysis is made by measuring analyte-line intensities from samples and standards. The intensity data is displayed on digital scalers or printed out.

11.4.3. Measurement along a Line

A line is scanned across the specimen by translating the specimen under a fixed electron beam or by deflecting the beam over a fixed specimen. Either type of scan can be continuous or incremental (stepwise).

The specimen can be translated for distances of the order 1 cm, manually or by motor drive, continuously with ratemeter–recorder output, or incrementally with scaler–printer readout. A motor-driven continuous scanner translates the specimen at 0.01–1 μm/s. A motor-driven step-scanner automatically translates the specimen in preselected increments of 0.1, 0.5, 1, 5, 10, or 50 μm, stopping at each point to scale and print out analyte-line counts and specimen current.

With only X (or Y) deflection, the electron beam scans a line across the specimen. The length of the scanned line can be varied from a point to perhaps 1000 μm (1 mm) by varying the amplitude of the X sweep coil current with the X size control (Figure 11.1). The line-scan rate can be varied from, say, 500/s to once every 10 s by varying the frequency of the sweep current. Cathode-ray tube display is required.

The concentration distribution profile of a selected analyte along a line across the specimen is mapped by setting an x-ray spectrometer at the analyte line, moving the specimen manually or with the step scanner in small increments, and scaling the intensity at each point. Profiles of two to four different analytes can be scanned simultaneously if two to four spectrometers are available. Subsequent line scans can be made along the same line for different elements or along different lines for the same elements. The intensity data is plotted manually on coordinates of intensity *versus* distance. A set of three such incremental line scans is shown in Figure 11.6C.

The same profiles are mapped much more rapidly and conveniently, but less precisely, by translating the specimen continuously with the motor drive while recording analyte-line intensity on the ratemeter–recorder. Profiles of two to four elements can be recorded simultaneously with a multipen recorder. A set of three such continuous line scans is shown in Figure 11.6B.

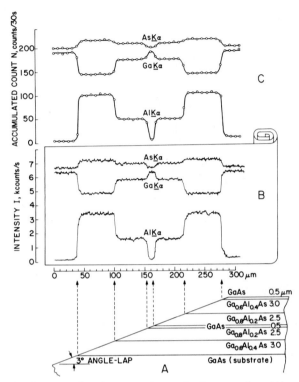

FIGURE 11.6. Continuous and stepwise line-scanning. (A)
Angle-lapped layered specimen of gallium arsenide (GaAs)
and gallium aluminum arsenide [(Ga,Al)As]. (B) Ratemeter–
recorder chart readout of continuous mechanical line scan
of A. (C) Plotted scaler readout of stepwise mechanical line
scan of A.

Finally, the concentration profile of a selected analyte along a line
can be displayed almost instantaneously by setting one of the x-ray spectrometers at the analyte line, allowing the electron beam to line-scan the
specimen, and displaying analyte-line intensity on the cathode-ray oscilloscope. The X-sweep coil deflects the electron beam in the cathode-ray
tube across the display screen in synchronism with the electron beam across
the specimen surface. The x-ray spectrometer ratemeter output then provides
the Y deflection of the cathode-ray tube beam, that is, deflects the beam
vertically in proportion to instantaneous analyte-line intensity. The result
is a bright-line cathode-ray trace of intensity *versus* distance, analogous to
the ratemeter–recorder charts or scaler plots described above. Such line
scans are shown in Figures 11.7 and 11.8B.

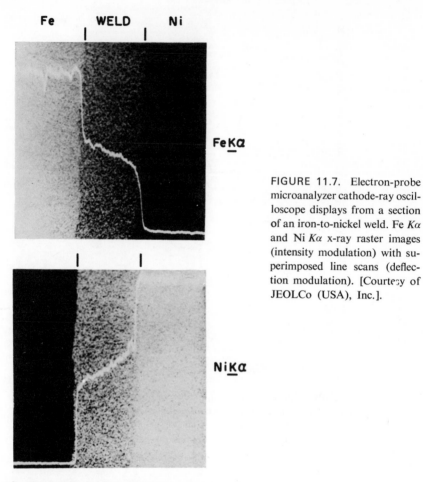

FIGURE 11.7. Electron-probe microanalyzer cathode-ray oscilloscope displays from a section of an iron-to-nickel weld. Fe $K\alpha$ and Ni $K\alpha$ x-ray raster images (intensity modulation) with superimposed line scans (deflection modulation). [Courtesy of JEOLCo (USA), Inc.].

11.4.4. Measurement over a Raster

A raster is scanned over the specimen by specimen translation or electron-beam deflection. The raster consists of repetitive line scans from top to bottom of the rectangular area. The individual lines may be continuous (*line raster*) or incremental (*point raster*). If the individual raster lines consist of specimen translations, the raster may have area up to ~ 1 cm square with ratemeter–recorder or scaler–printer readout for line or point rasters, respectively.

With both X and Y deflection, the electron beam scans a rectangular, usually square, line raster on the specimen. The X and Y size of the raster can be varied from a point to perhaps 1000 μm square by varying the amplitudes of the X and Y sweep currents with the X and Y size controls

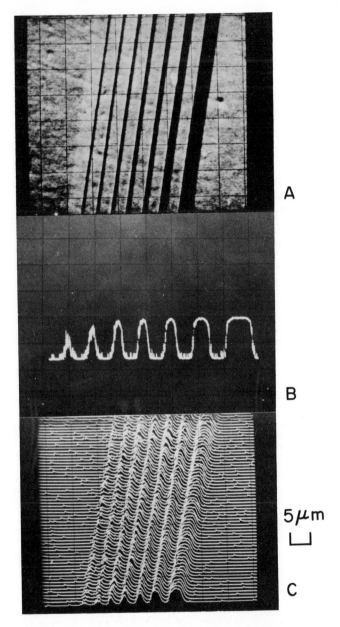

FIGURE 11.8. Electron-probe microanalyzer cathode-ray oscilloscope displays from a section of a layered gold-copper specimen in which the gold layers have thicknesses 0.2, 0.3, 0.6, 0.8, 1, 2, and 4 μm, respectively. [Courtesy of ETEC, Inc.]. (A) Specimen current raster image (intensity modulation). (B) Au $M\alpha$ x-ray line scan (deflection modulation). (C) Au $M\alpha$ x-ray raster image (deflection modulation of individual raster lines).

(Figure 11.1). The number of lines in the raster can be varied from ~100 to ~500 lines by varying the X sweep frequency. The raster scan rate can be varied from 1–150 s by varying the Y sweep frequency. Cathode-ray tube display is required.

With stepwise specimen translation, the action is as follows. A line is scanned incrementally from left to right (X direction); the specimen is then translated one increment downward (Y direction); another line is scanned, etc., until the entire point raster has been scanned. The X lines may be scanned left–right, right–left, left–right, etc., or left–right, left–right, left–right, etc. At each point, analyte-line counts are scaled, usually for a preset time. The multichannel analyzer, operated in multichannel scaler mode (Section 6.1.2), is extremely valuable for storing point-raster data. The accumulated analyte-line counts from successive raster points are stored in successive channels in the multichannel analyzer. Then, at the end of the scan, the data may be printed out in columns for manual plotting, or

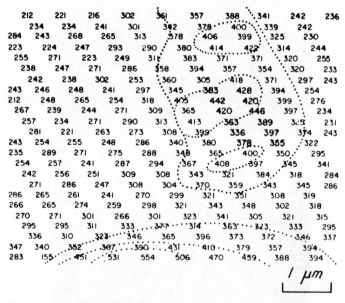

FIGURE 11.9. Electron-probe microanalyzer mechanical point raster: multichannel-scaler printout intensity topograph from an aluminum alloy having iron-rich inclusions. The numbers are Fe $K\alpha$ counts accumulated in the multichannel scaler channels. The isointensity lines were drawn in manually by joining points of substantially equal intensity; they delineate the iron-rich regions. [L. S. Birks, *Electron-Probe Microanalysis*, 2nd ed., Wiley–Interscience, New York, p. 81 (1971); courtesy of the author and publisher.]

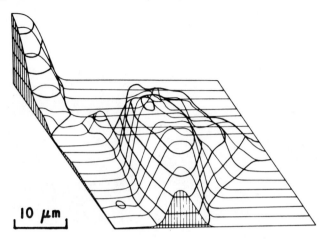

FIGURE 11.10. Electron-probe microanalyzer mechanical line raster from leaded brass (Cu–Zn–Pb). The figure shows successive line-scans of the Pb $L\alpha_1$ line read out on an X–Y recorder. The starting points of successive lines are displaced along an oblique line with respect to the horizontal (distance) axis. [K. F. J. Heinrich, *Advances in X-Ray Analysis* **7**, 385 (1964); courtesy of the author and Plenum Press.]

printed out in a two-dimensional array, each count in its correct relative position in the raster. Figure 11.9 shows such a two-dimensional printout, point-raster, intensity topograph.

With continuous specimen translation, the action is the same, except that the line translations are continuous rather than stepwise, and the data is read out on an X–Y recorder. Figure 11.10 shows such a line raster.

In the electron-beam scanning method, the X- and Y-sweep coils deflect the electron beam in a line raster on the specimen surface and cathode-ray tube display screen. The x-ray spectrometer pulse-height selector or ratemeter output then modulates the intensity (Z axis) of the cathode-ray beam in proportion to instantaneous analyte-line intensity, so the brightness at each incremental area on the display is proportional to analyte-line intensity—and thereby analyte concentration—at the corresponding incremental area on the specimen. Thus, the line-intensity and concentration distribution of a selected analyte over the raster area are displayed on the oscilloscope. Electron-beam, line-raster, cathode-ray tube displays are shown in Figures 11.7 and 11.11A.

When the output of the pulse-height selector is applied to the cathode-ray tube, the tube actually displays individual x-ray photons, each as a bright spot, giving a "snowy" display (Figures 11.7 and 11.11A). The

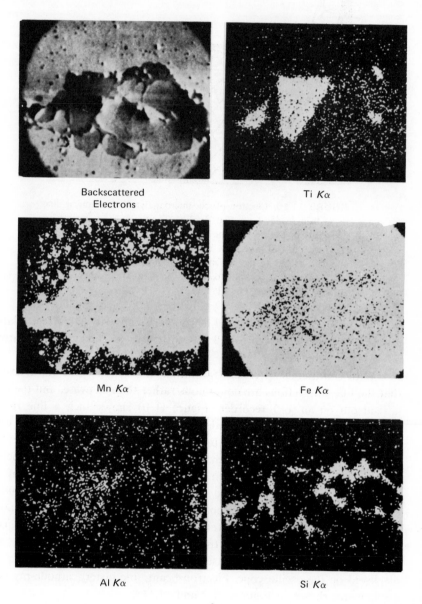

Backscattered
Electrons

Ti $K\alpha$

Mn $K\alpha$

Fe $K\alpha$

Al $K\alpha$

Si $K\alpha$

A

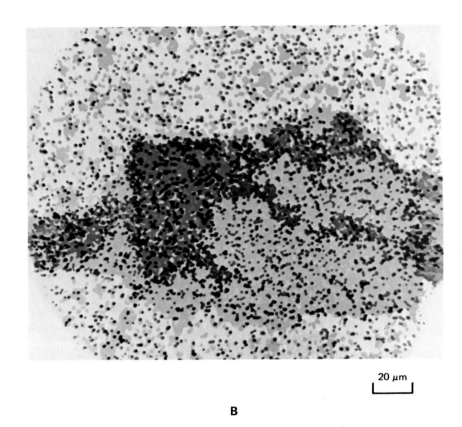

20 μm

B

FIGURE 11.11. Electron-probe microanalyzer displays of the same area of an inclusion in iron. [L. P. Salvage, *Metallography* **2,** 101(1969); courtesy of the author, Elsevier Publishing Co., and the International Microstructural Analysis Society.] (A) Cathode-ray oscilloscope raster images (intensity modulation) of backscattered electrons and $K\alpha$ x-rays of Ti, Mn, Fe, Al, and Si. (B) Indirect-method composite color photograph of all the x-ray images in A. The oscilloscope raster images in A were photographed on separate black-and-white negative films, each of which was then used to expose a different color diazochrome film to produce a color transparency. These color transparencies were then superimposed in register and the composite exposed onto a color film. The color code is as follows: red—Ti $K\alpha$; yellow—Mn $K\alpha$; light blue—Fe $K\alpha$; dark blue—Al $K\alpha$; and green—Si $K\alpha$. The dots represent individual x-ray photons; the Fe $K\alpha$ (light blue) and Mn $K\alpha$ (yellow) images appear continuous because the intensities of these lines are high.

"snow" fluctuates due to the random time distribution of x-rays. Suppose that a 300-line raster being scanned once every 2 s is displaying, say, Cu $K\alpha$ at average intensity 5000 counts/s. The raster may be regarded as consisting of $300 \times 300 = 90,000$ *picture elements*. Then, on an average, $(2 \times 5000)/90,000$ or one picture element in nine receives a pulse, that is, displays a bright spot. Of course, the spots are much more numerous in regions of the display corresponding to high copper concentration, less numerous in low-copper regions. Also, spots appear in the display even in regions corresponding to copper-free areas because of continuous background pulses. The image quality is improved by slower scanning or by integration of several scans by photographic recording or by a storage oscilloscope.

Another type of oscilloscope line-raster display combining the line and raster modes is described in the last paragraph of Section 11.4.1.3.

The discussions in Sections 11.4.2, 11.4.3, and 11.4.4 pertain to characteristic x-rays. However, absorbed, backscattered, and secondary electron currents are also read out and displayed in point, line, and raster modes. The backscattered-electron and specimen-current displays reveal the distribution of effective atomic number. The higher the effective atomic number of the incremental specimen area under the electron beam, the brighter is the corresponding point on the scattered-electron display, the dimmer is the point on the specimen-current display. All three types of electron display reveal surface topography. Secondary-electron displays have much higher resolution than x-ray, backscattered-electron, and specimen-current displays; this is because secondary electrons arise only within the surface area of electron-beam impact, whereas the other quanta arise within the wider and deeper electron-diffusion zone (Figure 11.2). Backscattered- and absorbed-electron line and raster scanning can be done at much higher frequency than x-ray scanning because the electrons are more numerous than x-rays by several orders.

11.4.5. Color Displays

Color photography can be used to aid in the visualization and interpretation of electron-probe raster displays. Different colors may be used to represent different chemical elements or different concentrations of the same element.

The first application can be realized in several ways. Successive raster images of the same specimen area can be displayed on the oscilloscope from, say, three x-ray spectrometers set for x-ray spectral lines of three different elements. Each image is photographed on the same color film, but through a filter of a different color. For example, the images of the shortest, intermediate, and longest wavelengths can be photographed through blue, green, and red filters—that is, in blue, green, and red light— respectively. In this way, a color picture of the specimen area is obtained in which different colors represent different elements, and variations in color density over the image represent variations in concentration of the element represented by that color. This technique may be termed the *direct method.*

Alternatively, the several oscilloscope images can be photographed on separate black-and-white positive films, then each film exposed in turn through a different color filter onto the same color film. A third technique is to photograph the oscilloscope images on separate black-and-white negative films, each of which is then used to expose a different color diazochrome film to produce a positive color transparency. These color transparencies are then superimposed in register, and the composite exposed onto a color film. Figure 11.11A shows the several black-and-white oscilloscope x-ray displays from an inclusion in iron, and Figure 11.11B shows the resulting color composite prepared by this technique. The two techniques described in this paragraph may be termed *indirect methods.*

Incidentally, the cathode-ray tube used for the direct method must have a phosphor having white—rather than yellow, green, or blue—luminescence; this does not necessarily apply to tubes used for the indirect methods.

Finally, the same type of color display can be obtained simultaneously by using the outputs of the three x-ray spectrometers to modulate the brightness of the red, green, and blue electron guns, respectively, of a color-television picture tube. The advantages of this method are that the picture is seen directly and can be adjusted to the operator's satisfaction prior to photographing it, and that errors arising from slight differences in size and orientation of the raster images and slight misalignments of superimposed images do not occur.

11.5. SPECIMEN CONSIDERATIONS

In x-ray secondary-emission (fluorescence) spectrometry, the depth of the surface layer contributing the measured x-rays is limited by the depth *from* which the analyte-line x-rays can emerge and is of the order 100 μm. In primary-emission spectrometry, the limit is imposed by the depth *to* which the electrons can penetrate with enough energy to exceed the analyte excitation potential and is of the order 1 μm. Consequently, absorption–enhancement effects are less severe and surface-texture effects more severe in electron-probe microanalysis than in fluorescence spectrometry. Microscopic topography can be detrimental in three ways. Incident electrons may be shielded from parts of the surface, and x-rays originating in parts of the surface may be shielded from the x-ray spectrometers. Also, microscopic slopes present local takeoff angles different from that of the specimen plane, and takeoff angle affects the x-ray intensity in the direction of the spectrometer.

The more severe surface-texture effects necessitate highly plane and polished specimen surfaces. For accurate quantitative analysis, or even for simple comparison of analyte-line intensities, all related specimens should be "potted" in the same plastic mold or cemented to the same mount, then polished to an identical surface plane and high polish. The final polish should be made with 0.25-μm diamond dust on a cloth lap. Other abrasives, such as aluminum oxide, silicon carbide, and cerium oxide, and lead- or tin-coated polishing wheels are avoided to prevent contamination of the specimen surface. Etches and electropolishes are avoided because they may alter surface composition and create topography.

Multiple-thin-layer specimens may be probed on their edges or angle-lapped to expose and, in effect, widen the several layers (Figure 11.6).

It is necessary to evaporate a carbon film 50–100 Å thick on electrically nonconductive specimens to prevent accumulation of electric charge.

It is usually difficult or impossible to observe features on highly polished surfaces under the light microscope. In such cases, the feature may be located as follows. Spacial variations in composition may be satisfactorily revealed in the scattered or absorbed electron raster displays. Alternatively, the specimen may be etched to reveal the features, then, under observation in a light microscope, scored, or pitted with a micro hardness tester to locate the features of interest. The specimen is then polished to remove the etch topography, but not the scoring or microhardness marks.

For any mode of measurement, the region of interest is selected by use of the light microscope and specimen translators. If observation at higher magnification is required (up to $5000\times$), the absorbed, scattered, or (preferably) secondary electron oscilloscope raster display may be viewed. The exact spot, line, or area to be studied may be placed by use of the light-microscope cross-hairs or, more precisely, the oscilloscope displays.

After the electron beam has bombarded the selected region for awhile, a visible dark spot, line, or square appears on the specimen viewed through the light microscope. The darkening results from carbonization, by the electron beam, of adsorbed residual oil vapors from the diffusion pump. These deposits may interfere with the x-ray measurements in one or more of three ways: (1) They reduce the energy of the incident electrons; (2) they absorb the emergent x-rays, although this effect is serious only for very long wavelengths; and (3) they contribute analyte-line intensity when carbon or other elements present in the deposit are to be determined. Otherwise, the deposits are beneficial in that they mark precisely the spot, line, or area investigated.

In electron-probe microanalyzers, the electron-beam diameter at the specimen plane is $\sim$1 μm. In scanning electron microscopes, the beam diameter is $\sim$200–1000 Å. The smaller the beam spot, the higher is the resolution, but the smaller is the maximum attainable beam current and therefore the smaller the attainable x-ray intensity. Larger beam diameters, up to $\sim$100 μm, may be used to increase the attainable beam current, average out surface texture and microheterogeneity, and reduce the rate of carbon deposition (see above).

Although most instruments are capable of operation up to 50 kV, 25 kV is a typical practical maximum operating potential. Lower potentials are used to: (1) limit the effective specimen thickness to that of the thinnest film in the group to be analyzed; (2) avoid excitation of underlying layers; (3) avoid excitation of interfering spectral lines; (4) improve spacial resolution by reduction of lateral electron scatter (Figure 11.2); and (5) reduce the magnitude of the atomic number and absorption intensity corrections (Section 11.6).

11.6. QUANTITATIVE ANALYSIS

In principle, in either primary or secondary x-ray emission spectrometry, one would expect the concentration $C_{A,X}$ (weight fraction) of analyte A in sample X to be given by

$$C_{A,X} = I_{A,X}/I_{A,A} = R_A \tag{11.1}$$

where $I_{A,X}$ and $I_{A,A}$ are analyte-line intensities from the sample and pure analyte, respectively. Although this equation may give high accuracy in favorable cases in both methods, it is not generally satisfactory in either. The reason is that the incident electron (or x-ray) beam and the emergent analyte-line x-rays are affected not only by the analyte, but also by the specimen matrix. The equation is usually more accurate in electron-probe microanalysis than in fluorescence spectrometry because the measured x-rays originate in a very thin surface layer, so the effect of the matrix is reduced.

When the equation does not give acceptable accuracy, the matrix effect must be dealt with in one of two ways, with calibration standards or by mathematical calculation.

The first of these methods is the more convenient and satisfactory. Analyte-line intensity is measured from the samples and from two or more standards of the same substance of known composition. The standards must be homogeneous on a submicrometer scale, and standards and samples must have identical fine surface finishes. An intensity-concentration relationship is established from the standard data and used to derive analyte concentrations for the samples.

When suitable standards are not available, resort is made to the mathematical methods. Essentially, the procedure is as follows.

1. Analyte-line intensity is measured from the samples and from a reference standard. Usually the standard is the pure analyte, but when its use is not feasible (gaseous elements, sodium, sulfur, bromine, etc.), a suitable compound or alloy having known composition is used. The more similar such standards are to the samples, the more effective they will be. Beam current must be constant for all measurements of any one analyte line, but specimen current must be allowed to vary with the mean atomic number of the specimen.

2. The measured intensities are corrected to increase their accuracy, that is, for instrument drift, dead time, background, etc.

3. The *relative intensity*—ratio of corrected analyte-line intensities from the sample and standard—is calculated for each sample.

4. The relative intensity is corrected for atomic number (Z), absorption (A), and fluorescence (F) effects to give analyte concentration.

a. Atomic number correction. As the average atomic number of the matrix decreases, the beam electrons penetrating the specimen lose energy more rapidly because light elements are more readily ionized; thus, the matrix absorbs more electrons, tending to reduce the number available for analyte-line excitation. If this were the only atomic number effect, a specified concentration of a specified analyte would give a lower analyte-line intensity from a low-Z matrix than from a high-Z matrix. However, as the average atomic number of the matrix decreases, the fraction of beam electrons that backscatter also decreases, making more electrons available for ionization—that is, for excitation—and tending to give a higher analyte-line intensity than in a high-Z matrix. These penetration and backscatter phenomena, which are also functions of beam acceleration potential, tend to compensate one another. A third atomic number effect is the ionization effect. The number of x-ray-producing ionizations excited by the electrons in unit mass of analyte is given by the ratio of the number of orbital electrons and mass of atoms of that element, that is, by the quotient of atomic number and weight, Z/A.

   b. *Absorption correction.* Analyte-line x-rays excited beneath the specimen surface undergo absorption in emerging from the specimen.

   c. *Fluorescence correction.* Analyte-line x-rays arise not only by primary excitation by the electron beam, but also by secondary excitation by spectral-line x-rays of other elements in the specimen and by the continuous x-rays. Such secondary excitation is efficient only for relatively intense spectral lines and continuum having wavelength shorter than, and relatively near, the analyte-line absorption edge.

   In general, analyte concentration C_A is given by

$$C_A = R_A(ZAF) \tag{11.2}$$

where R_A is the ratio of analyte-line intensities from the sample X and pure analyte A, $I_{A,X}/I_{A,A}$; and (ZAF) is the total correction for atomic number (Z), absorption (A), and fluorescence (F) effects. The atomic number correction contains factors for both electron backscatter and ionization-penetration. The fluorescence correction is usually limited to enhancement by lines of matrix elements, but may include enhancement by continuum. Incidentally, in the literature of electron-probe microanalysis, the symbol k is usually used for relative intensity rather than R.

   A comprehensive treatment of intensity corrections for electron-probe microanalysis is beyond the scope of this introductory chapter, but the chapter would not be complete without consideration of a typical correction procedure as an example. The original version of Colby's MAGIC (Microprobe Analysis Generalized Intensity Corrections) procedure provides an excellent example. In the following discussion, subscripts A and X, when used on r, s, $f(\chi)$, and (μ/ϱ), refer to pure analyte and sample, respectively.

   Equation (11.2) may be expanded as follows:

$$C_A = \frac{I_{A,X}}{I_{A,A}} \left(\frac{r_A}{r_X} \frac{s_X}{s_A} \right) \left[\frac{f(\chi)_A}{f(\chi)_X} \right] \left[\frac{1}{1 + \sum (I_j/I_e) + (I_{cont}/I_e)} \right] \qquad (j \neq A) \tag{11.3}$$

where the three bracketed terms are the atomic number, absorption, and fluorescence corrections, respectively, and are expanded below.

In the *atomic number correction*, r is the electron-backscatter coefficient and is the ratio of ionization produced to that which would be produced in the absence of backscatter; and s is the electron stopping power,

$$s_i = \frac{Z}{A} \log_e\left[1.17 \frac{(V + V_i)/2}{\bar{V}_i}\right] \tag{11.4}$$

where Z and A are atomic number and weight, respectively; V and V_i are electron acceleration potential and excitation potential, respectively; and $\bar{V}_i$ is the mean ionization potential.

In the *absorption correction*, $f(\chi)$ ("f of chi") is the fraction of the generated analyte-line photons that are actually emitted:

$$f(\chi)_i = \frac{1 + h}{[1 + (\chi/\sigma)]\{1 + h[1 + (\chi/\sigma)]\}} \tag{11.5}$$

where χ is the attenuation of the emergent analyte-line x-rays, h is an electron penetration term, and σ is the effective Lenard electron absorption coefficient as modified by Heinrich:

$$\chi_i = (\mu/\varrho)_{i,\lambda_A} \csc \psi \tag{11.6}$$

$$h_i = 1.2(A/Z^2)_i \tag{11.7}$$

$$\sigma_i = (4.5 \times 10^5)/(V^{1.67} - V_i^{1.67}) \tag{11.8}$$

where $(\mu/\varrho)_{i,\lambda_A}$ is the mass-absorption coefficient of element i for the analyte line λ_A; ψ is the x-ray emergence angle; and A, Z, V, and V_i are defined above.

In the *fluorescence correction*, the enhancement of the analyte line λ_A by spectral lines of matrix elements j is given by the ratio of analyte-line intensities excited by elements j, I_j, and directly by the electron beam, I_e, summed for all matrix lines that can significantly enhance λ_A:

$$\left(\frac{I_j}{I_e}\right)_{\lambda_A} = 0.5 P_{Aj} C_j \frac{r_A - 1}{r_A} \omega_j \frac{A_A}{A_j} \left(\frac{U_j - 1}{U_A - 1}\right)^{1.67} \frac{(\mu/\varrho)_{A,\lambda_j}}{(\mu/\varrho)_{X,\lambda_j}}$$

$$\times \left[\frac{\log_e(1 + u)}{u} + \frac{\log_e(1 + v)}{v}\right] \quad (j \neq A) \tag{11.9}$$

in which

$$U_A = V/V_A \quad \text{and} \quad U_j = V/V_j \tag{11.10}$$

$$u = [(\mu/\varrho)_{X,\lambda_A} \csc \psi]/(\mu/\varrho)_{X,\lambda_j} \tag{11.11}$$

$$v = \sigma/(\mu/\varrho)_{X,\lambda_j} \tag{11.12}$$

and where C_j is concentration (weight fraction) of enhancing element j; r_A is analyte absorption-edge jump ratio; ω is fluorescent yield; and A, Z, V, V_j, σ, χ, and ψ have the significance given above. The constant P_{Aj} has the value 1 for a $j\,K$ line enhancing an A K line or a $j\,L$ line enhancing an A L line, 4 for a K line enhancing an L line, and 0.25 for an L line enhancing a K line. The enhancement of λ_A by continuum $(I_{cont}/I_e)_\lambda$, may often be disregarded and is not considered here.

In all the foregoing equations, parameters for the sample X are calculated as follows:

$$r_X = \sum (C_i r_i) \tag{11.13}$$

$$s_X = \sum (C_i s_i) \tag{11.14}$$

$$f(\chi)_X = \sum [C_i(\mu/\varrho)_{i,\lambda_A}] \csc \psi \tag{11.15}$$

$$(\mu/\varrho)_{X,\lambda} = \sum [C_i(\mu/\varrho)_{i,\lambda}] \tag{11.16}$$

in which i represents the individual constituent elements—including the analyte—in the sample, and C is weight fraction.

There are simplified practical versions of some intensity correction systems, and tables of constants and other data for their solution are available. In many such cases, corrections for two-, three-, and possibly even four-component systems can be calculated manually on a desk calculator. However, in general, a computer is required for these calculations.

11.7. PERFORMANCE

Sensitivity performance of a modern commercial electron-probe microanalyzer is summarized in Table 11.2 for the $K\alpha$ lines of selected chemical elements from atomic number 4 to 29 (beryllium to copper). However, the detection limits given in the table apply only to specimens having effective atomic number similar to that of the analyte itself. The sensitivity may be much less than that given for a heavy matrix, and much greater for a light matrix. The data probably represents optimal performance. Background is relatively high because of the electron excitation, but peak-to-background ratios are reasonably high.

If the beam diameter and penetration are both ~ 1 μm, the effective specimen volume is ~ 1 μm³ or 10^{-12} cm³. For a density of 10 g/cm³, the effective specimen mass is 10^{-11} g, and, at the detection limit—say, 0.01 wt%—the contributing analyte mass is only 1 fg (10^{-15} g), compared with 1 ng (10^{-9} g) for x-ray fluorescence spectrometry. However, direct com-

TABLE 11.2. Performance of a Modern Electron-Probe Microanalyzer[a]

Element and atomic number	Specimen[b]	Crystal[c]	Line intensity[d] I_L $\left(\dfrac{\text{counts/s}}{\mu A}\right)$	Peak-to-background ratio, I_P/I_B	Limit of detectability[e] (wt%)
$_4$Be	Be	PbLig	6.0×10^2	50	1.7
$_5$B	B	PbSt	1.5×10^4	50	0.35
$_6$C	C	''	4.0×10^4	70	0.18
$_8$O	$SiO_2{}^b$	''	3.0×10^4	20	0.20
$_9$F	LiF^b	KHP	1.1×10^4	280	0.12
$_{11}$Na	$NaCl^b$	''	3.2×10^4	550	0.03
$_{12}$Mg	Mg	''	4.0×10^5	1500	0.01
$_{13}$Al	Al	''	8.0×10^5	1300	0.01
$_{14}$Si	Si	''	1.2×10^6	800	0.01
$_{22}$Ti	Ti	PET	1.8×10^6	700	0.01
$_{26}$Fe	Fe	LiF	2.5×10^5	460	0.03
$_{29}$Cu	Cu	''	1.5×10^5	230	0.05

[a] Courtesy of JEOLCo (USA), Inc.
[b] SiO_2 is 53% O; LiF is 73% F; NaCl is 39% Na.
[c] The symbols indicate, in order, lead lignocerate, lead stearate, potassium hydrogen phthalate, pentaerythritol, and lithium fluoride.
[d] All $K\alpha$ lines; acceleration potential 25 kV.
[e] Defined as $(3I_B^{1/2}/I_L)C$, where I_L and I_B are analyte-line and background intensities, respectively, and C is analyte concentration (wt%); $C = 100\%$, except as specified in note b above.

parison of the detection limits of electron-probe microanalysis and x-ray fluorescence spectrometry is not valid. A local concentration of 0.01 wt% (100 ppm) in the microanalyzer may correspond to an average concentration in the bulk sample of 1–0.001 ppm (1 ppm to 1 ppb).

11.8. APPLICATIONS

The electron-probe x-ray microanalyzer fills the great need for in-place nondestructive determination of composition and spacial variation of composition on a microscopic scale in analytical chemistry, metallurgy, ceramics, mineralogy, geology, and biology and medicine. The instrument is extremely valuable in materials research because the properties of materials depend not only on average composition, but also on local concentrations of elements in individual grains, at grain boundaries, etc.

Among the many classes of applications are the following: (1) analysis of individual crystallites or grains of phases in heterogeneous alloys, ceramics, and minerals; (2) analysis of microscopic precipitates, inclusions, and segregations in the same materials; (3) analysis of grain boundaries; (4) analysis of individual layers in sections of coated, plated, or corroded surfaces or of bonded surfaces; (5) mapping of concentration gradients at metal-to-metal or metal-to-ceramic interfaces, diffusion couples, diffusion zones, heterogeneous grain boundaries, etc.; (6) evaluation of microscale homogeneity; (7) mapping of distribution of elements over the surfaces of heterogeneous alloys, ceramics, minerals, etc.; (8) analysis of very small particles filtered from gases and liquids; (9) determination and distribution of natural and foreign elements in biological materials; (10) analysis of specific areas on small semiconductor electron devices and other very small parts; and (11) determination of composition and thickness of thin films.

Electron-probe microanalysis complements other methods of selected-area microanalysis. Selected-area accessories for x-ray fluorescence spectrometers sample a much larger area and deeper layer. Auger-electron spectrometry (AES), secondary-ion mass spectrometry (SIMS), and ion-scattering spectrometry (ISS) also sample somewhat larger areas, but much thinner layers—as thin as a monolayer.

11.9. COMPARISON WITH X-RAY FLUORESCENCE SPECTROMETRY

It must not be concluded that operation of the electron-probe micro-analyzer is as simple, convenient, rapid, and reliable as that of the x-ray fluorescence spectrometer. An experienced fluorescence spectrometer operator is likely to find the transition to electron-probe microanalysis very frustrating at first. The reverse transition should be a pleasant surprise!

The electron-probe microanalyzer is much more complex than the x-ray fluorescence spectrometer. A highly sophisticated and sometimes "temperamental" electron-optic column replaces the simple, stable, sealed-off x-ray tube. The requirement for inclusion of the specimen in the vacuum enclosure precludes liquid, slurry, and some types of biological specimens. Full-focusing curved-crystal spectrometers replace the nonfocusing flat-crystal spectrometers. Even the specimen stage is complex, with its several micrometer-driven movements and its requirements for highly reproducible positioning and freedom from backlash. For proper performance, the instrument must be precisely aligned so that the axes of the electron-optic column and light microscope and the x-ray paths to the several x-ray spectrometers all converge at a common point on the specimen plane.

In x-ray fluorescence spectrometry, the excitation conditions are the potential (kV) and current (mA) in the sealed-off x-ray tube. These conditions are very stable, accurately monitored on their respective meters, and readily reset as required, usually with selector switches. In electron-probe microanalysis, the excitation conditions are the acceleration potential (kV) and *electron-beam current* (μA, nA) in the electron-optic column. The potential is stable, accurately monitored on a meter, and readily reset, but the beam current may drift and is difficult to monitor. Beam current regulators are effective at the low currents used in scanning electron microscopes, but not at the higher currents used in electron-probe microanalyzers. The beam current may be reset by periodically directing it on a Faraday cup or on a fresh point of the same highly polished homogeneous metal, preferably a pure element, then resetting the *specimen current* to a constant value. The specimen current is readily monitored on a meter, but may vary, not only because of beam-current drift, but because of variation in mean atomic number from specimen to specimen and even from point to point on the same specimen. In quantitative analysis with calibration standards very similar to the samples, the specimen current may be reset at a constant value for each measurement. However, if mathematical atomic number correction is to be applied, it is not permissible to do this. Rather, the *beam* current must be maintained constant (see above).

In the x-ray fluorescence spectrometer, the simple act of placing a plane specimen in its compartment correctly locates the specimen surface. In the microanalyzer, the specimen plane is set by bringing the specimen surface into focus in the light microscope at highest magnification. Slight differences in focus may cause substantial differences in measured analyte-line intensity. If the specimen surface is not highly polished, intensity is likely to be different if high, intermediate, or low relief is focused. A highly polished, very clean, structureless surface may present no feature at all on which to focus, and other means are required to set the specimen plane. Even when the surface is favorable for focusing, it may be difficult to find the particular microscopic feature to be studied.

In electron-probe microanalysis, because of the small diameter of the electron-beam spot, problems of microheterogeneity and microtopography are severe. In x-ray fluorescence spectrometry, 30–100-μm finishes are usually satisfactory. In electron-probe microanalysis, 0.25-μm finishes or finer may be required, and it is useless to compare analyte-line intensities from surfaces having substantially different finishes. Very rough finishes may result in extremely low analyte-line intensity, even from pure analyte. Suitable calibration standards are difficult to obtain because of these micro-heterogeneity and microtopography problems.

Chemical effects on wavelength are the exception in x-ray fluorescence spectrometry, the rule in electron-probe microanalysis when the chemical state of the analyte varies among specimens.

The buildup of carbon at the beam spot can result in decrease in meas-ured intensity during the measurement. It is almost always necessary to make replicate measurements at different spots. When C $K\alpha$ is measured, it may be necessary to translate the specimen during measurement to expose clean surface continuously.

In an x-ray fluorescence spectrometer, once the exact 2θ angles for all analytes have been established, the goniometer may be reset to these angles any number of times. Repeaking is required only if the crystal or collimator is changed. This applies even when the instrument is peaked with pure analytes, then used to measure specimens having the analyte in combined form. In the electron-probe microanalyzer, the full-focusing spectrometers give extremely sharp peaks. These spectrometers are more difficult to "peak" than nonfocusing spectrometers, and slight departures from the peak cause greater decrease in measured intensity. Thermal effects on the analyzer crystal are also more severe. The peak setting is affected by—or, at a fixed setting, the measured intensity is affected by: (1) the focusing of the specimen plane in the light microscope; (2) differences in microfinish of the specimen surface; (3) differences in slope of the specimen surface, overall or locally; (4) differences in chemical state of the analyte; (5) drift in operating condi-tions of the electron-optic column; and (6) variation in position of the beam spot on the specimen plane. As a result, it may be necessary to peak all the x-ray spectrometers for each measurement. This is very difficult for low-intensity peaks because of fluctuations in the panel ratemeter and recorder.

It is evident that, even disregarding the requirement for resetting beam current, the procedure of locating an analytical feature, focusing it, placing it under the electron beam, peaking the several x-ray spectrometers, and scaling the intensities may be tedious and time-consuming.

The components of the x-ray spectrometers are relatively inaccessible so that changing crystals, slits, and detectors is inconvenient.

Nevertheless, in fairness it must be pointed out that electron-probe microanalysis is one of very few methods that can determine composition and spacial variation of composition at the micrometer level. One must expect and be willing to accept some inconvenience to obtain such information.

SUGGESTED READING

ADLER, I., *X-Ray Emission Spectrography in Geology*, Elsevier, Amsterdam (1966); Chaps. 8–11, pp. 164–240.

ANDERSEN, C. A., ed., *Microprobe Analysis*, Wiley–Interscience, New York, 571 pp. (1973).

BEAMAN, D. R., and J. A. ISASI, "Electron-Beam Microanalysis," *ASTM Special Technical Publication* **STP-506**, American Society for Testing and Materials, Philadelphia, 80 pp. (1972).

BERTIN, E. P., *Principles and Practice of X-Ray Spectrometric Analysis*, 2nd ed., Plenum Press, New York (1975); Chap. 21, pp. 903–945.

BIRKS, L. S., *Electron-Probe Microanalysis*, 2nd ed., Wiley–Interscience, New York, 190 pp. (1971).

REED, S. J. B., *Electron Microprobe Analysis*, Cambridge University Press, Cambridge, 400 pp. (1975).

TOUSIMIS, A. J., and L. MARTON, eds., *Electron-Probe Microanalysis*, Advances in Electronics and Electron Physics, Supplement 6, 450 pp. (1969).

Index